Innovative Decision-Making in Engineering:

The Role of Cognition, Heuristics, and Human Factors

Zian Shah Kabir
https://orcid.org/0000-0001-8710-5283
University of Technology Sydney, Australia

IGI Global
Scientific Publishing
Publishing Tomorrow's Research Today

Vice President of Editorial	Melissa Wagner
Director of Acquisitions	Mikaela Felty
Director of Book Development	Jocelynn Hessler
Production Manager	Mike Brehm
Cover Design	Jose Rosado

Published in the United States of America by
IGI Global Scientific Publishing
701 East Chocolate Avenue
Hershey, PA, 17033, USA
Tel: 717-533-8845 | Fax: 717-533-7115
Website: https://www.igi-global.com E-mail: cust@igi-global.com

Library of Congress Cataloging-in-Publication Data

LCCN: 2025034549 (CIP Data Pending)
ISBN13: 9798337357027
Isbn13Softcover: 9798337357034
EISBN13: 9798337357041

British Cataloguing in Publication Data
A Cataloguing in Publication record for this book is available from the British Library.

Table of Contents

Preface xiii

Acknowledgement xv

Introduction xvii

Chapter 1
Innovative Decision Making Through Artificial Intelligence in a Connected Healthcare System 1
Zian Shah Kabir, University of Technology Sydney, Australia

Chapter 2
Next-Generation Decision-Making for Human and Machine Intelligence 27
Javaria Ajoon, KPESE, Pakistan
Qazi Mudassar Ilyas, King Faisal University, Saudi Arabia

Chapter 3
Decision Making in Human-Computer Interaction 59
Vandana N. Revankar, St. Francis College, India
G. Nivashini, Karpagam Academy of Higher Education, India
P. Selvakumar, Department of Science and Humanities, Nehru Institute of Technology, Coimbatore, India
K. L. Meera, Christ Academy Institute of Advanced Studies, India
Undabatla Rambabu, Lakireddy Bali Reddy College of Engineering, India
M. R. Arun, Vel Tech Rangarajan Dr. Sagunthala R&D Institute of Science and Technology, India
T. C. Manjunath, Rajarajeswari College of Engineering, India

Chapter 4
Explaining Hallucination of Artificial Intelligence in Data-Driven Decision-Making 85
Zian Shah Kabir, University of Technology Sydney, Australia

Chapter 5
Aligned Minds, Efficient Machines: Integrating Neuromorphic Computing for Personalized AI 113
Ben Kennedy, Capitol Technology University, USA

Chapter 6
Analytical Tools in Decision Making: Decision Tree, Influence Diagram, and Stochastic Modeling 153
Samkit Shah, Parul University, India

Chapter 7
Impact of Artificial Intelligence on the Strategic Decision Making of Multinational Companies 183
Benjamin Macleod, University of Technology Sydney, Australia
Sancheeta Pugalia, University of Technology Sydney, Australia

Chapter 8
A Hybrid Decision-Making Program for Optimal Robot Selection Integrating Fuzzy Logic, Genetic Algorithms, and Reliability-Aware MCDM Under Industry 4 219
Nazim Ali Khan, Deenbandhu Chhotu Ram University of Science and Technology, India
Ajay Kumar, Department of Mechanical Engineering, Deenbandhu Chhotu Ram University of Science and Technology, India
Naseem Rao, Department of Electronics and Communication, Jamia Hamdard, New Delhi, India

Chapter 9
Cultural Influences on AI-Driven Decision-Making in Digital Commerce Systems 243
Chuhui Shao, University of Technology Sydney, Australia
Kyeong Kang, University of Technology Sydney, Australia

Chapter 10
Synergizing Unconventional Approaches for Hybrid Power Generation From Emerging Renewable Technologies 279
Md. Ziaul Islam, West Zone Power Distribution PLC, Bangladesh

Chapter 11
EEGNet-Based Emotion Recognition Using Multimodal EEG and ECG Signals: A Hybrid AI-Driven Model 307
T. Nithya, Velalar College of Engineering and Technology, India
V. Chandrasekaran, Velalar College of Engineering and Technology, India
S. Mahendrakumar, Velalar College of Engineering and Technology, India
N. V. Keerthana, Velalar College of Engineering and Technology, India
C. Vasuki, Nandha Engineering College, India
R. Loganathan, Paavai Engineering College, India
P. Sathyasutha, Gnanamani College of Technology, India
R. Maheshwari, Velalar College of Engineering and Technology, India

Chapter 12
Decision-Making in Emerging Technologies Backed by Cognitive Analytics for Accurate ESG Reporting and Organizational Resilience 339
Jose De Jesus Reyes-Sánchez, TECNM Instituto Tecnologico Superior de Fresnillo, Mexico

Compilation of References 379

About the Contributors 429

Index 437

Detailed Table of Contents

Preface xiii

Acknowledgement xv

Introduction xvii

Chapter 1
Innovative Decision Making Through Artificial Intelligence in a Connected Healthcare System 1
Zian Shah Kabir, University of Technology Sydney, Australia

Artificial intelligence (AI) has transformed the healthcare industry by improving diagnosis, recordkeeping, and predictive analysis. To provide an in-depth picture of artificial intelligence-enabled healthcare technology, this chapter centers on the emergence of AI capabilities in medical applications. This chapter develops an AI-powered innovative decision-making support framework that explains sequential steps to get an optimum level of healthcare outcomes through data input, perception analysis, decision intelligence, cognitive reasoning, explainable decision, and feedback loop. By leveraging innovative decision making and advanced artificial intelligence techniques, the chapter seeks to generate actionable and valuable insights that can improve patient outcomes with the least possible resources like health professionals, accommodation, and medical instruments. The chapter reveals considerable importance for hospital systems transitioning from reactive care to proactive that engages in innovative data-driven decision-making in a connected healthcare system.

Chapter 2
Next-Generation Decision-Making for Human and Machine Intelligence 27
Javaria Ajoon, KPESE, Pakistan
Qazi Mudassar Ilyas, King Faisal University, Saudi Arabia

Decision-making remains a defining challenge in both engineering and organizational domains. Human choices, while enriched by intuition and experience, are often undermined by bias, stress, and inconsistency. Machines, in contrast, deliver speed and analytical precision but struggle with context, ethics, and adaptability. This chapter examines how the decision-making landscape has shifted from human reliance to algorithmic models and now toward integrated human–machine systems. By outlining the cognitive processes behind human judgment, the computational methods that guide machine reasoning, and the design of collaborative frameworks, it highlights how hybrid approaches can achieve more resilient and transparent outcomes. Practical applications, ethical concerns, and emerging challenges are discussed, offering engineers and practitioners a roadmap for developing systems that balance efficiency with human values.

Chapter 3
Decision Making in Human-Computer Interaction 59
Vandana N. Revankar, St. Francis College, India
G. Nivashini, Karpagam Academy of Higher Education, India
P. Selvakumar, Department of Science and Humanities, Nehru Institute of Technology, Coimbatore, India
K. L. Meera, Christ Academy Institute of Advanced Studies, India
Undabatla Rambabu, Lakireddy Bali Reddy College of Engineering, India
M. R. Arun, Vel Tech Rangarajan Dr. Sagunthala R&D Institute of Science and Technology, India
T. C. Manjunath, Rajarajeswari College of Engineering, India

Decision making is a central cognitive process that plays a pivotal role in Human-Computer Interaction (HCI), influencing how users interact with systems, interpret information, and achieve their goals. At its core, decision making involves selecting the most appropriate course of action from among multiple alternatives based on the user's goals, available information, contextual cues, and cognitive resources. In HCI, every interaction—from clicking a button, selecting a menu option, navigating through an interface, to interpreting feedback—requires the user to make decisions, consciously or subconsciously. These micro and macro decisions accumulate to define the overall user experience and system effectiveness. Decision making in HCI is highly context-dependent. It varies across different domains (e.g., gaming, e-commerce, healthcare), task types (routine vs. novel), user expertise (novice vs. expert), and device characteristics (desktop, mobile, wearable). In routine interactions, decisions may be automated and habitual, relying on procedural memory.

Chapter 4
Explaining Hallucination of Artificial Intelligence in Data-Driven Decision-Making 85
Zian Shah Kabir, University of Technology Sydney, Australia

Artificial intelligence transforms a paradigm shift in delivering data-driven innovative knowledge, where hallucinated information can influence users' decision-making. It demands exploring the effects of hallucinations towards the trustworthiness and ethics of AI in decision-making. Thus, this chapter aims to explain the implications of hallucinations across data-driven decision-making. The chapter conceptualizes a framework for explaining a process flow of mitigating AI-driven hallucinations that engages critical flaws in promoting innovative and strategic decision making for professional practices. Theoretically, the framework explains the effects of hallucinations and how it impacts decision making, following multiple theories and models. We present a metatheoretical framework to study the hallucinations of LLMs-based AI, which will allow researchers and practitioners to capture a multidisciplinary theoretical lens to investigate the complexities of AI-driven hallucinations in the data-driven innovative decision-making process.

Chapter 5
Aligned Minds, Efficient Machines: Integrating Neuromorphic Computing for Personalized AI 113
Ben Kennedy, Capitol Technology University, USA

AI systems exhibit robust reasoning abilities yet face challenges in personalization and sustainability, as cognitive augmentation technologies often show rigidity and high energy use that limit effectiveness. This paper introduces a framework that integrates psychometric alignment, modular cognitive agents, and neuromorphic computing to deliver scalable, personalized, energy-efficient augmentation. The Synthetic Cognitive Augmentation Network Alignment Questionnaire (SCANAQ) captures users' cognitive inclinations (Tate, 2024a) and informs the Synthetic Cognitive Augmentation Network Using Experts (SCANUE), a modular system modeled on prefrontal-cortex subregions (Tate, 2024b). The Spiking Transformer Augmenting Cognition (STAC) pipeline targets neuromorphic edge deployment: STAC V1 demonstrates a hybrid fine-tuning approach, and STAC V2 provides an ANN-to-SNN conversion pathway that yields spiking-compatible networks (Tate, 2025a; Tate, 2025b). Future work includes empirical hardware validation and standardized neuromorphic benchmarking (Davies et al., 2021; Ostrau et al., 2022).

Chapter 6
Analytical Tools in Decision Making: Decision Tree, Influence Diagram, and Stochastic Modeling.. 153
Samkit Shah, Parul University, India

Uncertainty in decision-making is common in fields like management, healthcare, finance, and engineering. To manage complexity, tools such as Decision Trees, Influence Diagrams, and Stochastic Models are widely used. This chapter explains how these tools help structure choices, show variable relationships, and simulate uncertain outcomes. A case study on pandemic vaccine distribution illustrates their combined use in solving complex problems. Ethical concerns, such as over-reliance on models, are also addressed. The study promotes a hybrid approach to improve clarity, strategy, and resilience in uncertain situations.

Chapter 7
Impact of Artificial Intelligence on the Strategic Decision Making of Multinational Companies 183
Benjamin Macleod, University of Technology Sydney, Australia
Sancheeta Pugalia, University of Technology Sydney, Australia

Artificial Intelligence (AI) is increasingly being leveraged into strategic decision-making processes across multinational companies, reshaping global industries. The rapid development and adoption led by multinational companies raises various questions on the actual capabilities and limitations of AI, what specific tasks can effectively be improved and associated risks to consider. This chapter focuses on the impacts of AI on global industries through a comparative case study examining four multinational corporations, comparing their unique implementation of AI across five criteria.The findings support the ability for AI to fundamentally address inherent decision-making challenges related to accuracy and timeliness of data, automation, while also highlighting areas sensitive to over-reliance on these automated processes. The conclusion provides a framework for businesses and policymakers to responsibly implement AI, including recommendations for improving AI governance, investing in employee upskilling, and ensuring ethical AI deployment to enhance long-term strategic outcomes.

Chapter 8
A Hybrid Decision-Making Program for Optimal Robot Selection Integrating Fuzzy Logic, Genetic Algorithms, and Reliability-Aware MCDM Under Industry 4 219
Nazim Ali Khan, Deenbandhu Chhotu Ram University of Science and Technology, India
Ajay Kumar, Department of Mechanical Engineering, Deenbandhu Chhotu Ram University of Science and Technology, India
Naseem Rao, Department of Electronics and Communication, Jamia Hamdard, New Delhi, India

Productivity and performance improvement in Industry 4.0 revolves around optimal robotics selection. Relying on traditional robot selection approaches often results in improperly defined criteria and ignored benchmarks, such as a robot's reliability. This work solves the problem with a new approach by creating a Hybrid Decision Model Process (HDMP), which fuses together Fuzzy Logic, Genetic Algorithms, and principles of Reliability Engineering with multiple Multicriteria Decision Making (MCDM) techniques like TOPSIS, COPRAS, and ARAS. The HDMP has taken into account the essentials of a candidate's selection criteria, such as payload capacity, reach, accuracy, operational speed, power usage, price, and reliability, which includes MTBF and other failure metrics. Consistent and accurate results were obtained using eight MCDM tools along with Spearman's rank correlation coefficient

SRC for model validation. This approach offers an alternative for enhancement in the robot selection process by focusing on productivity, efficiency, and operational dependability throughout the framework of Industry 4.0.

Chapter 9
Cultural Influences on AI-Driven Decision-Making in Digital Commerce Systems 243
Chuhui Shao, University of Technology Sydney, Australia
Kyeong Kang, University of Technology Sydney, Australia

This chapter explores how cultural diversity shapes consumer interactions with AI-driven digital commerce systems. While AI technologies revolutionize online shopping through personalized recommendations and intelligent assistants, their effectiveness varies significantly across different cultural backgrounds. This chapter proposes a theoretical framework examining how cultural values influence trust, acceptance, and usage patterns of AI commerce systems. This research introduces the Cultural-Adaptive Decision-Making (CADM) framework to guide the development of culturally-sensitive AI commerce systems. This framework provides theoretical design principles for creating adaptive interfaces that respond to cultural preferences while maintaining system efficiency. The study offers conceptual insights for global commerce companies seeking to optimize AI implementations across multicultural markets, ultimately enhancing user experience and business performance through AI design. This work provides a new perspective and possibilities for AI-driven commerce innovation.

Chapter 10
Synergizing Unconventional Approaches for Hybrid Power Generation From Emerging Renewable Technologies 279
Md. Ziaul Islam, West Zone Power Distribution PLC, Bangladesh

The global energy transition needs solutions that go beyond traditional solar and wind sources, which have issues with reliability and location. This chapter focuses on hybrid power generation systems that utilize unconventional renewable resources such as footstep power generators, thermoelectric harvesters, noise-to-power converters and other unorthodox approaches to energy storage. Each of these technologies has limitations in terms of location and output. But when these are combined into hybrid systems, they become more resilient, stable and scalable. Strategies include mapping regional energy needs, designing modular systems with smart controllers and using AI management to optimize different energy flows. Future energy independence will be shaped by IoT, advanced materials and community driven manufacturing enabling flexible systems. By focusing on collaboration across fields and human centered design, the chapter argues that hybridizing unconventional renewables is not just a supplementary option but a crucial strategy for sustainable and decentralized power systems.

Chapter 11
EEGNet-Based Emotion Recognition Using Multimodal EEG and ECG Signals: A Hybrid AI-Driven Model 307
T. Nithya, Velalar College of Engineering and Technology, India
V. Chandrasekaran, Velalar College of Engineering and Technology, India
S. Mahendrakumar, Velalar College of Engineering and Technology, India
N. V. Keerthana, Velalar College of Engineering and Technology, India
C. Vasuki, Nandha Engineering College, India
R. Loganathan, Paavai Engineering College, India
P. Sathyasutha, Gnanamani College of Technology, India
R. Maheshwari, Velalar College of Engineering and Technology, India

Physiological signal-based emotion recognition has emerged as a significant field of study with applications in interactive system interfaces, affective computing and healthcare. The DREAMER dataset to explore the effectiveness of EEGNet, a compact convolutional neural network designed for EEG-based interfaces between brain and computers in emotion classification. EEGNet can attain superior emotion recognition accuracy by learning directly from raw EEG data, thus outperforming traditional feature-based methods. The experimental results demonstrate that EEGNet attains accuracy for valence classification and arousal classification, significantly surpassing conventional machine learning approaches. This paper demonstrates the usefulness of EEGNet for creating EEG-based emotion identification systems as a versatile and effective solution for real-world application. The possible future concepts under this research include multimodal data integration and subject independent classification, which enhance deep learning in physiological signal processing.

Chapter 12
Decision-Making in Emerging Technologies Backed by Cognitive Analytics for Accurate ESG Reporting and Organizational Resilience 339
Jose De Jesus Reyes-Sánchez, TECNM Instituto Tecnologico Superior de Fresnillo, Mexico

The research explores how the synergy between cognitive analytics and emerging technologies reconfigures corporate decision-making processes to generate more accurate ESG reporting and strengthen organizational resilience. Through a qualitative analysis of emblematic cases in logistics, manufacturing and health, it is shown that technological modularity shortens learning cycles and, when integrated with transformational leadership and transparent data governance, increases supply chain traceability and reduces technical debt. The findings show that competitive value arises from converting the abundance of information into actionable knowledge, legitimizing iterative experimentation and aligning financial incentives with verifiable socio-environmental metrics.

Compilation of References 379

About the Contributors 429

Index 437

Preface

The human-computer interaction is engaging uncertainties in engineering practices that demands a new dimension on how decision-making is processed, controlled, and executed. The edited book "*Innovative Decision-Making in Engineering: The Role of Cognition, Heuristics, and Human Factors*" explores this growing paradigm—where technology, human and machine intelligence, social, cultural, and data-driven insight are amalgamated to reaffirm the decision-making across industries. This edited book gets together diverse research ideas to examine how cognitive processes, human factors, cultural factors, data-driven innovative models, artificial intelligence (AI), and strategic contexts impact engineering and organizational values.

ORGANIZATION OF BOOK

This book is comprised of thirteen chapters that delivers distinct values of innovative decision-making. The opening chapters, "*Innovative Decision-Making through Artificial Intelligence in a Connected Healthcare System*" and "*Next-Generation Decision-Making for Human and Machine Intelligence,*" establish a critical understanding of how AI enhances human cognitive capabilities towards data-driven decision-making, particularly in connected healthcare and complex adaptive systems. These chapters establish the shade for the integration of cognition, human and machine intelligence, and computation capabilities in modern engineering practices.

Subsequent deliberations, containing "*Decision-Making in Human-Computer Interaction*" and "*Explaining Hallucination of AI in Data-Driven Decision-Making,*" explore a new direction of explaining interoperability, user experience, and the accountability of AI-driven insights. In "*Aligned Minds, Efficient Machines: Integrating Neuromorphic Computing for Personalized AI*" and "*Analytical Tools in Decision-Making (Decision Tree, Influence Diagram, and Stochastic Modelling),*" authors extant research frameworks in explaining analytical methods and personalized service mechanisms. These chapters contribute to make a clear understanding of decision-making through computational techniques, machine intelligence with human cognitive functions, emphasizing both algorithmic accuracy and judgemental reasoning.

The chapter volume extends into the industrial, social, ethical, and cultural dimensions of decision-making in engineering practices. Chapters such as *"Impact of Artificial Intelligence on the Strategic Decision-Making of Multinational Companies"* and *"A Hybrid Decision-Making Program for Optional Robot Selection Integrating Fuzzy Logics, Genetic Algorithms, and Reliability-Aware MCDM under Industry 4.0"* deliver empirical and methodological insights into how advanced technologies transform from monolithic to industry 4.0 by incorporating AI-driven processes and innovative decision-making frameworks. On the other hand, *"Cultural Influences on AI-Driven Decision-Making in Digital Commerce System"* underlines the significance of social, ethical, cultural, and contextual heuristics in decision-making to enhance the design and implementation methods in complex adaptive systems.

The final chapters—*"Synergizing Unconventional Approaches for Hybrid Power Generation from Energy Renewable Technologies," "EEGNet-Based Emotion Recognition Using Multimodal EEG and EEG Signals: A Hybrid AI-Driven Model,"* and *"Decision-Making in Emerging Technologies Backed by Cognitive Analysis for Accurate EEG Reporting and Organization Resilience"*—exemplify the frontier of multidisciplinary research where smart technologies create an enabling actor for decision-making, particularly in renewable energy, organizational resilience, human-brain interface, and neuroscience that congregate through cognitive, technological, and data-driven innovation.

This collaborative effort brings out a holistic approach to explain the decision-making process in engineering practices through cognition, heuristics, human factors, ethical, social, and cultural values, where advanced technologies, data-driven analytics, and computational capabilities play critical roles. It is my deterministic believe that this chapter volume will serve as both a scholarly reference and a source of inspiration for researchers, professionals, industry practitioners, regulators, governments, and students seeking to understand, design, implement the next generation decision support systems through human-machine interaction capabilities and data-driven decision-making.

Dr Zian Shah Kabir

Editor

Innovative Decision-Making in Engineering: The Role of Cognition, Heuristics, and Human Factors

IGI Global Publishers

Acknowledgement

The development and successful completion of "Innovative Decision-Making in Engineering: The Role of Cognition, Heuristics, and Human Factors" has been made possible through tremendous efforts, valuable insights, and commitment of many persons and organizations. I would like to express my gratitude to all authors who scholarly dedicate their time and efforts to deliver valuable works form the intellectual foundation of knowledge. Their research outcomes, innovative ideas, and contribution to developing knowledge in the emerging context of decision-making in engineering practices that have made this book as a deep and intense resource with multidisciplinary fields.

Special thanks are conveyed to the reviewers and editorial board members of IGI Global for their valuable review comments, constructive feedback, suggestions, and guidance throughout the journey. Their academic rigor and research knowledge have confirmed that the chapters mounted in this volume enrich to the highest standards of scholarly excellence.

I am also truly grateful to the research communities and organizations that have collaborated in the study of cognitive and data-driven decision-making. Their valuable supports and collegial engagement have significantly augmented the presentations in this work.

A special thanks is conveyed to the technical experts, development editors, and coordinators at IGI Global for their professionalism, guidance, and supports during every stage of production. Their dedication and contribution have been influential in transforming this collaborative effort into a cohesive, organized, and impactful publication.

Finally, I like to acknowledge the unwavering and valuable support of colleagues, mentors, and family members who have encouraged me to dedicate for the successful accomplishment of this project. Their encouragement and peer influence develop my social norms in extending the editorial scope for decision-making of human-machine interaction through data-driven innovation.

I hope that this edited book will serve as immense values to the continuing discourse on cognition, heuristics, social, cultural, and human factors in engineering decision-making, inspiring further research and managerial applications in this rapidly evolving field of study.

Dr Zian Shah Kabir

Editor

Innovative Decision-Making in Engineering: The Role of Cognition, Heuristics, and Human Factors

IGI Global Publishers

Introduction

This book addresses critical aspects of decision-making that are emerging field of interests to researchers, professionals, industry experts, and organizational leaders who are steering the integration of cognitive science, advanced technologies, and human-cantered design through digital transformation to achieve the Industrial 4.0. As information systems become heterogenous, the capability to take judgemental and ethical decisions at the convergence of human and machine intelligence has become a challenging factor in engineering practices.

Collectively, the editor and contributing authors of *Innovative Decision-Making in Engineering: The Role of Cognition, Heuristics, and Human Factors* represent a comprehensive piece of research work on how decision-making processes are evolving in the age of human-machine interaction. This edited volume shapes the growing body of knowledge that connects human factors, cognitive, heuristics, social cultural, ethical, and advanced technologies, offering practical and theoretical perspectives in decision-making. The volume accumulates innovative research findings in the context of cognitive science, emotional and machine intelligence, and organizational capabilities, contemplating a holistic approach in engineering practices.

The chapter volume advances the conceptual foundation by exploring the embryonic dynamics of decision-making in complex adaptive systems and outlining how human-machine interaction engage computational intelligence to deliver data-driven insights. The chapters explore diverse and emerging topics, including AI-driven healthcare decision support systems, neuromorphic computing for personalized recommendations, and value co-creation through data-driven innovation. The volume also expands into strategic and innovative use case scenarios—such as decision-making in multinational corporations, fuzzy logic-based robotics, and cognitive frameworks—explaining how analytical tools enhance to judgmental and intuitive decisions.

I decidedly recommend this book to researchers, professionals, industry practitioners, governments, regulators, and students interested in knowing how the blend of cognitive science and innovative technologies is transforming the decision-making process in engineering and other perspectives. It contributes to the growing discourse on how human emotional feelings and machine intelligence can co-create values in decision-making to create more resilient, sustainable, equitable, and adaptive information systems. This collection of converged and multidisciplinary knowledge makes it supportive to both technical and non-technical readers, allowing them to apply its theoretical and managerial implications to their own professional practices.

This volume pursues as a precious contribution to contemporary knowledge and practice, advancing a forward-thinking perspective on decision-making through cognitive, human and machine intelligence, and innovative data-driven insights in the Fourth Industrial Revolution.

Zian Shah Kabir
University of Technology Sydney, Australia

Chapter 1
Innovative Decision Making Through Artificial Intelligence in a Connected Healthcare System

Zian Shah Kabir
https://orcid.org/0000-0001-8710-5283
University of Technology Sydney, Australia

ABSTRACT

Artificial intelligence (AI) has transformed the healthcare industry by improving diagnosis, recordkeeping, and predictive analysis. To provide an in-depth picture of artificial intelligence-enabled healthcare technology, this chapter centers on the emergence of AI capabilities in medical applications. This chapter develops an AI-powered innovative decision-making support framework that explains sequential steps to get an optimum level of healthcare outcomes through data input, perception analysis, decision intelligence, cognitive reasoning, explainable decision, and feedback loop. By leveraging innovative decision making and advanced artificial intelligence techniques, the chapter seeks to generate actionable and valuable insights that can improve patient outcomes with the least possible resources like health professionals, accommodation, and medical instruments. The chapter reveals considerable importance for hospital systems transitioning from reactive care to proactive that engages in innovative data-driven decision-making in a connected healthcare system.

DOI: 10.4018/979-8-3373-5702-7.ch001

1. INTRODUCTION

Researching the effects of artificial intelligence (AI) on clinical decision-making and healthcare delivery practices is the key concern of this chapter. Increasing patient loads, staffing shortages, clinical decision making, cultural barriers to adoption, clinical integration, and the need to make more fast and accurate clinical decisions have stretched traditional triage systems to their limits. In response, AI is being looked to as a potential game-changer—offering the possibility to make faster, consistent, and accurate than conventional approaches in a connected healthcare system (Rahman et al., 2024). Meanwhile, AI has become an uprising technology which can uphold the decision-making process and utilize minimal resources in delivering optimum healthcare services. This chapter addresses the computational power of AI that can assist in providing complex healthcare interventions and reinforce better curing applications.

The chapter explores how AI technology transforms into a connected healthcare landscape, achieving a higher level of accuracy in health monitoring, detection, and diagnosis while maintaining transparency in the decision-making process (Mikhailov, 2024). AI provides a way to modernize features that do not need human intervention, and that can support healthcare professionals to derive an innovative decision making in delivering better treatment and diagnosis by providing advanced understanding of health conditions and reasoning process. Not only do we shed light on its transformative potential, but this chapter also offers healthcare organizations through valuable insights on how best to utilize AI for improving patient outcomes and operational efficiency considering trust, acceptance, usability, system integration and cultural adoption.

Despite significant technological advancements in healthcare informatics, current medical decision support systems face multifaceted challenges that can limit their effectiveness and technological adoption. These challenges are particularly evident in complex diagnostic scenarios where multiple data types - including electronic health records, medical imaging, and real-time patient monitoring - must be analyzed simultaneously for accurate clinical decision-making. This chapter is dedicated to exploring a healthcare-specific innovative decision-making framework through AI technology. It seeks to identify use case scenarios and challenges specific to a connected healthcare industry, providing an in-depth perspective on how AI technology integrates into various clinical workflows. This goal seeks to comprehend the nuances involved with adopting AI technology which are essential for healthcare organizations operating across various specialties.

Healthcare providers increasingly confront unprecedented challenges in clinical decision-making, driven by an exponential surge in medical data complexity and volume. The recent report of Future Business Insights indicates that the world health market is projected to rise to US $ 504.17 billion by 2032, with a CAGR of 44.0% (Fortune Report, 2025). This critical situation has catalyzed the urgent need for adopting innovative decision-making processes, particularly those powered by AI, enhancing diagnostic accuracy and clinical efficiency while reducing medical errors.

This chapter addresses these fundamental challenges through a novel approach: the systematic integration of advanced AI technologies with explainable and innovative decision support frameworks in medical systems. This chapter focuses on three key innovations: the development of a hybrid AI-powered decision-making based regulatory and policy framework that combines deep learning with explainable AI methods for enhanced diagnostic accuracy to enhance user experience, and create more usability; the implementation of real-time processing capabilities for diverse medical data types through system integration; and the creation of transparent, interpretable decision support interfaces that build trust while maintaining high performance standards.

The significance of this chapter extends beyond technological innovation to address critical healthcare challenges, particularly for cultural barriers to AI adoption and clinical integration through diversified nature of healthcare. This chapter explains an AI-powered innovative decision-making support framework that effectively processes complex medical data while providing clearer and actionable insights, reducing medical errors through innovative decision making. The framework can improve patient outcomes through early and accurate diagnosis, increase healthcare delivery efficiency by reducing decision time, address the effects of cultural background, ethical and moral values, and demonstrate the practical implementation of explainable AI in clinical settings. Our approach uniquely combines state-of-the-art machine learning algorithms with natural language processing and intelligent data analytics, providing healthcare professionals with reliable, transparent, and timely decision support that integrates seamlessly with existing clinical workflows.

The chapter is structured to provide a comprehensive examination of AI integration in delivering innovative decision making to a connected healthcare industry. Following this introduction, this chapter presents an extensive background analysis of current literature that contextualizes our work within the latest healthcare technology trends. Then, the chapter explains an AI-powered innovative decision-making support framework that includes multifaceted constructs. The recommendation section details our novel contributions, followed by the significance and innovation section. After that, we conclude with anticipated results for future development.

2. BACKGROUND

2.1 Decision Making Transformation

Decision-making was initially conceived as a cognitive mechanism for humans in which they select the best choice from possible alternatives based on specific criteria (Busemeyer & Townsend, 1993). Decision-making is a rational thought process that identifies cost and benefit analysis through information gathering from different sources. However, decisions may not be always rational where different uncertainties like cognitive prejudices can affect human decision-making through overconfidence, insightful factors, or visual illusion (Blanco, 2017). Some of these uncertainties may arise when delivering autonomous driving, smart logistics through robotic arms, remote surgery using IoT, crop disease detection through AI, and large-scale military applications.

In the AI era, it supports innovation by automating tasks and providing real-time insights to enhance decision-making by analyzing large volumes of data (Jarrahi, 2018). Predictive analysis through advanced machine learning algorithms analyzes historical data to enable proactive decision-making and streamline complex adaptive processes, reduce costs, and optimize resources. AI is advancing healthcare, retail, logistics, transport, education, and mining by incorporating machine intelligence that simulates human intelligence. It can resolve critical issues effectively and efficiently using heuristic data and perceived experiences. However, the decision-making framework considering AI was unexplored in the context of a connected healthcare system (Mikhailov, 2024; Ryffel et al., 2018).

In the realm of digital transformation within healthcare, there are various types of uncertainties associated with heterogenous environment that require data-driven decision-making framework (Batko & Ślęzak, 2022). However, AI-powered decision-making framework in a connected healthcare can be developed by introducing clinical and diagnostic through intelligence-enabled wearable medical devices (Barnes & Zvarikova, 2021), predictive diagnostics for mental health (Chancellor & De Choudhury, 2020), automated approach for depression detection through facial expressions (Gilanie et al., 2022), and healthcare techniques on diagnostics (Kalra et al., 2024). Therefore, it is essential to take innovative decision making through AI by addressing all possible uncertainties for critical infrastructures like the healthcare industry.

In the earlier stage, decision making was derived as a mental process in executing a course of action after gathering observations from the respective environment. In system engineering, the process is explained through sequential steps: determining the problem, analyzing the situation, exploring and assessing options, and making decisions. With the evolution of advanced technologies, the decision-making process is not only related to human intelligence, but also it engages machine intelligence. Therefore, decision making is a fundamental process that involves human-machine interaction, socio-technological impacts, uncertainties, ethics, and moral judgement.

2.2 Artificial Intelligence

Artificial Intelligence (AI) is a technology where a system or machine performs tasks as human intelligence through a combination of algorithms, iterative improvements, access, and data processing (Alimadadi et al., 2020). Machine learning is a subset of AI which is an algorithm that makes a system in absorbing data and acting gradually without being explicitly programmed. These features are applied in the field of healthcare in many innovative ways.

The European Union AI strategy derived the term AI as systems that show intelligent behavior by analyzing their environment and taking measures–with some degree of autonomy–to achieve desired goals (European Commission, 2021). AI is not just replacing human intelligence—it supports systems capabilities of engaging machine intelligence with rationalities through external data and enhances decision-making with minimal human efforts. However, there is a scope to extend research on developing AI-driven decision-making considering uncertainties.

2.3 The Role of AI in Healthcare

AI is reshaping the healthcare industry by enabling predictive modelling, personalized medicine, and improved operational efficiency (Chancellor & De Choudhury, 2020). As healthcare systems increasingly rely on data-driven decision-making, AI has the potential to enhance diagnostic accuracy (Kalra et al., 2024; Barnes & Zvarikova, 2021), optimize resource allocation (Rahman et al., 2024), personalized treatments (Melnykova et al., 2019), and reduce hospital readmissions. However, despite growing interest and proven benefits, the implementation of AI remains uneven due to technical, cultural, regulatory, ethical, and organizational challenges. This chapter reviews the existing literature examining the role of AI in healthcare, the challenges to its adoption, and current research gaps, particularly in the context of a connected healthcare system for patients with feedback processes.

AI has become an integral part of modern healthcare by transforming clinical decisions considering critical uncertainties. The strength of AI lies in integrating diverse data sources, such as electronic health records (EHRs), and real-time sensor outputs, to support accurate diagnoses and treatment personalization (Ristevski & Chen, 2018). Personalized medicine is a key application area where AI can play a crucial role in modernizing automation in the healthcare industry. Further, integrated frameworks that can combine dynamic clinical measurements (e.g., vital signs) with static patient attributes (e.g., genetic markers) to deliver individualized treatments (Melnykova et al., 2019). In a broader operational context, hospitals can get measurable benefits and innovative intervention mechanisms from AI adoption into existing healthcare systems. However, an integrated AI-powered decision-making framework is required to explain connected healthcare. For instance, Batko and Ślezak (2022) found that 47.6% of Polish healthcare institutions use structured data analytics to improve clinical workflows, leading to a 33% reduction in bed shortages. These results are consistent with Gartner's 2021 findings, which reported up to 40% efficiency gains in hospitals using predictive analytics for resource management. However, the AI adoption in decision-making framework for a connected healthcare system is still unexplored.

2.4 Role of AI in Healthcare Decision-Making

AI has emerged as a transformative force in healthcare delivery, offering innovative solutions for interventions, disease prevention, diagnosis, treatment, and remote patient monitoring. The role of AI in healthcare is critical to addressing predictive and prospective analytics that engage in a systemic process in decision making. The decision making not only lies with the system-generated outcomes but also supports health professionals in taking their innovative decision making. For example, manual data input and traditional diagnostic procedures are both time-consuming and error-prone, affecting medical efficiency and potentially having adverse effects on patient outcomes. AI tools like machine learning and predictive analytics, on the other hand, can automate repetitive processes and facilitate data-driven decision-making, which not only lowers errors but also improves resource allocation, speeds up treatment, and increases the overall operational effectiveness of healthcare facilities (Mikhailov, 2024; Chen et al., 2023). However, there are still important questions about how AI can fit in with current healthcare processes, especially when it comes to juggling the roles of AI systems and human clinical physicians (Kalra et al., 2024). Furthermore, it is imperative to assess if AI is applicable across many medical specialties or if its benefits are restricted to domains like pathology and radiology.

Although the healthcare industry accommodates AI, still there is a fundamental challenge in taking human decision-making accurately, considering critical and high-pressure scenarios (Barnes & Zvarikova, 2021). Critics claim that, while AI can imitate certain scenarios, it may not fully represent the contextual judgments made by skilled clinicians when caring for patients (Kalra et al., 2024). Furthermore, the possibility of over-reliance on AI technology has prompted concerns since trainers may rely on automated systems and miss the importance of developing critical thinking abilities of health professionals. On the other hand, inadequate pre-trained data with uncertainties can misguide users by delivering hallucinated information. Moreover, data privacy and ethical values might have consequences for AI-driven decision-making processes in a connected Heathcare system.

Recent developments in AI-based Clinical Decision Support Systems have shown remarkable potential in enhancing clinical accuracy and efficiency through rational decision-making (Mikhailov, 2024; Ryffel et al., 2018). AI has demonstrated the potential to significantly improve the precision and suitability of clinical decision support systems while attracting widespread interest from medical practitioners. However, advanced AI models like federated learning and chain-of-thoughts can be integrated to develop an integrated AI-powered decision-making framework to achieve a connected healthcare system (Mohri et al., 2019; Wei et al., 2022).

2.5 Current State of AI in Complex Decision-Making

Large Language Models (LLMs) have enabled the deployment of advanced AI applications by simulating complex decision-making processes that require advanced understanding and logic for a particular use case scenario. The rapid advancement of AI has highlighted the importance of language models' capabilities in determining complex reasoning processes. While researchers continue to improve the capabilities of LLMs in the pursuit of artificial general intelligence, SLM (Small Language Models) research can make machine intelligence more accessible, affordable, and efficient for critical tasks (Lu et al., 2024). Further, the substantial computational functionalities of those models can enable the AI systems through complex reasoning mechanisms that can handle diversified queries by reducing system complexities, external cognitive biases, and AI-driven hallucinations. This chapter addresses how innovative decision making can overcome those uncertainties and challenges by accommodating complex reasoning tasks using advanced AI models.

However, the optimization of small language models through chain-of-thought (CoT) reasoning can enhance a promising direction for innovative decision making through advanced AI capabilities (Wei et al., 2022). In clinical applications, the CoT can make reasonable decisions that can resolve difficult tasks to accommodate a sequence of intermediate steps in challenging situations. Previous studies indicate that specialized models using chain-of-thought could effectively pursue excessive medical records that can enhance the overall performance in complex decision making. Their work showed how domain specialization combined with explicit reasoning paths could create more efficient solutions for specific tasks in a connected healthcare system.

Recent research has focused on making small language models more efficient through domain specialization. Arefeen et al. (2024) demonstrated that domain-specific optimization could reduce API costs by 67.81% while maintaining comparable performance to larger models. For example, CoT is improving the accuracy of large models in mathematical reasoning that can enhance reasoning abilities, transferability, performance optimization, and adaptation processes. It demonstrates that explicit reasoning pathways may lead to significant gains in a model's performance. Prior findings suggest that integrated AI-based problem-solving systems may be able to complete multi-step solutions for large scale projects such as software engineering programming tasks, digital therapeutics, smart logistics, and medical diagnostics (Yang et al., 2023).

Moreover, the development of a hybrid AI architecture develops innovative decision making that can combine deep learning with explainable AI (XAI) methods for enhanced diagnostic accuracy in a connected healthcare. The hybrid model enhances the implementation of real-time processing capabilities for diverse medical data types, and the creation of transparent and interpretable decision support interfaces that can build trust while maintaining high performance standards. XAI has emerged as a crucial part of the AI-based clinical decision-making process. It can process versatile nature of medical data, identify patterns of disease, provide evidence-based recommendations, and support interactive clinical decision making for interventions, diagnosis, and treatment. Further, Holzinger emphasizes that beyond XAI can create an enabling environment to make information systems more trusted in connected healthcare (Holzinger et al., 2020).

In contrast, recent innovations have accelerated the adoption of Federated Learning (FL) in healthcare. To address data heterogeneity, FedProx and personalized FL (Rajendran et al., 2024) tailor global models to local contexts, enhancing relevance for diverse cohorts. Communication efficiency, a barrier to scalability, has been improved via gradient quantization (Konečný et al., 2016) and adaptive aggregation, reducing bandwidth usage by 60%. NVIDIA Clara exemplifies practical deployment, enabling hospitals to collaborate in real time (Maier-Hein et al., 2022). Equity-driven techniques like Fair-FL (Mohri et al., 2019) and contribution-aware weighting can support the critical needs in favor of high-resource healthcare institutions. For security, hybrid frameworks combining FL with differential privacy (DP) (Wei et al., 2022) or Secure Multi-Party Computation (SMPC) (Ryffel et al., 2018) safeguard against health data breaches. Future directions include federated reinforcement learning for dynamic and innovative decision-making and regulatory standardization for accountability. These innovations can collectively bridge the gap between theoretical FL and real-world healthcare equity.

2.6 Human-Factors, Ethics, and Cultural Effects

Nowadays, human-centered design considers human factors, uncertainties, ethics, tasks, context, social, cultural, and other externalities. Human factors may involve various constructs from cognitive and perceptual engagements, including representative biases, overconfidence, poor judgment, and negligence (Dror, 2020). Naturally, the human brain processes information through cognitive mechanisms and focuses on it with the support of memory capacity. The information processing may be described through sequential steps: step 1 - individuals may focus on information that is eye-catching but irrelevant while executing tasks; step 2 -individual may be misguided when salience features in the content are absent during the information processing using memory before the task performance (Jamieson & Skraaning, 2017). But illusions of control, bad apples (question of competencies), blind spots, predetermination, and overconfidence may engage in decision-making through cognitive biases. Therefore, human factors may play a crucial role related to the processing of quality information in designing AI-powered decision-making framework for a connected healthcare system that can address visual presentation, emotional content, conciseness, imageability, temporal, and spatial proximity.

In contrast, professional and organizational codes of conduct act as a driving force to maintain ethical values which create an enabling environment in taking rational decision-making. Even though professionals inherit virtues and norms from their respective societies, ethical code of conduct can make a binding upon them to follow honesty, compliance, fairness, loyalty, respect for public interest, environment and sustainable growth. Professionals can maintain their ethical codes before taking decisions that may avoid undesirable options in complex situations.

On the other hand, culture can also impact information system adoption. Hofstede initially explained cultural values through a few characteristics such as uncertainty avoidance, power distance, masculinity-femininity, individualism/collectivism, and time-orientation (Hofstede, 2011). People from different cultural backgrounds may experience different perceptions and behavioral intentions in adopting technologies (Leidner & Kayworth, 2006). In the context of cultural values, the same technology or information system can be adopted in different ways across cultures and result in different benefits.

National cultures with low uncertainty avoidance have different degrees of perception and technological adoption with characteristics. For example, people from United States are habituated with technological advancements, entrepreneurships, and ready to adopt changes, people from Sweden are flexible and curious to accept new ideas, and higher resistance to uncertainties, and people from United Kingdom are supportive to experimentation and reluctant to rules. On the contrary, high uncertainty avoidance cultures have higher perception and resistance in adopting technologies. People from Japan consider risk analysis and in-detail planning before technological adoption, people from South Korea follow structured processes and predictive analysis, and people from Mexico have preferences to follow guidelines and curious for risks and uncertainties.

3. MAIN FOCUS OF THE CHAPTER

Despite significant technological advancements in healthcare informatics, current medical decision support systems (MDSS) face multifaceted challenges that limit their effectiveness and adoption. Traditional MDSS, while valuable, increasingly struggle with three critical limitations: processing the exponentially growing volume of diverse medical data with existing clinical workflows, and there is a challenge to providing transparent, explainable recommendations that healthcare providers can trust (Kalra et al. 2024, Rahman et al., 2024). These challenges are particularly evident in complex diagnostic scenarios where multiple data types - including electronic health records, medical imaging, and real-time patient monitoring - must be analyzed simultaneously for accurate clinical decision-making.

3.1 Example 01: Decision Making during COVID-19

AI has emerged as a transformative force in tackling pandemic situations during COVID-19, offering innovative solutions for disease prevention, diagnosis, treatment, and patient monitoring (Alimadadi et al., 2020). Recent developments in AI-based Clinical Decision Support Systems have shown remarkable potential in enhancing clinical decision-making accuracy and efficiency. AI has demonstrated the potential to significantly improve innovative decision making while attracting widespread interest from medical practitioners.

AI-enabled wearable medical devices and Internet of Things (IoT) applications have demonstrated particular promise in healthcare delivery. These systems enable real-time continuous remote monitoring of patients' vital signs, with research showing their effectiveness in COVID-19 prevention, screening, and treatment (Barnes & Zvarikova, 2021). The Internet of Medical Things (IoMT) furthers developments in sensing, actuation, and connectivity to considerably enhance the quality and accessibility of healthcare services.

Data-driven innovative decision-making is a key competitive advantage for organizations, particularly for medical organizations. By incorporating data into their processes, organizations can better respond to users' demands, optimize operations, and enhance user experiences. For instance, clinics can analyze patients' historical records and extract disease patterns to refine treatment strategies, manage appointments, and improve healthcare services. Leading healthcare institutions can use AI to personalize patients' experiences and predict their healthcare requirements, ensuring they deliver treatments that fulfil patients' preferences (Mikhailov, 2024). Effective AI-driven healthcare can help hospitals improve patient care by analyzing a large volume of data. This enables health providers to identify trends, enhance diagnostic accuracy, and develop evidence-based treatment strategies, directing better healthcare outcomes, and increased optimum level of productivity.

3.2 Example 02: Mental Health Disorder

A machine learning approach, a subset of AI, can facilitate the process of early detection of mental disorders through personalized treatment procedures, improved prognosis, and monitoring the progress of mental status. It can implement an emotional coping mechanism and prevent the rise of mental health conditions through AI and identify different diagnosis challenges in several ways by using its accurate data analytics. It is evident that AI has already made great improvements in early detection of mental illness such as anxiety and depression. Previous research explains that while deep learning models show high predictive accuracy, they often lack interpretability, making it challenging to validate how diagnostic conclusions are reached—a crucial consideration for both clinicians and patients (Zhou et al., 2020).

Machine learning tools, capable of identifying subtle behavioral or linguistic patterns associated with mental health conditions, can play an essential role in early detection. For instance, Gilanie et al. (2022) demonstrated that facial micro-expressions, detectable through computer vision techniques, can reveal underlying depressive states before they become clinically observable, suggesting a path to real-time, non-invasive monitoring for early diagnosis for mental health disorders. Similarly, natural language processing (NLP) models can analyze language patterns in social media posts that can uncover symptoms of depression or anxiety based on subtle shifts in tone, word choice, or sentiment, allowing for early warning signs to be detected even outside clinical settings (Calvo et al., 2020). Therefore, AI-driven models can support health professionals in making innovative decisions through predictive and prospective analytics.

Islam et al. (2024) applied Random Forests and Support Vector Machines (SVMs) as a supervised learning to predict anxiety, mental depression, and classify mental health states from multimodal health data. Furthermore, Convolutional Neural Networks and Recurrent Neural Networks as deep learning mechanisms were executed to detect mood disorders by analyzing smartphone data that is more complex, such as text and speech rhythms (Olawade et al., 2024). Moreover, previous research indicates that Natural Language Processing (NLP) can be effectively utilized for markers that can be used to detect mental health problems with demonstrated accuracy to identify linguistic mental health issues by analyzing unstructured text data from social media posts (Chancellor & De Choudhury, 2020).

Decision-making in healthcare systems is challenging for patients and health professionals due to feedback delays, time constraints, and cognitive load. Understanding their decision-making processes in such tasks is critical for updating pre-training data. Learning innovative decision-making involves acquiring cause, reasoning, and effect of knowledge and recognizing patterns from embedded information and past experiences to develop a robust decision-making process in relevant contexts. Innovative decision-making involves making multiple, sequential preferences that adapt to evolving conditions over time. These conditions may occur from previous decisions, inaction, or environmental changes.

3.3 Example 03: Federated Learning for Healthcare

Traditional centralized AI models often perpetuate disparities by training data skewed toward well-resourced populations, leading to inaccurate predictions for marginalized groups. FL mitigates this by enabling decentralized collaboration, allowing underrepresented institutions to contribute data without compromising privacy. For instance, a study demonstrated that FL-based COVID-19 prediction models reduced diagnostic disparities by incorporating data from underserved communities. Furthermore, equity-driven aggregation methods like Fair- FL (Mohri et al., 2019) minimize worst-case errors across demographic subgroups, ensuring models generalize equitably. Such approaches align with regulatory mandates for equitable care delivery, as FL inherently promotes inclusivity by design. By decentralizing data ownership, FL ensures that AI-driven decisions reflect global patient diversity, fostering trust in automated clinical tools.

3.4 Example 04: Activity Mapping (Patient Management)

A connected healthcare system can be designed considering interactions, interrelationships, and interdependence among associated actors. Activity mapping is a critical task to develop system mapping to achieve the overall goal of performing multiple actions (Gopal & Clarke 2015). In the context of a healthcare system, we can explain how an integrated system can include critical management from accident to hospital release using an embedded and adaptive healthcare system. In figure 1, the patient attendance immediately after the accident is addressed as a first step in the activity process. Then, we include police and paramedics as parallel activities in the diagram. One of the critical elements in the activity process is hospital admission and treatment integrating nurses and doctors within an embedded healthcare system. Finally, patient release activity is included in the diagram to create a connected healthcare system.

The activity map identifies all possible actors considering a use case scenario (health management system), which can explain a visual representation of the key activities of actors within or external organizations that influence the system, as well as their working relationships to get optimum productivity from the system.

Figure 1. Activity Diagram of Healthcare System

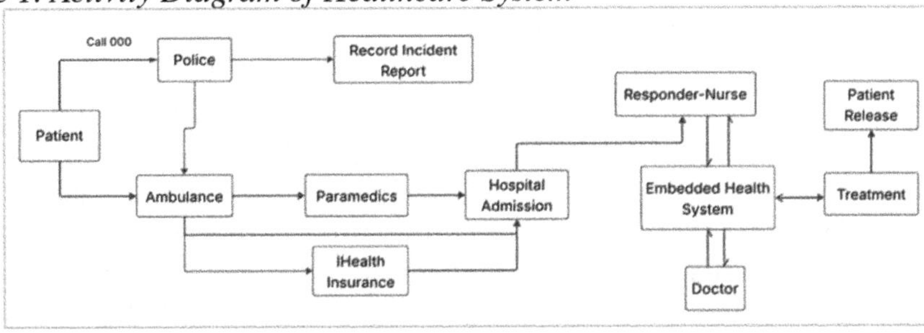

4. DECISION MAKING MODEL

Activity Theory was rooted in psychology (Vygotsky & Cole, 1978) and human computer interaction (HCI) (Nardi, 1996) that presents a rich lens to analyze how AI systems enhance human-machine driven decision-making in integrated healthcare. The theory explains a holistic way to cognize how people think, learn, and interact with critical and complex adaptive systems, particularly in a connected healthcare context. In innovative decision-making, AI models act as a critical role in mediating a correlation between human and decision challenges with uncertainties in socio-technical systems. This chapter describes how AI enhances decision-making in a connected healthcare system by explaining data interpretation, scenario analysis, counterfactual reasoning, and meaningful decisions.

In another context, the Decision Field Theory (DFT) cognitively grounded a model that depicts how decision-making can evolve through human preferences, social factors, and machine functionalities over time, considering critical uncertainties in complex situation (Busemeyer & Townsend, 1993). The theory can be applied to explain the process flow of getting valuable insights through innovative AI-driven decision-making. This chapter follows the DFT to apply in a connected healthcare as a complex adaptive system and develops an AI-powered decision-making framework to address contextual influences on innovative decision-making out of traditional settings (Hancock et al., 2024).

The proposed framework adopts human-machine centric modeling, explainable decisions, contextual awareness, personalized deliberations, and interactive decision support systems, following a dynamic stochastic process through cognitive, heuristics, and human factors. By embedding the DFT within AI decision-support tools, this chapter explains how an adaptive healthcare system can deliver psychologically aligned AI-driven innovative decision-making through human machine interactions—leading to achieve more trust, ethics, cultural values, satisfaction, personalization, and robust service delivery.

In the advent of AI models, Chain of Thought (CoT) and Federated Learning are also crucial to explain innovative decision-making in connected healthcare. The CoT can support adopting intermediate steps, reflecting human cognitive engagements through complex reasoning, transparency, and error detection processes (Wei et al., 2022), particularly for connected healthcare. In the context of healthcare, the CoT can execute as a diagnostic assistant to evaluate diagnostic criteria, compose treatment strategies, and create flagging alarms (Arefeen et al., 2024).

On the other hand, Federated Learning can allow multiple healthcare entities to collaboratively train a shared model without swapping raw patient data. The FL can deploy decentralized optimization and privacy enhancement techniques to ensure privacy compliance, multi-site generalization, and robustness in overfitting versatile data sources from different perspectives (Rajendran et al., 2024). The proposed AI-powered decision-making framework (as shown in Figure 2) describes explainable and collaborative intelligence through data-efficient learning mechanisms and adaptive environments.

Figure 2. AI-powered innovative decision-making support framework

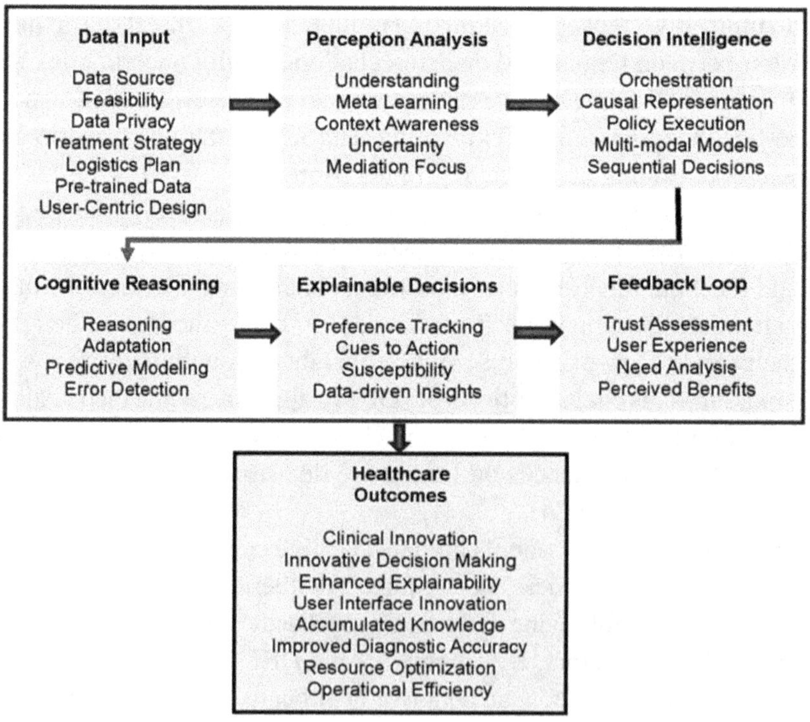

This chapter explains an AI-powered innovative decision-making framework that can leverage AI to support health professionals and medical organizations in delivering innovative, adaptive, and faster decisions, particularly in complex scenarios with uncertainties. The framework integrates multiple theories and models, and explains human and machine judgement, data modeling, user experience through cultural values, data privacy, and process mechanisms to deliver predictions, valuable insights, recommendations, and actionable learning.

The AI-powered decision-making support framework includes data analytics, intelligent sensing, causal understanding, and human-AI collaboration. It moves beyond static decision trees or dashboards to a complex adaptive system that learns, reasons, and supports critical decision-making in connected healthcare.

System mapping is a critical process to identify associated actors and their supporting roles to enhance systems thinking for service and value co-creation through strategic development (Gopal & Clarke 2015). Following the AI-powered decision-making framework, the chapter draws out a system mapping to enhance service productivity that explains how stakeholders can apply the principles of systems design following visual mapping.

The system map (as shown in Figure 3) acts as a visualization tool that describes the activities of actors and explains the sequences of AI-powered decision-making in connected healthcare.

Figure 3. System Mapping for AI-powered Innovative Decision-Making

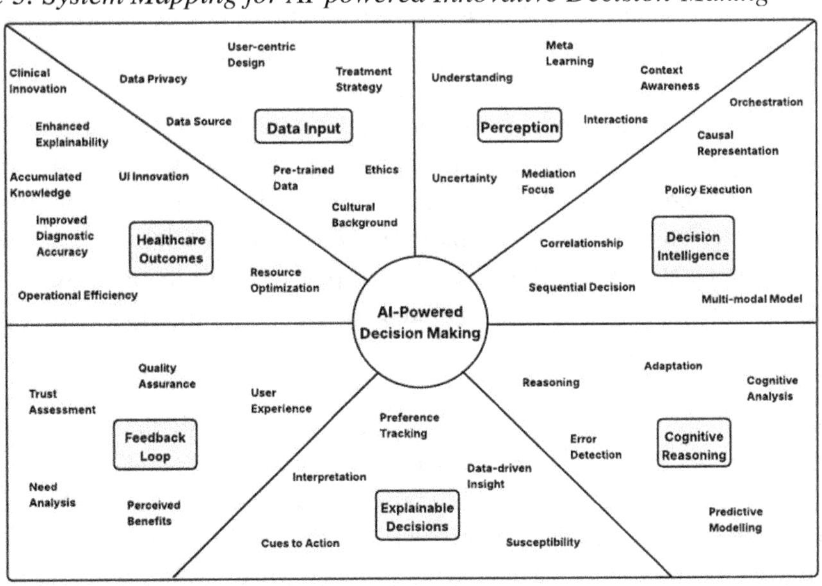

5. SIGNIFICANCE AND INNOVATION

Even though existing studies have demonstrated great potential for AI in detecting disease, critical gaps remain between AI-powered innovative decision making towards performance and clinical usability. As discussed in the previous sections, many high-performing models provide little insight into how decisions are made. This proposal is significant in directly addressing that challenge by combining a strong predictive model with an explainability and user feedback loop that ultimately directs healthcare outcomes.

The research into AI-powered medical decision support systems stands as a beacon of significance and innovation, illuminating new pathways in healthcare delivery. This section addresses the significance and innovation of the research problem, identifies its beneficiaries, and outlines the innovative dimensions it adds to the healthcare landscape while providing a strong justification for the significance and innovativeness of the research by drawing upon context as its basis for exploring AI's far-reaching implications in healthcare.

The chapter offers clear benefits to multiple stakeholders in the healthcare industry. Clinicians will gain a tool that not only flags early signs of cognitive decline, but also explains its reasoning in an understandable way, which can improve diagnostic confidence, and minimizes the black-box nature of AI models discussed in the earlier sections. Patients and health professionals can benefit from less invasive and earlier detection that helps prompt intervention, differing from traditional methods. Furthermore, AI researchers will gain a replicable approach to combining AI in a healthcare context, while technology and digital health organizations may adapt these findings towards developing advanced screening tools.

The innovation of this chapter lies not only in its technical contribution, but also in its emphasis on usability, transparency, and ethical deployment. One key innovation is the integration of explainability directly into the framework development process, which moves beyond traditional benchmarks to ensure that the AI's reasoning is interpretable. Another distinctive aspect lies in the inclusion of clinician feedback, which introduces a user-centered evaluation framework that enhances trust and practical value.

This chapter also introduces the use of spontaneous speech as a measurable, non-invasive digital biomarker, which makes it more suitable for widespread screening purposes. Moreover, by using open-access datasets and aiming for low-cost deployment, this chapter supports scalable and equitable diagnostic solutions in connected healthcare. These elements all together make the proposal a socially impactful and technologically innovative approach to AI adoption in taking innovative decision making.

The framework is proposed for a healthcare domain-specific context. The chapter has identified the most important parameters, metrics, and components for AI that can influence innovative decision making in connected healthcare. This creates the opportunity to create domain-specific mitigation strategies in the healthcare industry.

The significance of this chapter extends beyond academic contributions to direct impacts on healthcare practices. As highlighted in the model section, AI may lead to supporting health professionals in delivering treatments, while AI-driven healthcare systems can make critical innovative decision making, ensure patient safety, and potentially secure public trust in healthcare systems. The proposed framework addresses these key issues by providing a set of evaluation criteria enabling healthcare AI developers and clinical health professionals to select the most appropriate mitigation techniques for their specific contexts.

The chapter lies in its innovative decision making not only for individual implementation techniques, but also in how these techniques can interact and apply to healthcare-specific challenges. Through this integrated approach, the framework can provide a foundation for developing safer, fairer healthcare AI systems that improve patient outcomes and enhance the quality of clinical, innovative decision-making.

As AI applications in healthcare continue to expand, the importance of taking innovative decision-making demands to ensure robust and connected healthcare. We believe the framework proposed in this chapter can serve as a standard model for healthcare professionals and AI practitioners, ensuring both safety and performance to drive an equitable, fair, and accurate system for future healthcare.

AI-powered medical decision support systems' transformative potential lies in taking innovative decision making, and they have ushered in an entirely new era of healthcare knowledge, technical capability, and clinical practices that is reflected in this research's significance. As such, the significance of this chapter explains the enhancement of Clinical decision-making that can Improve diagnostic accuracy by analyzing vast amounts of medical data through advanced AI algorithms, and these AI systems improve diagnostic precision and reduce human error.

6. LIMITATIONS AND RECOMMENDATIONS

There are specific challenges and limitations of AI-based approaches. The most important one is about ethical treatment and data privacy that is related to accessing personal and sensitive patient data. Patients and their guardiancies should be fully informed, and health professionals should get prior consent to deliver AI-driven assessment. Further, reliability and accuracy issues of AI algorithms are also critical aspects that mainly depend on quality of data. For instance, inaccurate data may deliver hallucinated information if input training data is incomplete or somehow biased. There may have been an incorrect diagnosis by AI and machine learning algorithms since mental health disorders are complex. Further recommendations are essential to enhance training and validation of AI algorithms with different datasets, implement accountable AI systems, and integrate human supervision with AI-based health-related interventions.

This chapter proposes comprehensive recommendations that can assist healthcare organizations, medical professionals, and policymakers in capitalizing on the potential of AI while mitigating associated challenges. These actionable insights can guide all stakeholders to accommodate AI technology in making well-informed innovative decisions.

1. Regular monitoring and evaluation of AI System Performance: Healthcare organizations shall implement measures for continuous assessment of AI system performance in clinical settings to get innovative decision making. This requires regularly tracking diagnostic accuracy, system reliability, and user satisfaction while recognizing emerging capabilities. Regular performance monitoring and validation can facilitate these evaluations, enabling healthcare providers to maintain optimal system performance.
2. Careful Integration of AI Models: Before implementing AI systems, organizations shall conduct careful evaluation of various explainable, predictive, and federated AI approaches. This can involve analyzing the effectiveness of different models, assessing their impact on user trust, and understanding their integration requirements. Healthcare organizations can select appropriate explanation methods to assess the system requirements that can enhance clinical innovative decision-making while maintaining patient and health professional confidence.
3. Optimizing AI Systems for Clinical Workflows: This chapter highlights the transformative potential of AI in improving clinical efficiency by enhancing innovative decision making. We recommend healthcare organizations to explore an efficient way that can customize AI solutions into their specific clinical requirements, developing tailored interfaces, preferences, and workflows to enhance existing clinical practices and therapeutic interventions.
4. Robust Security and Compliance Protocols: Security and compliance considerations are vital for AI-driven innovative decision making in a connected healthcare system. Organizations shall maintain comprehensive security protocols, data protection, access control, and regular audit trails that can make an innovative decision-making process while ensuring compliance with healthcare regulations. They can also execute continuous monitoring to maintain high security standards and regulatory compliance.

5. Specialized Clinical Implementation Strategies: As AI applications vary across medical specialties, we advise organizations to develop implementation strategies tailored to specific clinical domains that can enhance their decision making. This involves identifying particular use cases and challenges within each special area, and crafting adoption plans accordingly. By aligning AI capabilities with clinical requirements, healthcare organizations can optimize the AI's potential for improving patient care.
6. Continuous Professional Development: As AI technology rapidly evolves, we recommend organizing training programs regularly for healthcare professionals. Clinical staff shall have to develop their adequate knowledge and expertise on AI capabilities, explanation methods, and best practices through structured training programs and regular updates. A well-trained clinical workforce is better positioned to effectively utilize AI tools while maintaining high standards of patient care.
7. Develop Ethical Governance Frameworks: Establish standardized protocols for privacy-preserving techniques (e.g., differential privacy, homomorphic encryption) to align with GDPR and HIPAA requirements. Institutional review boards should audit fairness of metrics, such as demographic parity and equal opportunity, to prevent algorithmic bias.
8. Real-time Support: Enables immediate decision support during critical clinical scenarios, improving response times and patient outcomes through instantaneous analysis of medical images and patient data.
9. Pattern Recognition: Advanced capability to identify subtle patterns and correlations in patient data that might be missed by human observation, particularly in complex medical imaging scenarios.

7. CONCLUSION

This chapter offers insight into the community into how AI-driven systems can accommodate innovative decision making into a connected healthcare that provides effective support to patients, offering on-demand assistance.

All in all, this research demonstrates the massive potential that exists for AI to improve healthcare efficiency and physician training, etc. By integrating AI technologies such as machine learning, virtual reality simulation, etc., it is possible to significantly optimize the healthcare process, improve patient management, and improve treatment processes to enhance operational efficiency. However, the study also highlights the ethical, operational, and technical challenges that must be addressed in order for AI to be widely used in the healthcare sector, and therefore recommends that healthcare organizations adopt pilot projects, enhance AI knowledge training for healthcare professionals, and develop personalized AI learning plans for relevant personnel to balance AI-assisted learning with human supervision.

The proposed framework may assist clinicians and other health professionals to reduce the burden on correct data requirements and inefficient technology in taking innovative decision making in connected healthcare. Apart from detection and diagnosis of diseases, AI can offer remote digital therapeutic treatments through advanced data analytics that can enhance decision making. All these benefits contribute to society's ability to deliver connected healthcare and promote awareness of health conditions.

The proposed framework can be used for more quantitative validation methods by using more real-world case studies or creating contextual simulations. This could help to further identify or validate the effectiveness of specific mitigation techniques. Future research work can also assess how AI-driven innovative decision making can impact overall healthcare systems. As human oversight is essential for decision-making in healthcare, understanding how AI input can cause decision making will be vital for the success of AI implementation in healthcare facilities.

REFERENCES

ACS Code of Professional Ethics. (2023). Source: https://www.acs.org.au/memberships/professional-ethics-conduct-and-complaints.html

Alimadadi, A., Aryal, S., Manandhar, I., Munroe, P. B., Joe, B., & Cheng, X. (2020). Artificial Intelligence and Machine Learning to Fight COVID-19. *Physiological Genomics, 52*(4), 200–202. DOI: 10.1152/physiolgenomics.00029.2020 PMID: 32216577

Arefeen, M. A., Debnath, B., & Chakradhar, S. (2024). Lean Context: Cost-efficient domain-specific question answering using LLMs. *Proceedings of the 2024 Conference on Empirical Methods in Natural Language Processing*, 2301-2312.

Barnes, R., & Zvarikova, K. (2021). Artificial Intelligence-enabled Wearable Medical Devices, Clinical and Diagnostic Decision Support Systems, and Internet of Things-based Healthcare Applications in COVID-19 Prevention, Screening, and Treatment. *American Journal of Medical Research (New York, N.Y.), 8*(2), 9–22. DOI: 10.22381/ajmr8220211

Batko, K., & Ślęzak, A. (2022). The use of Big Data Analytics in healthcare. *Journal of Big Data, 9*(1), 3–3. DOI: 10.1186/s40537-021-00553-4 PMID: 35013701

Blanco, F. (2017). Cognitive bias. In Vonk, J., & Shackelford, T. (Eds.), *Encyclopedia of animal cognition and behavior*. Springer., DOI: 10.1007/978-3-319-47829-6_1244-1

Busemeyer, J. R., & Townsend, J. T. (1993). Decision field theory: A dynamic-cognitive approach to decision making in an uncertain environment. *Psychological Review, 100*(3), 432–459. DOI: 10.1037/0033-295X.100.3.432 PMID: 8356185

Calvo, R. A., Milne, D. N., Hussain, M. S., & Christensen, H. (2020). Natural language processing in mental health applications using non-clinical texts. *Natural Language Engineering, 23*(5), 649–685. DOI: 10.1017/S1351324916000383

Chancellor, S., & De Choudhury, M. (2020). Methods in predictive diagnostics for mental health on social media: A survey. In *Proceedings of the 26th ACM SIGKDD International Conference on Knowledge Discovery & Data Mining* (pp. 3473-3474). ACM.

Chen, R. J., Wang, J. J., Williamson, D. F. K., Chen, T. Y., Lipkova, J., Lu, M. Y., Sahai, S., & Mahmood, F. (2023). Algorithmic fairness in artificial intelligence for medicine and healthcare. *Nature Biomedical Engineering, 7*(6), 719–742. DOI: 10.1038/s41551-023-01056-8 PMID: 37380750

Dror, I. E. (2020). Cognitive and human factors in expert decision making: Six fallacies and the eight sources of bias. *Analytical Chemistry, 92*(12), 7998–8004. DOI: 10.1021/acs.analchem.0c00704 PMID: 32508089

European Commission. (2021). *Artificial intelligence*. European Commission – Shaping Europe's Digital Future. https://digital-strategy.ec.europa.eu/en/policies/artificial-intelligence

Gilanie, G., Hassan, M. U., Asghar, M., Qamar, A. M., Ullah, H., Khan, R. U., Aslam, N., & Khan, I. U. (2022). An automated and real-time approach of depression detection from facial micro-expressions. *Computers, Materials & Continua, 73*(2), 2514–2530. DOI: 10.32604/cmc.2022.028229

Gopal, S., & Clarke, T. (2015). *System mapping: A guide to developing actor maps.* FSG.

Hancock, T. O., Hess, S., Choudhury, C. F., & Tsoleridis, P. (2024). Decision field theory: An extension for real-world settings. *Journal of Choice Modelling, 52*, 100495. DOI: 10.1016/j.jocm.2024.100495

Hofstede, G. (2011). Dimensionalizing Cultures: The Hofstede Model in Context. *Online Readings in Psychology and Culture, 2*(1). Advance online publication. DOI: 10.9707/2307-0919.1014

Holzinger, A., Goebel, R., Fong, R., Moon, T., Müller, K. R., & Samek, W. (2020). xxAI - Beyond Explainable Artificial Intelligence. In: Holzinger, A., Goebel, R., Fong, R., Moon, T., Müller, KR., Samek, W. (eds) xxAI - Beyond Explainable AI. xxAI 2020. Lecture Notes in Computer Science, vol 13200. Springer, Cham. DOI: 10.1007/978-3-031-04083-2_1

Islam, M. M., Hassan, S., Akter, S., Jibon, F. A., & Sahidullah, M. (2024). A comprehensive review of predictive analytics models for mental illness using machine learning algorithms. *Healthcare Analytics, 6*, 100350. DOI: 10.1016/j.health.2024.100350

Jamieson, G. A., & Skraaning, G.Jr. (2017). Levels of Automation in Human Factors Models for Automation Design: Why We Might Consider Throwing the Baby Out with the Bathwater. *Journal of Cognitive Engineering and Decision Making*, 1555343417732856.

Jarrahi, M. H. (2018). Artificial intelligence and the future of work: Human-AI symbiosis in organizational decision making. *Business Horizons, 61*(4), 577–586. DOI: 10.1016/j.bushor.2018.03.007

Kalra, N., Verma, P., & Verma, S. (2024). Advancements in AI based healthcare techniques with FOCUS ON diagnostic techniques. *Computers in Biology and Medicine*, *179*, 108917–108917. DOI: 10.1016/j.compbiomed.2024.108917 PMID: 39059212

Leidner, D. E., & Kayworth, T. (2006). Review: A Review of Culture in Information Systems Research: Toward a Theory of Information Technology Culture Conflict. *Management Information Systems Quarterly*, *30*(2), 357–399. DOI: 10.2307/25148735

Lu, Z., Li, X., Cai, D., Yi, R., Liu, F., Zhang, X., Lane, N. D., & Xu, M. (2024). Small language models: Survey, measurements, and insights. ArXiv (Cornell University). https://doi.org//arxiv.2409.15790.DOI: 10.48550

Maier-Hein, L., Eisenmann, M., Sarikaya, D., März, K., Collins, T., Malpani, A., Fallert, J., Feussner, H., Giannarou, S., Mascagni, P., Nakawala, H., Park, A., Pugh, C., Stoyanov, D., Vedula, S. S., Cleary, K., Fichtinger, G., Forestier, G., Gibaud, B., & Speidel, S. (2022). Surgical data science - from concepts toward clinical translation. *Medical Image Analysis*, *76*, 102306. DOI: 10.1016/j.media.2021.102306 PMID: 34879287

Melnykova, N., Shakhovska, N., ml, M. G., & Melnykov, V. (2019). Using big data for formalization of the patient's personalized data. *Procedia Computer Science*, *155*, 624–629. DOI: 10.1016/j.procs.2019.08.088

Mikhailov, V. (2024). Revolutionizing Healthcare: The Use of Artificial Intelligence in Clinical Care. *International Journal of Advanced Engineering Technologies and Innovations*, *10*(2), 718–735. https://ijaeti.com/index.php/Journal/article/view/528

Mohri, M., Sivek, G., & Suresh, A. T. (2019, May). Agnostic federated learning. In *the International Conference on Machine Learning* (pp. 4615-4625). PMLR.

Nardi, B. A. (1996). Activity theory and human-computer interaction. *Context and consciousness: Activity theory and human-computer interaction, 436*(1), 7-16.

Olawade, S. A., Zheng, Y., & Samuel, M. A. (2024). Mental health detection and treatment using artificial intelligence: A review of techniques and applications. *Journal of Advanced Health Informatics*, *14*(2), 120–134.

Rahman, A., Debnath, T., Kundu, D., Islam, S., Airin Afroj Aishi, Sadia Sazzad, Sayduzzaman, M., & Band, S. S. (2024). Machine learning and deep learning-based approach in smart healthcare: Recent advances, applications, challenges and opportunities. 11(1), 58–109. .DOI: 10.3934/publichealth.2024004

Rajendran, S., Pan, W., Sabuncu, M. R., Chen, Y., Zhou, J., & Wang, F. (2024). Learning across diverse biomedical data modalities and cohorts: Challenges and opportunities for innovation. *Patterns (New York, N.Y.)*, *5*(2), 100913. Advance online publication. DOI: 10.1016/j.patter.2023.100913 PMID: 38370129

Report, F. (2025). AI in Healthcare Market Size & Global Trends. https://www.fortunebusinessinsights.com/industry-reports/artificial-intelligence-in-healthcare-market-100534?utm_source=chatgpt.com

Ristevski, B., & Chen, M. (2018). Big data analytics in medicine and healthcare. *Journal of Integrative Bioinformatics*, *2017*(0030), 1–5. DOI: 10.1515/jib-2017-0030 PMID: 29746254

Ryffel, T., Trask, A., Dahl, M., Wagner, B., Mancuso, J., Rueckert, D., & Passerat-Palmbach, J. (2018). A generic framework for privacy preserving deep learning. arXiv preprint arXiv:1811.04017.

Vygotsky, L. S., & Cole, M. (1978). *Mind in Society: Development of Higher Psychological Processes*. Harvard University Press.

Wei, J., Wang, X., Schuurmans, D., Bosma, M., Ichter, B., Xia, F., Chi, E., Le, Q., & Zhou, D. (2022). Chain-of-thought prompting elicits reasoning in large language models. *Advances in Neural Information Processing Systems*, *35*, 24824–24837.

Yang, X., Chen, A., PourNejatian, N., Shin, H. C., Smith, K. E., Parisien, C., Compas, C., Martin, C., Costa, A. B., & Flores, M. G. (2023). A large language model for electronic health records. npj. *Digital Medicine*, *5*(1).

Zhou, X., Jin, K., Shang, Y., & Guo, G. (2020). Visually interpretable representation learning for depression recognition from facial images. *IEEE Transactions on Affective Computing*, *11*(3), 542–555. DOI: 10.1109/TAFFC.2018.2828819

KEY TERMS AND DEFINITIONS

Innovative Decision Making: This is a cognitive state of humans and machines to innovate and prioritize decision making for complex adaptive systems. In healthcare, gather information from all possible areas and gain approval in advance to speed up the process, reduce resistance to change, and avoid personal responsibility for worst decisions.

AI-Driven Healthcare System: AI-driven healthcare is a robust system that can enhance diagnostic reasoning capabilities and improve predictive analytics for patient status, enabling more accurate and timely medical decisions while reducing computation complexities.

Artificial intelligence Models: The development of AI models in healthcare systems has introduced many transformative opportunities that enhance innovative decision making. AI models include explainable and predictive methods such as Large Language Models (LLMs), Explainable AI (XAI). Chain of Thought (CoT), Federated Learning (FL) models, Small Language Models (SLM),

Explainability and Trust: Explainability and trust concerns play a critical role in assessing user requirements and confidence in connected healthcare, which prioritizes innovative decision making. It aims to gain a thorough understanding of AI implementation challenges and solutions, while helping healthcare organizations navigate the complex terrain of AI adoption safely while maintaining clinical trust and regulatory compliance.

Chapter 2
Next-Generation Decision-Making for Human and Machine Intelligence

Javaria Ajoon
https://orcid.org/0009-0003-7543-1876
KPESE, Pakistan

Qazi Mudassar Ilyas
https://orcid.org/0000-0003-4238-8093
King Faisal University, Saudi Arabia

ABSTRACT

Decision-making remains a defining challenge in both engineering and organizational domains. Human choices, while enriched by intuition and experience, are often undermined by bias, stress, and inconsistency. Machines, in contrast, deliver speed and analytical precision but struggle with context, ethics, and adaptability. This chapter examines how the decision-making landscape has shifted from human reliance to algorithmic models and now toward integrated human–machine systems. By outlining the cognitive processes behind human judgment, the computational methods that guide machine reasoning, and the design of collaborative frameworks, it highlights how hybrid approaches can achieve more resilient and transparent outcomes. Practical applications, ethical concerns, and emerging challenges are discussed, offering engineers and practitioners a roadmap for developing systems that balance efficiency with human values.

DOI: 10.4018/979-8-3373-5702-7.ch002

1. INTRODUCTION

Decision-making involves identifying available options, comparing them, assessing outcomes, and taking actions to achieve goals in a given situation. Optimal decision-making is crucial for problem-solving, resource management, and handling uncertainty. Human beings usually rely on experience, intuition, and heuristics for effective decision-making. Although these shortcuts help individuals make effective decisions, pioneering work by Tversky and Kahneman in the 1970s showed how intuition and heuristics can also bias our decision-making (Kahneman, 1982; Tversky & Kahneman, 1973). They introduced two heuristics, availability and anchoring, and demonstrated how they can bias our decision-making process. As shown in Figure 1, availability refers to similar situations or examples related to the current decision situation. As individuals are usually influenced more by recent, vivid, or dramatic events, they are prone to skipping a detailed similarity analysis demanded by the current situation, and their decisions are skewed by such events, possibly resulting in incorrect or suboptimal decisions. Anchoring refers to an arbitrary number in the mind of a decision maker from their environment, that distorts their subsequent judgements, as this anchor (arbitrary number) is usually irrelevant in the current decision situation. This brief discussion shows how intuition and heuristics can result in suboptimal decisions in situations that require more in-depth and objective analysis.

Figure 1. Cognitive heuristics and decision bias

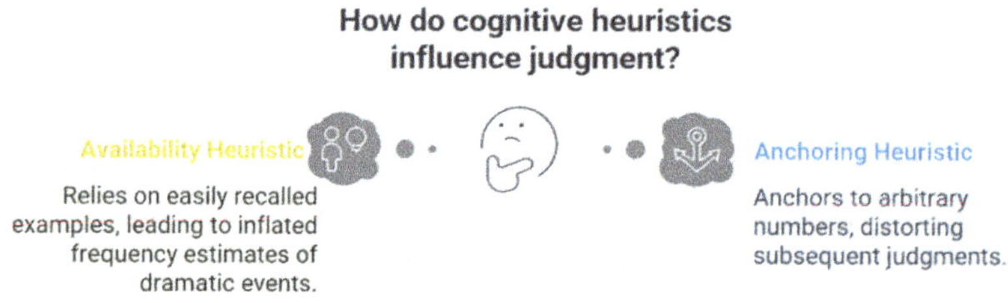

The recent advancements in machine intelligence and large-scale data processing have opened new doors for decision-making based on logical reasoning, statistical inference, and optimization. These techniques are highly scalable and efficient compared to human capabilities. However, they lack context-awareness and explainability required for high-stakes situations. These limitations can lead to critical errors if the decisions are made without required human oversight.

Hybrid decision-making techniques and systems can merge the two decision paradigms mentioned above i.e human and machine, to exploit their strengths while overcoming their limitations. These techniques combine the flexibility, intuition, experience, and heuristics of human beings with the speed, scale, and precision of machines for optimal decision making. This synergic effect is especially useful in engineering domains, such as control systems, safety-critical systems, and project planning under uncertain conditions.

In recent years, several researchers have studied and proposed hybrid decision-making techniques based on the integration of human cognitive capabilities with machine architectures. Wilson et al. identify human-enabled data representations, shared goals, and distributed responsibility of agents as key enablers of such systems (Wilson et al., 2025). Rooy introduces the concept of Type 1 and Type 2 thinking, where AI systems augment human cognitive strengths in the decision-making process (Rooy, 2024). The author studied the governance-level human-machine decision systems and stressed the need for a balanced integration of the two components.

These new trends suggest a strategic shift towards a next-generation decision-making paradigm, characterized by:

1. **Mutual adaptation**: systems that adapt to human behavior and vice versa (Nikolaidis et al., 2017)
2. **Cognitive augmentation**: decision-support systems that augment human reasoning with facts and figures, mitigate bias, and help overcome pitfalls in heuristics, such as availability and anchoring (Miller & Dorneich, 2006).
3. **Explainable decisions**: hybrid decision-making systems that can explain their outcomes, thus improving user confidence and trust (Malamuthu et al., 2025a).

This chapter explores how next-generation hybrid systems will transform the decision-making process through the effective collaboration of humans and machines. The objectives of the chapter are as follows:

- Understand how humans use mental shortcuts, intuition, and past experience to make choices.
- Highlight how human factors, such as emotions, feelings, stress, fatigue, and bias, hamper their decision-making abilities.
- Develop an understanding of how machines make decisions through algorithms, data processing, and computational methods.
- Study the evolution of decision-making with the integration of human and machine intelligence.
- Discuss how machines can be designed to overcome human limitations for improved decision making.
- Present a clear structure to educate engineers and practitioners on building better hybrid decision-making systems.

- Provide real-world examples of hybrid decision making in engineering fields, like aviation, robotics, and healthcare.
- Encourage human-machine collaboration to solve complex engineering problems.

This chapter will provide a clear understanding of cognitive science and automated decision-making processes. Our goal is to provide a clear roadmap to engineering professionals for combining both forms of intelligence for improved decision-making.

2. BACKGROUND

2.1. Bounded Rationality, Heuristics, and the Cognitive Miser

Herbert Simon proposed the theory of bounded rationality, which implies that, because of limited information and cognitive abilities, human beings often tend to satisfice rather than optimize their decision (Raab & Gigerenzer, 2015; Simon, 1997). These decision constraints were elaborated in the General Problem Solver framework proposed by Simon and Newell (Tversky & Kahneman, 1974). The decision-makers often conserve their cognitive effort through various heuristics, such as availability, anchoring, and representativeness. However, as stated above, these heuristics introduce errors in the decision-making process. This behavior is also reflected in the "cognitive miser" framework proposed by Fiske and Taylor (Fiske & Taylor, 1991). Contrary to the popular opinion of heuristics being an erroneous decision-making model, the ecological rationality model proposed by Gigerenzer and Gaissmaier suggests that in many real-life situations, heuristics are a better choice compared to more complex and expensive decision strategies (Gigerenzer & Gaissmaier, 2011).

Figure 2. Foundations of decision-making

2.2. Noise, Bias, and Emotional Influences in Human Judgment

Gilovich et al. and Kahneman et al. differentiate between two key factors influencing human judgment, bias and noise (Gilovich et al., 2013; Kahneman et al., 2022). While bias is caused by systematic errors in judgment, noise represents random and unpredictable variability across decisions. For instance, Dlugos et al.'s analysis showed that 50% of clinicians' decisions exhibited pattern noise (Dlugos et al., 2025).

Human choices and judgements are also influenced significantly by emotional biases. Finucane et al. introduced the affect heuristic, which shows how mood influences risk and benefit assessments (Finucane et al., 2000). Similarly, the somatic marker hypothesis proposed by Damasio et al. describes the role of bodily feedback on human decision-making (Damasio et al., 1997). A research study by Englich and Soder suggests negative mood enhances anchoring bias in individuals (Englich & Soder, 2009). Another study by Bodenhausen et al. shows individuals are prone to pay more attention to pessimistic information because of negativity bias (Bodenhausen et al., 2000). These studies prove how human emotions not only bias their judgment but also amplify decision inconsistency.

2.3. Stress, Fatigue, and Decision-Making Under Strain

Stress, fatigue, and strain significantly affect human decision-making prowess, resulting in suboptimal decisions in critical situations. Porcelli and Delgado argue that, under acute stress, human beings rely more on their intuition rather than analytical processing because of reduced prefrontal cortex activity (Porcelli & Delgado, 2017). Wemm and Wulfert also showed that stress reduces loss aversion and impairs executive tasks for most individuals (Wemm & Wulfert, 2017).

Fatigue, whether physical or mental, clouds individuals' judgments and leads to suboptimal decisions. For example, a study by Scott et al. reported that sleep-deprived and fatigued nurses later regretted their decision made in such conditions as they rated those clinical judgements poor (Scott et al., 2014). Similarly, Danziger et al. studied judicial proceedings and the behavior of judges under fatigue (Danziger et al., 2011). The study divided the judges' day into two sessions separated by food breaks. It concluded that the favorable decisions by judges gradually dropped from about 65% at the beginning of each session to zero by the end. Another interesting study by Vohs et al. found that making decisions also drains the energy of individuals, thus deteriorating the quality of decisions made late in the day after several decisions have been made earlier in the day (Vohs et al., 2008).

2.4. Cognitive Architecture and Human–Machine Interaction

The pioneering work by Newell et al. in cognitive science has inspired the development of cognitive architecture, which views the human problem-solving process as a series of logical steps guided by a rule-based system (Newell et al., 1972). Since then, researchers have been trying to replicate this process in machines (Langley et al., 2009).

The modern cognitive architectures, such as adaptive control of thought-rational (ACT-R) (Kim & Nam, 2020) and SOAR (Hanford et al., 2009), are based on learning, memory, and perception modules to simulate complex human behavior that cannot be achieved by a simple rule-based process (Anderson et al., 2004). The recent developments in the domain of mutual adaptation mechanisms, which strive to achieve a stronger coordination between humans and machines, have resulted in significant progress in tasks such as robot motion planning (Dragan et al., 2013). Predictive models take this concept further, enabling machines to predict human intentions and errors before they occur and adjust their response accordingly. For instance, Griffiths et al. proposed a resource-rational analysis framework based on predictive models to design AI systems that account for bounded rationality and adapt their responses accordingly (Griffiths et al., 2015).

These recent developments are enabling new cognitive architectures to improve human-machine coordination, enabling the development of intelligent systems that are not reactive but anticipatory and adaptive. These contemporary architectural designs find several applications in the complex domains of manufacturing, autonomous vehicles, and assistive technologies.

2.5. Machine Decision-Making: Algorithms and Explainability

Contemporary machine decision systems employ data-driven models, such as neural networks and probabilistic classifiers, to achieve their goals. However, these are black box solutions whose outputs cannot be easily explained to human beings. This reduces the transparency and trustworthiness of such systems. Explainable artificial intelligence (XAI) research endeavors to explain the machine's decision to human beings. Arrieta et al. present an excellent survey of recent XAI opportunities and challenges (Arrieta et al., 2019). To this end, Salih et al. discuss strengths and weaknesses of two widely used methods, SHapley Additive exPlanations (SHAP) and local interpretable model agnostic explanation (LIME) (Salih et al., 2025).

Researchers are experimenting with hybrid models. Tahir et al. developed a hybrid XAI model by integrating SHAP and LIME for autonomous vehicles (Tahir et al., 2024). The study achieved a high explainability without sacrificing the performance of the systems. Similarly, Malamuthu et al. developed a hybrid deep learning and rule-based machine decision system to achieve improved explainability without any performance degradation (Malamuthu et al., 2025b).

A comprehensive overview of theories, limitations, and system designs discussed above is given in Table 1.

Table 1. Summary of foundational theories, human decision-making limitations, machine augmentation strategies, and representative academic references

Limitation / Factor	Effect on Decision	Mechanism	Mitigation Approach	Representative Studies
Bounded rationality	Satisficing, suboptimal choices	Cognitive constraints, limited optimization	Decision satisficing, structured options	(Simon, 1997)
Heuristics & biases	Systematic deviations from normative reasoning	Mental shortcuts (e.g., anchoring)	Heuristic awareness, environment-fit design	(Tversky & Kahneman, 1974)
Ecological rationality	Effective simple rule application	Adaptive use of heuristics	Contextual heuristic deployment	(Gigerenzer & Gaissmaier, 2011)
Affect heuristic	Skewed risk-benefit evaluations	Emotion-based judgment	Emotional regulation reminders	(Finucane et al., 2000)
Somatic marker hypothesis	Emotion-guided decision making	Physiological (somatic) feedback	Biofeedback-informed support systems	(Damasio et al., 1997)
Mood-based anchoring	Amplified anchoring under negative mood	Mood-influenced heuristic bias	Mood-aware prompts, re-anchoring strategies	(Englich & Soder, 2009)
Negativity bias	Overweighting negative outcomes	Attentional bias toward negative info	Balanced framing techniques	(Bodenhausen et al., 2000)
Acute stress	Reliance on intuition, impaired planning	Prefrontal cortex suppression	Stress-aware pacing, decision automation	(Porcelli & Delgado, 2017)
Reduced risk sensitivity	Lower loss aversion under pressure	Dampened affective response	Alert thresholds, risk visualization	(Wemm & Wulfert, 2017)
Impaired executive control	Poor task performance under stress	Prefrontal impairment	Task redistribution, real-time coaching	(Vohs et al., 2008)
Fatigue-induced noise	Decision inconsistency and variability	Neural resource depletion	Mandatory breaks, adaptive deadlines	(Scott et al., 2014)
Inconsistent risk preferences	Erratic risk-reward choices	Cognitive depletion	Consistency checks, decision support	(Porcelli & Delgado, 2017)
Decision fatigue	Impulsivity and poor self-control	Ego-depletion from repeated choices	Choice architecture, decision hygiene	(Danziger et al., 2011)
Burnout	Emotional exhaustion, impaired judgment	Chronic stress and resource depletion	Monitoring, role rotation	(Michailidis & Banks, 2016)
Cognitive architectures	Modeled human reasoning and adaptation	Forward/inverse decision modeling	Interactive agent frameworks	(Langley et al., 2009)

continued on following page

Table 1. Continued

Limitation / Factor	Effect on Decision	Mechanism	Mitigation Approach	Representative Studies
Cognitive augmentation	Support under stress via UI adaptations	Adaptive interfaces	Stress/fatigue detection, system prompts	(Griffiths et al., 2015)
Explainable AI	Improved trust and comprehension	Transparent model output	SHAP/LIME, hybrid explanation tools	(Arrieta et al., 2019)
Hybrid decision frameworks	Enhanced joint human–machine decision outcomes	Mixed-initiative control & feedback loops	Shared authority, transparency systems	(Tahir et al., 2024)

3. HUMAN INTELLIGENCE IN DECISION-MAKING

After discussing the foundations of human cognition, decision-making, and the associated challenges, this section explores how these theories are applied to solve engineering problems and make optimal choices in complex environments. We discuss the importance of tacit knowledge and its application in engineering judgment. The role of cognitive diversity and collaborative decision-making in teams is also elaborated. This section also explores the complex cognitive phenomena, such as metacognition, intuition vs analysis in creative problem solving. Finally, we highlight the ethical challenges in decision-making as they have a bearing on society and individuals.

3.1. Tacit Knowledge and Human Judgement

Although the engineering solutions are usually based on exact formulae and scientific laws, engineering judgement also relies on tacit knowledge of seasoned engineers. This kind of knowledge, which is acquired through experience, failure, and success case studies accumulated over a long period, cannot be articulated easily. However, these insights play a key role in the decision-making process of experienced engineers. Polanyi & Sen argue that individuals usually possess more knowledge than they can express (Polanyi & Sen, 2009). This is particularly true for complex engineering decisions, where seasoned professionals rely heavily on intuition and pattern matching in addition to the well-defined and codified body of knowledge. For instance, a veteran civil engineer might be able to identify any discrepancies in a structural design, and this judgment is not based on measurable data but accumulated perceptual and procedural knowledge. Engineering judgment is especially useful in cases with incomplete, ambiguous, or noisy data.

Figure 3. Human intelligence in decision-making

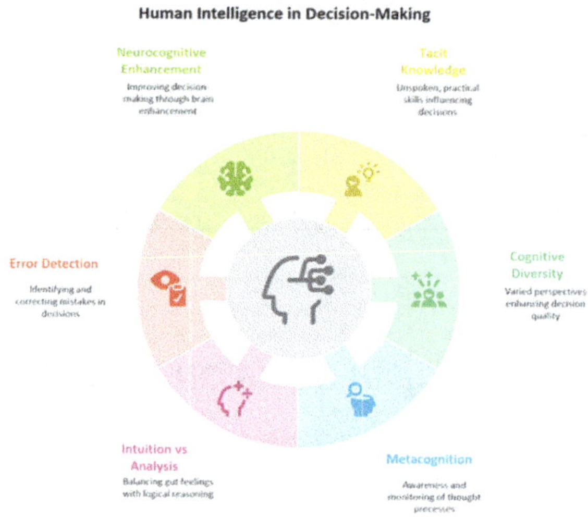

3. 2. Cognitive Diversity and Collaborative Decision-Making

Isolated decision-makers rarely suffice in moderate and complex engineering problems. As the theory of cognitive diversity suggests, teams composed of diverse engineers with different decision styles, cognitive abilities, disciplinary training, and heuristic styles make better decisions (Lamm et al., 2012). A study by Page showed that a group of diverse individuals usually outperforms like-minded experts if they can overcome communication barriers (Page, 2008). This diversity might be exhibited in the form of varying cognitive approaches, risk management behaviors, problem decomposition approaches, or cognitive biases, all of which help in achieving this synergic effect of optimal decision-making.

While collaborative decision-making generally improves the overall outcomes, it is crucial for experts to avoid some pitfalls. While diversity is encouraged, decision-makers need to agree on common goals, constraints, and decision criteria to avoid misalignment. Developing shared mental models in early phases of problem-solving helps develop this common understanding, crucial for encouraging constructive dissent to avoid conformity and groupthink (Malan, 2010).

3.3. Metacognition and Decision Monitoring

In addition to the basic attributes required for decision-making mentioned above, including knowledge, experience, intuition, and heuristics, another quality crucial to effective decision-making is metacognition. Metacognition refers to one's ability to think about and reflect on one's own decisions. Metacognitive strategies allow engineers to assess their own judgements and identify situations when they are over-confident, when they need more information, or when they should defer a decision.

Bastian and Zucchella argue that metacognition significantly improves decision-making by reducing overconfidence and hindsight bias of engineers (Bastian & Zucchella, 2022). The study suggests several strategies and tools to improve metacognition, such as checklists, reflective prompts, and design retrospectives. In high-stakes engineering decisions, such as deciding whether to shut down a nuclear reactor, abort a rocket launch, or approve a skyscraper design, the ability to pause and reflect is usually the difference between success and disaster.

Another related decision-making technique is decision latency, which involves deliberately delaying a decision till more information becomes available. Decision latency is also helpful in emotion regulation to make unbiased decisions devoid of undue stress on decision-makers. These attributes are usually undervalued in today's fast-paced engineering projects, but they are crucial to making stable decisions free of bias, overconfidence, or insufficient data.

3.4. Intuition vs Analysis in Creative Problem Solving

Although well-structured and straightforward engineering problems can often be solved by procedural logic and heuristics, complex and ill-structured engineering challenges require intuitive thinking and out-of-the-box solutions. Thus, creative synthesis and lateral thinking set exceptional engineers apart from the ordinary. Studies on innovation and engineering problem-solving show that such engineers move fluidly between analytical and creative modes, especially in early project phases (Pahl et al., 2007). Several case studies have shown that breakthroughs come from an intuitive-style problem-solving approach, rather than data analysis and procedural knowledge. For instance, engineers at Apple came up with the power button/switch placement using such an approach (Norman & Euchner, 2023). Similarly, Tesla reimagined vehicle architecture through out-of-the-box thinking (World Design Council, 2025).

3.5. Error Detection and Ethical Reasoning

Many engineering decisions involve crucial trade-offs that may have serious ramifications regarding the lives of individuals and communities. The ability to detect and correct errors during the design phase might be the difference between an engineering masterpiece and a disaster. This ability requires not only technical knowledge but also cognitive courage to reject poor design and develop safe and secure designs. The best engineering professionals not only possess the ability to identify issues in others' work but also review their own decisions critically and make necessary revisions.

Error detection depends on the project environment, feedback loops, and ethical reasoning. Studies on moral cognition suggest that engineers with sound moral values and ethical training are more likely to resist flawed directives and voice concerns about unsafe decisions (Bairaktarova & Woodcock, 2017). This calls for education programs that inculcate strong character in the engineers and mentoring avenues to produce socially responsible practitioners.

Moreover, ethical dilemmas often require non-binary decisions, balancing cost, performance, safety, and sustainability. Such complex decisions require high cognitive flexibility and ethical reasoning, which algorithms cannot be entrusted with.

3.6. Neurocognitive Enhancement to Decision-Making

The latest technologies in decision-making and decision-support, such as neurofeedback, brain-computer interfaces, and wearable cognitive monitors, are transforming the decision-making landscape in the engineering discipline, especially complex systems (Shmorgun & Lamas, 2014). These technologies can monitor the brain activity of decision makers in real time and detect intervals of fatigue, lack of concentration, and mental overload. These triggers are then used to optimize the presentation of information to the user. For instance, an electroencephalogram (EEG) headband can be used in control room operations to detect mental fatigue and adjust the data stream or suggest a rest period accordingly (Fan et al., 2015).

Such engineering advancements are giving rise to closed-loop engineering systems, where engineers' mental state is directly used to make optimal decisions. These systems can reduce the probability of suboptimal decisions by offloading highly demanding tasks, suggesting periods of breaks, or modifying decision priority in favor of simpler decisions. This emergent behavior signals a shift from traditional decision-making to a decision-augmentation paradigm where the boundary between the role of humans and machines in decision-making is blurred.

However, this integration of biological and artificial intelligence raises serious ethical concerns about privacy, accountability, and informed consent. For example, how is human decision performance data stored? Who has access to this data? If a decision is changed because of neurofeedback, is the human being still accountable for any damages caused by this decision?

Looking ahead, this fusion of human and machine intelligence in decision-making opens exciting opportunities for more stable and resilient engineering decisions. The ability of machines to process large amounts of data in real-time and reduce error rates, coupled with the cognitive and creative abilities of engineers, possesses tremendous potential for solving complex engineering problems and making excellent choices in such systems. Ultimately, machines should support human decision-making to enhance their capacity to handle complex situations, reduce stress, and uncertainty in high-stakes environments.

4. MACHINE INTELLIGENCE IN DECISION-MAKING

As technology continues to improve, intelligent systems are starting to handle more decision-making tasks, especially in fields like construction, supply chains, and areas where safety is a top concern. These were once jobs handled mainly by people. But humans can mess up, they get exhausted, emotional, or forget important details. Machines do not deal with those issues. They run on logic, data trends, and quick processing. They can scan through vast amounts of information without losing focus or slowing down. Because of that, machines are being used more often in situations that require rapid and accurate decisions. Still, there are challenges to face, like whether humans can rely on the outcome machines produce, how understandable those outcomes are, and what happens when the data they use is incomplete or wrong. This section covers the main systems and techniques machines use to make decisions, along with the real-life problems that come with letting technology take the lead.

4.1. Algorithmic Decision Models

Machines make choices by following detailed, step-by-step sets of instructions known as algorithms. These are structured methods based on rules, logic, or math, designed to solve specific problems. Algorithmic decision-making models work well in environments where inputs and the outcome are well defined and predictable. Techniques such as decision trees, linear programming, and rule-based models have become common tools for tackling complex problems, largely because they offer an organized framework for analysis (Russell & Norvig, 2021).

Among these approaches, linear programming is particularly notable, as it is widely applied in supply chain management to streamline operations and support informed decision-making. It helps in making decisions where cost, time, labor, and resources need to be balanced to ensure efficient delivery and production. Since the variables are quantifiable and the constraints are defined clearly, this technique proves to be highly effective.

Decision trees are another useful tool that simplifies complex issues by breaking them into a series of yes/no questions. This method moves step by step instead of analyzing everything at once, which makes the decision process logical and clear. For instance, in a modern irrigation system, some farmers use smart systems to decide when and where to water their crops. The system might ask itself questions like: Is the soil too damp? Has it rained recently? What time of day is it? What sort of plant is growing here? By asking a series of such questions, it can be determined if watering is needed at that moment. It saves water, avoids overwatering, and makes sure that each part of the field gets the necessary ingredients.

Rule-based systems, on the other hand, rely on predefined "if-then" conditions. These are often used in industrial settings for tasks like checking whether a machine follows operational rules or confirming that a construction design meets the required codes (Mitchell, 1997). Let us consider an example of a small factory that packs snacks into bags. A simple system is in place to check each pack before it moves forward. If a bag feels too light, it is simply removed. If the seal is broken or inappropriate, the system sends it to a separate tray. It does not try to learn or improve; rather, it sticks to the same predefined checks every time.

Despite all these factors, these models continue to have some limitations. When the data is unclear or the environment changes rapidly, algorithmic systems can struggle. They are not built to handle uncertainty or unexpected inputs. Another key disadvantage is that they do not learn on their own and follow the same rules until a human updates them. If the system needs to respond to something new, it will not adapt unless it is reprogrammed (Jarrahi, 2018).

4.2. Machine Learning and Pattern Recognition

Machine learning (ML) helps machines go beyond fixed rules. Instead of being told exactly what to do, they learn patterns from data. This is very useful in engineering environments, where real-time data from sensors, user behavior, or environmental changes must be considered (Mitchell, 1997).

There are three major types:

Supervised Learning: Systems learn from examples with known answers (e.g., predicting part failure based on past breakdowns).

Unsupervised Learning: The system finds hidden patterns without needing labeled data (e.g., identifying usage patterns in smart grids).

Reinforcement Learning: Machines learn by trying things out and learning from success or failure (used in robotics and adaptive traffic systems).

Deep learning, which is part of machine learning, uses artificial neural networks to handle complex tasks like image classification or speech recognition (LeCun et al., 2015). For example, drones can scan bridge surfaces and detect cracks using deep learning models trained on images of damaged infrastructure.

But these models are "black boxes." They make accurate predictions, but even developers may not understand exactly how they reached a decision. This becomes a problem in areas like healthcare or aviation, where transparency is critical.

Figure 4. Comparison of different decision models, contrasting algorithmic approaches, machine learning techniques, and rule-based systems

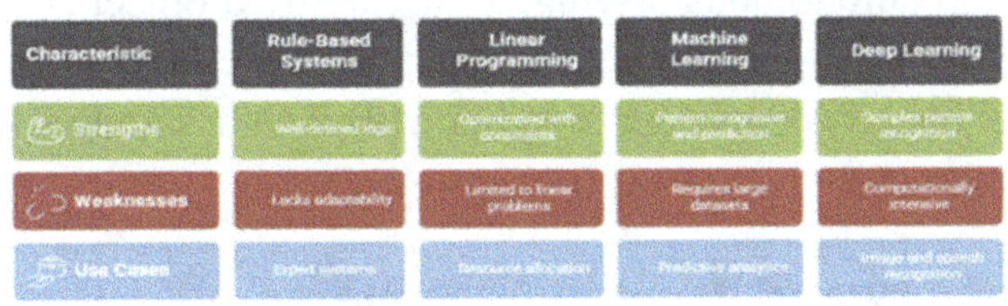

4.3. Decision Support Systems in Engineering

Decision Support Systems (DSS) assist engineers by combining their knowledge with computational tools to analyze data and support better decision-making, especially in complex or stressful situations. For instance, when paired with Building Information Modeling (BIM), DSS allows structural engineers to test different design options while considering various limitations. In manufacturing settings, these systems help by issuing safety warnings and improving machine efficiency. Similarly, smart grid DSS monitors the flow of energy and suggests ways to reduce consumption (Turban et al., 2011).

Modern decision support systems (DSS) increasingly use machine learning models that evolve and improve with experience. By drawing on feedback and past results, they refine their recommendations over time. This adaptability has led to their growing use in areas such as large-scale infrastructure planning, traffic management, and disaster risk assessment.

4.4. Limitations and Risks of AI-Based Decisions

While machine intelligence offers many benefits, it also comes with notable risks:

Bias in data: Machine models learn from historical data, which may reflect existing social or operational biases. For instance, biased hiring data can result in discriminatory recruitment decisions (Barocas et al., 2019).

Overfitting: A model trained too closely on past data may perform poorly when faced with new situations.

Automation bias: Users may accept AI recommendations without question, overlooking potential errors.

Systemic risk: In connected systems, a single erroneous decision can cascade into larger failures.

For example, in autonomous driving, a misclassified object could lead to fatal accidents. Hence, continuous monitoring and human oversight are critical. Redundancy checks, fallback systems, and simulation testing are standard safety practices in such environments.

4.5. Explainability and Trust in AI Systems

Trust plays a crucial role in how people engage with intelligent systems. For users to feel confident, they must be able to see the reasoning behind a system's decisions. This becomes particularly important in high-risk areas such as healthcare, construction, and defense, where the consequences of errors can be severe.

Explainable Artificial Intelligence (XAI) tools like SHAP (Shapley Additive Explanations) and LIME (Local Interpretable Model-Agnostic Explanations) are designed to make complex "black-box" models easier to understand. SHAP works by estimating the contribution of each input feature to the final decision, while LIME creates simplified, local models around individual predictions to explain how they were generated.

In engineering, such tools are often applied to identify faults in maintenance systems and to provide justification for modifications in automated designs. Tahir et al. introduced a hybrid framework that integrates SHAP and LIME, enabling real-time interpretability for autonomous systems (Tahir et al., 2024).

Another growing approach is the use of counterfactual explanations. For example, "If the temperature had been five degrees lower, the machine would not have shut down", which can be particularly valuable in training processes and audit activities (Wachter et al., 2017). Figure 5 compares these methods, highlighting their working principles, applications, and limitations.

Figure 5. Overview of decision explanation methods, comparing SHAP, LIME, and counterfactual approaches for improving transparency and interpretability in AI systems

Characteristic	SHAP	LIME	Counterfactual	Hybrid (SHAP+LIME)
What It Does	Explains feature importance	Builds interpretable local models	Offers "what-if" reasoning	Combines both strengths
Where It's Used	Finance, health, maintenance	Diagnostics, design audit	Safety reviews, legal compliance	Self-driving, robotics
Limitation	High computation for large models	Sensitive to input noise	It can be hard to generate for deep models	Complex to deploy

5. HUMAN–MACHINE COLLABORATIVE DECISION-MAKING

Hybrid decision-making rests on the recognition that humans and machines work best as partners with complementary strengths, rather than as competing problem-solvers. Humans offer insight, empathy, and ethical judgment, while machines contribute unwavering computational capacity, expansive memory, and the skill to uncover patterns within vast data streams. When thoughtfully integrated, these capabilities can lead to decisions that are faster, more precise, and more ethically sound than those either could produce on their own (Wilson et al., 2025)

5.1. Complementary Strengths of Humans and Machines

Humans have a unique capacity to make sense of unclear situations, pick up on social nuances, and weigh ethical considerations when outcomes are uncertain. These abilities grow out of real-world experience, cultural background, and the tacit knowledge we accumulate but often cannot fully express (Polanyi & Sen, 2009). Machines, on the other hand, can plough through enormous volumes of structured and unstructured data in moments, spot patterns too subtle for the human eye, and carry out repetitive tasks tirelessly, unaffected by fatigue or distraction (Miller & Dorneich, 2006).

Table 2 outlines the distinct advantages and limitations of humans and machines, showing how their differences can balance each other in joint decision-making. Figure 6 summarizes the strengths and weaknesses of humans and machines in decision-making.

Table 2. Comparative strengths and weaknesses of humans and machines in decision-making

Capability	Humans Strengths	Machines Strengths	Humans Limitations	Machines Limitations
Context Understanding	Deep cultural, ethical, and situational awareness	Fast, consistent interpretation of structured context	Susceptible to bias, fatigue, and inconsistent judgment	Lacks cultural grounding, no lived experience
Creativity	Novel framing, lateral thinking, adaptability	Generative creativity in bounded domains	Cognitive limits, stress impact innovation	Dependent on training data; lacks true abstraction
Data Handling	Pattern recognition in familiar settings	Rapid large-scale analysis and recall	Limited working memory, prone to overload	Requires clean, relevant data
Adaptability	Flexible reasoning under novel conditions	Fast parameter adjustments within defined constraints	Fatigue reduces adaptability	Struggles outside the trained scope
Ethical Reasoning	Balances human, societal, and environmental factors	Can enforce coded ethical rules	Vulnerable to pressure and moral disengagement	Constrained by predefined moral codes

Figure 6. Comparative strengths and limitations of humans and machines in decision-making

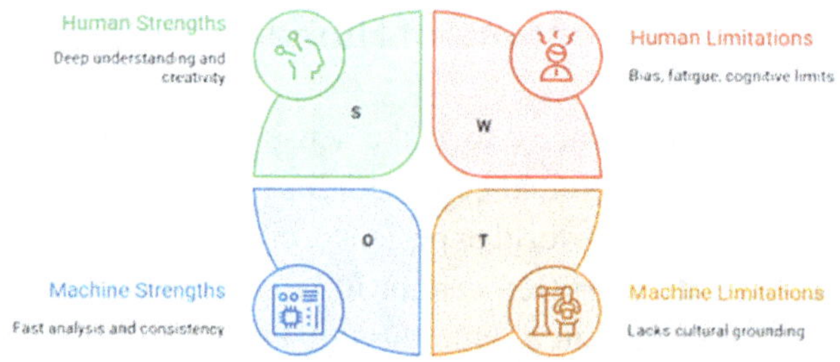

When these abilities are deliberately integrated, the weaknesses of one are offset by the strengths of the other because machines reduce human bias and inconsistency. In contrast, humans provide ethical and contextual framing that machines lack.

5.2. Interaction Models and Interfaces

The quality of a human–machine partnership is shaped largely by how the two communicate. Interaction models define this exchange of control and information.

- **Direct Control**: The human directs every action, with the machine acting purely as a tool.
- **Supervisory Control:** The machine operates independently within set limits but defers to humans for exceptions (Sheridan & Verplank, 1978).
- **Mixed-Initiative Systems:** Either human or machine can initiate action, and control shifts dynamically according to context.

Recent developments in multimodal interaction, which integrate voice, gesture, haptic feedback, and even neural inputs, are making collaboration more fluid and intuitive (Jiang et al., 2025). These richer channels reduce misunderstanding and keep the human engaged even when much of the heavy lifting is automated.

5.3. Shared Mental Models and Situational Awareness

Collaboration depends on a shared understanding of the problem, the goal, and the current situation, which researchers refer to as Shared Mental Models (SMMs) (Nikolaidis et al., 2017). If humans and machines picture the state of the world differently, coordination breaks down. Alignment can be achieved through:

1. **Transparent Information Flow:** Systems should show not only outcomes but also how those outcomes were reached, as promoted in explainable AI research (Arrieta et al., 2019).
2. **Common Vocabulary:** Interfaces and prompts must use language familiar to the operator's domain.
3. **Context Sensitivity:** Machines should adjust the level of detail according to stress, workload, or urgency to avoid cognitive overload (Griffiths et al., 2015).

Aviation offers a clear example: modern cockpits now give pilots predictive warnings rather than static readouts, aligning machine outputs with the pilot's mental model of the aircraft's condition (Salih et al., 2025).

5.4. Cognitive Load Management in Hybrid Teams

Even the most skilled operator has a limit to how much information they can process. Hybrid systems must manage cognitive load carefully to prevent overload, especially during emergencies.

Machines can help by:

- **Filtering and Prioritizing Data:** Showing only information that matters at the moment (Griffiths et al., 2015).
- **Automating Repetitive Subtasks:** Freeing human attention for creative or ethical decision elements.
- **Adjusting Task Pacing:** Slowing down or batching decisions when humans show signs of fatigue.

In healthcare, adaptive decision-support systems have been shown to improve diagnostic accuracy by presenting prompts according to physician workload rather than in a fixed sequence.

5.5. Adaptive Autonomy and Decision Delegation

Adaptive autonomy is about letting the level of automation change as the situation demands, sometimes giving more control to the machine, other times putting it back in human hands. This shift depends on how well the human is performing, how complex the situation is, and how critical the decision might be (Parasuraman et al., 2000). At its core, it comes down to three ideas:

- **Watching performance closely**, keeping an eye on how both the human and the machine are doing in real time, so control can shift when one is better suited to handle the task.
- **Knowing when to hand over control**, having clear points or conditions, learned through experience or set in advance, that signal it's time to change who is in charge.
- **Balancing trust**, making sure the human does not rely too much on the system, but also doesn't dismiss its advice without reason.

Trust is fragile and must be handled with care. If a machine's decisions seem unclear or inconsistent, people will be quick to ignore even good suggestions. On the other hand, if the system always works without explaining itself, the human can stop thinking critically and lose valuable skills. Building the right balance means designing systems that explain their reasoning, understand the context they are working in, and training operators to work comfortably with them.

One clear example is in autonomous underwater vehicles (AUVs). Here, adaptive autonomy allows the vehicle to handle steady navigation on its own, but the human steps in when something unusual or dangerous happens. In these moments, the system needs to recognize when to step back, and the human needs to recognize when to let the system take the lead.

5.6. Case Study – Human–Machine Collaboration in Search-and-Rescue Robotics

Search-and-rescue (SAR) operations represent one of the most demanding contexts for testing human–machine decision-making. These missions often occur in high-stress environments marked by incomplete information, unstable terrain, and extreme time constraints. Collaborative decision-making between human responders and robotic systems introduces new possibilities for operational safety, efficiency, and ethical accountability.

In a simulated urban disaster exercise, a multi-disciplinary research team deployed semi-autonomous ground robots equipped with LiDAR(Light Detection and Ranging), thermal, and acoustic sensors to identify survivors in collapsed structures. Humans supervised the mission through an interactive control interface, setting broad objectives while allowing the robots to autonomously navigate, scan, and report potential life signs. The hybrid architecture used a hierarchical decision process, robots handled local perception and obstacle avoidance.

At the tactical level, robots made fast probabilistic judgments using reinforcement learning models trained on historical rescue data. At the operational level, human supervisors integrated sensor reports, validated confidence levels, and issued adaptive commands. At the strategic level, decision-making was collaborative, balancing human intuition about risk with machine precision in localization and mapping. This distributed cognition framework reduced overall search time and improved situational awareness.

Trust emerged as the central determinant of success. Early trials revealed human hesitation to rely on autonomous detection, particularly when the robots flagged ambiguous thermal readings. Transparency mechanisms, such as on-screen explanations of the robot's confidence intervals and sensor rationale helped establish calibrated trust. Conversely, human oversight corrected overconfidence when robots misclassified metallic surfaces as heat signatures. Through iteration, both agents adapted, forming a shared mental model of the environment.

The trial's results showed tangible benefits: response time decreased by 28 percent, and false detection rates fell substantially compared with fully manual operations.

Human cognitive load was reduced as robots managed environmental scanning, allowing operators to focus on ethical prioritization and real-time planning. This hybrid model demonstrated how human judgment and machine autonomy could complement rather than compete with one another.

Ethical and regulatory oversight were embedded throughout. Every robotic action and decision was logged for traceability, supporting post-mission accountability. Sensitive imagery from onboard cameras was anonymized in compliance with privacy standards inspired by the EU AI(European Union Artificial Intelligence) Act. Human participants provided informed consent for workload monitoring, ensuring that the system's data governance aligned with both operational and ethical best practices.

This case demonstrates that hybrid decision-making in robotics is not merely a technical challenge but a social and moral one. The success of the SAR operation relied on the integration of human empathy, contextual awareness, and ethical reasoning with machine-level precision and endurance. Together, they achieved an equilibrium of autonomy and accountability which is a model for the next generation of responsible engineering systems.

6. CHALLENGES AND FUTURE DIRECTIONS

Blending human judgment with machine intelligence holds tremendous promise, yet making it work in practice is far from simple. The obstacles are layered, stretching beyond algorithms and hardware into the realms of human psychology, ethics, workplace dynamics, and trust. Building systems that people will embrace means ensuring they are not only powerful and precise but also open, flexible, and fair in how they operate. Below are some core challenges that must be addressed before hybrid decision-making can reach its full potential.

- **Data Quality and Model Transparency**

Even the most sophisticated system can only work with the material it is fed. If the data is outdated, incomplete, or shaped by past biases, the machine will quietly carry those flaws into its outputs. And because many advanced models still operate like black boxes, even their creators can not always explain every decision in a way that makes sense to a non-technical audience. For someone on the receiving end, "the system says so" is not enough; it feels like being asked to follow advice without understanding the reason. That's not a great recipe for trust.

- **Limited Adaptability and Clumsy Handover**

Adaptability is another stubborn hurdle. These systems perform brilliantly when the world behaves as expected, but reality is rarely that tidy. A strange set of sensor readings, an emergency no one saw coming, or a one-off problem in a supply chain can leave the machine scrambling. In theory, this is when control shifts back to the human. In practice, that handover can be clumsy because the system may still be speaking in technical shorthand, and the human, having been in "observer mode," may need a few moments to get fully back into the driver's seat. Sometimes those moments do not exist.

- **Human Over-reliance and Distrust**

Humans, for their part, have their own traps. When a machine consistently gets things right, we start to rely on it more than we should, letting our sharpness dull over time. That over-reliance is dangerous when the unexpected happens. On the other side, if the machine makes one high-profile mistake, some people will never trust it again, no matter how good it is afterward. Both extremes, blind faith and complete rejection, undo the very point of building a hybrid system.

- **Accountability and Ethical Risks**

Then there is the question no one can dodge: when a joint decision goes wrong, who owns the outcome? Is it the coder, the operator, the manager, or the whole organization? The answer is often murky. And with new areas like neuroadaptive decision-support, where systems actually monitor brain activity to tailor their responses, new ethical issues arise. Who has access to that personal data? How long is it kept? Could it be used for purposes no one agreed to? On a broader scale, there is the real risk of a "decision-making gap" between those with access to advanced hybrid systems and those without, creating a new kind of inequality that extends far beyond technology itself.

- **Organizational and Cultural Resistance**

Inside organizations, culture can be just as much of a roadblock as technical limits. People who have built careers on their judgment may not welcome machine advice, especially if it is delivered in unfamiliar terms. Others might resist because they see the system as a step toward replacing them. Overcoming that kind of resistance takes more than proving the machine's accuracy. It requires steady trust-building, honest communication, and systems that prove their worth under real conditions, not just in controlled trials. The following figure provides a clear and well-structured explanation of the challenges.

Figure 7. Key challenges in hybrid human–machine decision-making, including issues of trust, accountability, adaptability, and organizational acceptance

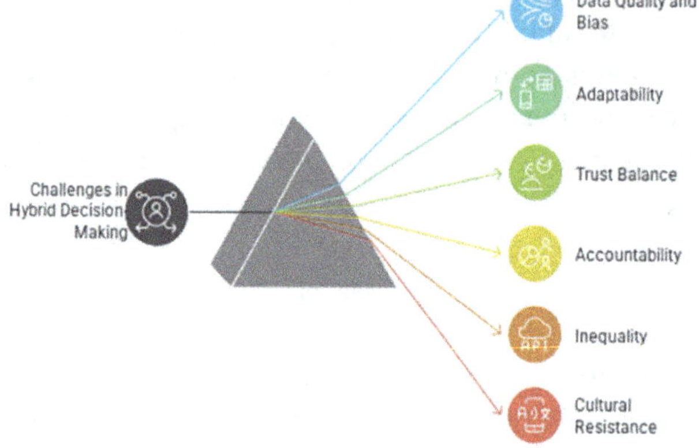

Looking forward, progress will depend on more than faster processors or bigger datasets. It will mean building systems that can explain themselves differently to different people using the right level of detail, in the right tone, at the right moment. It will mean adding "trust meters" that notice when someone is leaning too heavily on the system's judgment, or ignoring it entirely, and then adjusting how the system interacts. It will mean putting ethics and accountability into the blueprint from day one, rather than patching them in after deployment. And it will mean making sure humans stay sharp, treating the system as a partner, not a crutch. Done right, these changes could move hybrid decision-making from a promising idea to something people trust with the decisions that matter most.

7. CONCLUSION

The move from decisions made solely by people to decisions shaped by human-machine teams is one of the most significant shifts in how we think about problem-solving. People bring context, empathy, and an instinctive sense for the "why" behind a choice. Machines bring tireless processing, lightning-fast analysis, and the ability to see patterns we would otherwise miss. The real advantage comes when these strengths are not just sitting side by side but actively reinforcing one another.

That kind of partnership does not happen by accident. It takes careful design, training, and the willingness to revisit and refine how people and machines interact over time. Humans will always be prone to fatigue, bias, and moments of poor judgment. Machines will always be limited by the data they've seen and the rules they have been given. When each side is set up to cover the other's weaknesses, the results can be stronger than either could achieve alone.

If we get this right, hybrid systems won't just make us faster; they will make us better by helping us reach decisions that are fairer, clearer, and more aligned with the values we want to uphold. However, this will only happen if we view technology not as a replacement, but as an extension of human capability. It means engineers, policymakers, and front-line users working together, staying open to change, and keeping people's needs at the center of design.

In the end, the aim is not to hand everything over to algorithms, or to cling to human judgment as if it is infallible. It is to create a working relationship where the precision of machines meets the conscience of people, and where the choices we make together can withstand both scrutiny and the real-world impact they carry.

REFERENCES

Anderson, J. R., Bothell, D., Byrne, M. D., Douglass, S., Lebiere, C., & Qin, Y. (2004). An Integrated Theory of the Mind. *Psychological Review, 111*(4), 1036–1060. DOI: 10.1037/0033-295X.111.4.1036 PMID: 15482072

Arrieta, A. B., Díaz-Rodríguez, N., Ser, J. D., Bennetot, A., Tabik, S., Barbado, A., García, S., Gil-López, S., Molina, D., Benjamins, R., Chatila, R., & Herrera, F. (2019). *Explainable Artificial Intelligence (XAI): Concepts, Taxonomies, Opportunities and Challenges toward Responsible AI* (No. arXiv:1910.10045). arXiv. https://doi.org//arXiv.1910.10045DOI: 10.48550

Bairaktarova, D., & Woodcock, A. (2017). Engineering Student's Ethical Awareness and Behavior: A New Motivational Model. *Science and Engineering Ethics, 23*(4), 1129–1157. DOI: 10.1007/s11948-016-9814-x PMID: 27752963

Barocas, S., Hardt, M., & Narayanan, A. (2019). *Fairness and Machine Learning: Limitations and Opportunities*. MIT Press. https://fairmlbook.org

Bastian, B., & Zucchella, A. (2022). Entrepreneurial metacognition: A study on nascent entrepreneurs. *The International Entrepreneurship and Management Journal, 18*(4), 1775–1805. DOI: 10.1007/s11365-022-00799-1

Bodenhausen, G. V., Gabriel, S., & Lineberger, M. (2000). Sadness and Susceptibility to Judgmental Bias: The Case of Anchoring. *Psychological Science, 11*(4), 320–323. DOI: 10.1111/1467-9280.00263 PMID: 11273392

Damasio, A. R., Everitt, B. J., & Bishop, D. (1997). The somatic marker hypothesis and the possible functions of the prefrontal cortex. *Philosophical Transactions of the Royal Society of London. Series B, Biological Sciences, 351*(1346), 1413–1420. DOI: 10.1098/rstb.1996.0125 PMID: 8941953

Danziger, S., Levav, J., & Avnaim-Pesso, L. (2011). Extraneous factors in judicial decisions. *Proceedings of the National Academy of Sciences of the United States of America, 108*(17), 6889–6892. DOI: 10.1073/pnas.1018033108 PMID: 21482790

Dlugos, K. V., Mazwi, M., Lao, R., & Honjo, O. (2025). Noise is an underrecognized problem in medical decision making and is known by other names: A scoping review. *BMC Medical Informatics and Decision Making, 25*(1), 86. DOI: 10.1186/s12911-025-02905-z PMID: 39962496

Dragan, A. D., Lee, K. C. T., & Srinivasa, S. S. (2013). Legibility and predictability of robot motion. *2013 8th ACM/IEEE International Conference on Human-Robot Interaction (HRI)*, 301–308. DOI: 10.1109/HRI.2013.6483603

Englich, B., & Soder, K. (2009). Moody experts—How mood and expertise influence judgmental anchoring. *Judgment and Decision Making*, *4*(1), 41–50. DOI: 10.1017/S1930297500000693

Fan, X., Zhou, Q., Liu, Z., & Xie, F. (2015). Electroencephalogram assessment of mental fatigue in visual search. *Bio-Medical Materials and Engineering*, *26*(1, suppl), S1455–S1463. DOI: 10.3233/BME-151444 PMID: 26405908

Finucane, M. L., Alhakami, A., Slovic, P., & Johnson, S. M. (2000). The affect heuristic in judgments of risks and benefits. *Journal of Behavioral Decision Making*, *13*(1), 1–17. DOI: 10.1002/(SICI)1099-0771(200001/03)13:1<1::AID-BDM333>3.0.CO;2-S

Fiske, S. T., & Taylor, S. E. (1991). *Social Cognition*. McGraw-Hill Humanities/ Social Sciences/Languages.

Gigerenzer, G., & Gaissmaier, W. (2011). Heuristic Decision Making. *Annual Review of Psychology*, *62*(1), 451–482. DOI: 10.1146/annurev-psych-120709-145346 PMID: 21126183

Gilovich, T., Griffin, D., & Kahneman, D. (Eds.). (2013). *Heuristics and Biases: The Psychology of Intuitive Judgment*. Cambridge University Press.

Griffiths, T. L., Lieder, F., & Goodman, N. D. (2015). Rational Use of Cognitive Resources: Levels of Analysis Between the Computational and the Algorithmic. *Topics in Cognitive Science*, *7*(2), 217–229. DOI: 10.1111/tops.12142 PMID: 25898807

Hanford, S. D., Janrathitikarn, O., & Long, L. N. (2009). Control of Mobile Robots Using the Soar Cognitive Architecture. *Journal of Aerospace Computing, Information, and Communication*, *6*(2), 69–91. DOI: 10.2514/1.37056

Jarrahi, M. H. (2018). Artificial intelligence and the future of work: Human–AI symbiosis in organizational decision making. *Business Horizons*, *61*(4), 577–586. DOI: 10.1016/j.bushor.2018.03.007

Kahneman, D. (1982). *Judgment under uncertainty: Heuristics and biases*. Cambridge University. DOI: 10.1017/CBO9780511809477

Kahneman, D., Sibony, O., & Sunstein, C. R. (2022). *Noise: A Flaw in Human Judgment*. Little, Brown Spark.

Kim, N., & Nam, C. S. (2020). Adaptive Control of Thought-Rational (ACT-R): Applying a Cognitive Architecture to Neuroergonomics. In Nam, C. S. (Ed.), *Neuroergonomics: Principles and Practice* (pp. 105–114). Springer International Publishing., DOI: 10.1007/978-3-030-34784-0_6

Lamm, A. J., Roberts, T. G., Irani, T. A., Snyder, L. J. U., & Brendemuhl, J. (2012). The Influence of Cognitive Diversity on Group Problem Solving Strategy. *Journal of Agricultural Education, 53*(1), 18–30. DOI: 10.5032/jae.2012.01018

Langley, P., Laird, J. E., & Rogers, S. (2009). Cognitive architectures: Research issues and challenges. *Cognitive Systems Research, 10*(2), 141–160. DOI: 10.1016/j.cogsys.2006.07.004

LeCun, Y., Bengio, Y., & Hinton, G. (2015). Deep learning. *Nature, 521*(7553), 436–444. DOI: 10.1038/nature14539 PMID: 26017442

Malamuthu, B. K., Balakrishnan, T. P., Kumar, R. P. N., Venkataramanaiah, B., & Malathy, V. (2025a). Explainable AI for Decision-Making: A Hybrid Approach to Trustworthy Computing. *International Journal of Computational and Experimental Science and Engineering, 11*(2), 2. Advance online publication. DOI: 10.22399/ijcesen.1684

Malamuthu, B. K., Balakrishnan, T. P., Kumar, R. P. N., Venkataramanaiah, B., & Malathy, V. (2025b). Explainable AI for Decision-Making: A Hybrid Approach to Trustworthy Computing. *International Journal of Computational and Experimental Science and Engineering, 11*(2), 2. Advance online publication. DOI: 10.22399/ijcesen.1684

Malan, R. (2010). *The role of shared mental models of strategic thinking in the development of organisational strategy* [PhD Thesis]. University of Southern Queensland.

Michailidis, E., & Banks, A. P. (2016). The relationship between burnout and risk-taking in workplace decision-making and decision-making style. *Work and Stress, 30*(3), 278–292. DOI: 10.1080/02678373.2016.1213773

Miller, C. A., & Dorneich, M. (2006). From Associate Systems to Augmented Cognition: 25 Years of User Adaptation in High Criticality Systems. . . *Arlington.*

Mitchell, T. M. (1997). *Machine Learning*. McGraw-Hill.

Newell, A., & Simon, H. A. (1972). *Human problem solving* (Vol. 104). Prentice-hall Englewood Cliffs.

Nikolaidis, S., Hsu, D., & Srinivasa, S. (2017). Human-robot mutual adaptation in collaborative tasks: Models and experiments. *The International Journal of Robotics Research, 36*(5–7), 618–634. DOI: 10.1177/0278364917690593 PMID: 32855581

Norman, D., & Euchner, J. (2023). Design for a Better World: A Conversation with Don Norman. *Research Technology Management, 66*(3), 11–18. DOI: 10.1080/08956308.2023.2183015

Page, S. (2008). *The Difference: How the Power of Diversity Creates Better Groups, Firms, Schools, and Societies - New Edition*. Princeton University Press., DOI: 10.1515/9781400830282

Pahl, A. K., Newnes, L., & McMahon, C. (2007). A generic model for creativity and innovation: Overview for early phases of engineering design. *Journal of Desert Research, 6*(1/2), 5. DOI: 10.1504/JDR.2007.015561

Parasuraman, R., Sheridan, T. B., & Wickens, C. D. (2000). A model for types and levels of human interaction with automation. *IEEE Transactions on Systems, Man, and Cybernetics. Part A, Systems and Humans, 30*(3), 286–297. DOI: 10.1109/3468.844354 PMID: 11760769

Polanyi, M., & Sen, A. (2009). *The tacit dimension. With a new foreword by Amartya Sen*. The University Of.

Porcelli, A. J., & Delgado, M. R. (2017). Stress and decision making: Effects on valuation, learning, and risk-taking. *Current Opinion in Behavioral Sciences, 14*, 33–39. DOI: 10.1016/j.cobeha.2016.11.015 PMID: 28044144

Raab, M., & Gigerenzer, G. (2015). The power of simplicity: A fast-and-frugal heuristics approach to performance science. *Frontiers in Psychology, 6*. Advance online publication. DOI: 10.3389/fpsyg.2015.01672 PMID: 26579051

Rooy, D. V. (2024). Human–machine collaboration for enhanced decision-making in governance. *Data & Policy, 6*, e60. DOI: 10.1017/dap.2024.72

Russell, S., & Norvig, P. (2021). *Artificial Intelligence: A Modern Approach* (4th ed.). Pearson.

Salih, A., Raisi-Estabragh, Z., Galazzo, I. B., Radeva, P., Petersen, S. E., Menegaz, G., & Lekadir, K. (2025). A Perspective on Explainable Artificial Intelligence Methods: SHAP and LIME. *Advanced Intelligent Systems, 7*(1), 2400304. DOI: 10.1002/aisy.202400304

Scott, L. D., Arslanian-Engoren, C., & Engoren, M. C. (2014). Association of sleep and fatigue with decision regret among critical care nurses. *American Journal of Critical Care: An Official Publication. American Journal of Critical Care, 23*(1), 13–23. DOI: 10.4037/ajcc2014191 PMID: 24382613

Sheridan, T. B., & Verplank, W. L. (1978). *Human and Computer Control of Undersea Teleoperators*. MIT Man-Machine Systems Laboratory. DOI: 10.21236/ADA057655

Shmorgun, I., & Lamas, D. (2014). *Towards a Design Space for Ubiquitous Computing*. Springer International Publishing., DOI: 10.1007/978-3-319-07788-8_6

Simon, H. A. (1997). *Models of Bounded Rationality: Empirically Grounded Economic Reason*. The MIT Press., DOI: 10.7551/mitpress/4711.001.0001

Tahir, H. A., Alayed, W., Hassan, W. U., & Haider, A. (2024). A Novel Hybrid XAI Solution for Autonomous Vehicles: Real-Time Interpretability Through LIME–SHAP Integration. *Sensors (Basel)*, *24*(21), 21. Advance online publication. DOI: 10.3390/s24216776 PMID: 39517672

Turban, E., Sharda, R., & Delen, D. (2011). *Decision Support and Business Intelligence Systems* (9th ed.). Pearson.

Tversky, A., & Kahneman, D. (1973). Availability: A heuristic for judging frequency and probability. *Cognitive Psychology*, *5*(2), 207–232. DOI: 10.1016/0010-0285(73)90033-9

Tversky, A., & Kahneman, D. (1974). Judgment under Uncertainty: Heuristics and Biases. *Science*, *185*(4157), 1124–1131. DOI: 10.1126/science.185.4157.1124 PMID: 17835457

Vohs, K. D., Baumeister, R. F., Schmeichel, B. J., Twenge, J. M., Nelson, N. M., & Tice, D. M. (2008). Making choices impairs subsequent self-control: A limited-resource account of decision making, self-regulation, and active initiative. *Journal of Personality and Social Psychology*, *94*(5), 883–898. DOI: 10.1037/0022-3514.94.5.883 PMID: 18444745

Wachter, S., Mittelstadt, B., & Russell, C. (2017). Counterfactual explanations without opening the black box: Automated decisions and the GDPR. *SSRN*, *31*(2), 841–887. DOI: 10.2139/ssrn.3063289

Wemm, S. E., & Wulfert, E. (2017). Effects of Acute Stress on Decision Making. *Applied Psychophysiology and Biofeedback*, *42*(1), 1–12. DOI: 10.1007/s10484-016-9347-8 PMID: 28083720

Wilson, B., Lakshmanan, K., Dix, A., Rahat, A., & Roach, M. (2025). A Crossroads for Hybrid Human-Machine Decision-Making. In Meo, R., & Silvestri, F. (Eds.), *Machine Learning and Principles and Practice of Knowledge Discovery in Databases* (Vol. 2134, pp. 316–322). Springer Nature Switzerland., DOI: 10.1007/978-3-031-74627-7_24

World Design Council. (2025, July 14). *Case Study: How Tesla Used Design Thinking to Redefine the Automotive Experience*. https://worlddesigncouncil.org/blog/case-study-how-tesla-used-design-thinking-to-redefine-the-automotive-experience/

ADDITIONAL READINGS

Al Humdan, E., Smith, J., & Zhao, L. (2025). Enhancing transportation agility through neuroadaptive AI decision support systems. *Transportation Research Interdisciplinary Perspectives, 17*, 100190. DOI: 10.1016/j.trip.2025.100190

Bhattacharyya, S., Valeriani, D., Cinel, C., Citi, L., & Poli, R. (2017). Walking improves group decision-making performance with collaborative BCI. In T. O. Zander & S. Fairclough (Eds.), *Current Research in Neurotechnology: Proceedings of NAT'17* (pp. 1–10). Springer. (No DOI)

Birbaumer, N. (2024). "Your Thoughts are (Were) Free!": Brain–Computer Interfaces, neurofeedback, and the future of mind-reading. *Applied Psychophysiology and Biofeedback, 49*, 96–104. DOI: 10.1007/s10484-024-09648-z PMID: 38874845

Daly, J. J. (2015). Brain–computer interface: Current and emerging technologies. *Journal of Neuroengineering and Rehabilitation, 12*, 45. DOI: 10.1186/s12984-015-0050-5

Gu, X., Cao, Z., Jolfaei, A., Xu, P., Wu, D., Jung, T.-P., & Lin, C.-T. (2020). EEG-based brain-computer interfaces: A survey of signal sensing, computational intelligence, and translational applications. *arXiv preprint.* https://doi.org//arXiv.2001.11337DOI: 10.48550

Jiang, R., Fairclough, S., & Zander, T. O. (2025). Designing for cognitive resilience in human-AI teams: A neuroadaptive approach. *Journal of Cognitive Engineering and Decision Making, 19*(1), 1–15. DOI: 10.1177/1555343425123456

Jiang, X., Gao, X., & Rao, R. P. N. (2021). Interface, interaction, and intelligence in generalized brain–computer interface evolution. *Trends in Cognitive Sciences, 25*(8), 710–722. DOI: 10.1016/j.tics.2021.05.002 PMID: 34120845

Jiang, Y. (2024). Application strategies of brain–computer interface in computer engineering education. *IEEE Transactions on Education, 59*(4), 319–326. DOI: 10.1109/TE.2016.2558168

Jiang, Y., & Nam, C. S. (2015). A collaborative BCI for ALS patients in assistive robotics domains. [nature.com]. *Proceedings of the Human Factors and Ergonomics Society Annual Meeting, 59*, 716–720. DOI: 10.1177/1541931215599019

Onciul, R., & [others]. (2025). Artificial Intelligence and Neuroscience: Transformative potential of neuroadaptive education systems. *Journal of Personalized Learning Analytics, 14*(2), 550. DOI: 10.3390/jpla14020550 PMID: 39860555

Papakostas, C., Troussas, C., Krouska, A., & Sgouropoulou, C. (2025). NAMI: A Neuro-Adaptive Multimodal Architecture for Wearable Human-Computer Interaction. *Preprints*. DOI: 10.20944/preprints202501.0031.v1

Poli, R., Valeriani, D., & Cinel, C. (2016). Enhancement of group perception via a collaborative brain-computer interface. *IEEE Transactions on Biomedical Engineering*, *63*(5), 1238–1248. DOI: 10.1109/TBME.2016.2598875 PMID: 28541187

Saha, S. (2021). Progress in Brain–Computer Interface: Challenges and emerging opportunities. *Frontiers in Human Neuroscience*, *15*, 110. DOI: 10.3389/fnhum.2021.00310

Yadav, H., & Maini, S. (2023). Electroencephalogram-based brain-computer interface: Applications, challenges, and opportunities. *Multimedia Tools and Applications*, *82*(30), 15653–15680. DOI: 10.1007/s11042-023-15653-x PMID: 37362726

Zhou, X., & Liao, P.-C. (2022). Weighing votes in human-machine collaboration for hazard recognition: Inferring decision confidence from EEG wavelets. *Frontiers in Neuroinformatics*, *16*, 8. DOI: 10.3389/fninf.2022.00008

Chapter 3

Decision Making in Human-Computer Interaction

Vandana N. Revankar
https://orcid.org/0009-0001-4342-8408
St. Francis College, India

G. Nivashini
Karpagam Academy of Higher Education, India

P. Selvakumar
https://orcid.org/0000-0002-3650-4548
Department of Science and Humanities, Nehru Institute of Technology, Coimbatore, India

K. L. Meera
https://orcid.org/0009-0008-8276-1669
Christ Academy Institute of Advanced Studies, India

Undabatla Rambabu
https://orcid.org/0000-0003-2731-1610
Lakireddy Bali Reddy College of Engineering, India

M. R. Arun
https://orcid.org/0000-0002-8291-2956
Vel Tech Rangarajan Dr. Sagunthala R&D Institute of Science and Technology, India

T. C. Manjunath
https://orcid.org/0000-0003-2545-9160
Rajarajeswari College of Engineering, India

ABSTRACT

Decision making is a central cognitive process that plays a pivotal role in Human-Computer Interaction (HCI), influencing how users interact with systems, interpret information, and achieve their goals. At its core, decision making involves selecting the most appropriate course of action from among multiple alternatives based on the user's goals, available information, contextual cues, and cognitive resources. In HCI, every interaction—from clicking a button, selecting a menu option, navigating

DOI: 10.4018/979-8-3373-5702-7.ch003

through an interface, to interpreting feedback—requires the user to make decisions, consciously or subconsciously. These micro and macro decisions accumulate to define the overall user experience and system effectiveness. Decision making in HCI is highly context-dependent. It varies across different domains (e.g., gaming, e-commerce, healthcare), task types (routine vs. novel), user expertise (novice vs. expert), and device characteristics (desktop, mobile, wearable). In routine interactions, decisions may be automated and habitual, relying on procedural memory.

THE ROLE OF DECISION MAKING IN HCI

Decision making is a central cognitive process that plays a pivotal role in Human-Computer Interaction (HCI), influencing how users interact with systems, interpret information, and achieve their goals. At its core, decision making involves selecting the most appropriate course of action from among multiple alternatives based on the user's goals, available information, contextual cues, and cognitive resources. In HCI, every interaction—from clicking a button, selecting a menu option, navigating through an interface, to interpreting feedback—requires the user to make decisions, consciously or subconsciously. These micro and macro decisions accumulate to define the overall user experience and system effectiveness. Decision making in HCI is highly context-dependent. It varies across different domains (e.g., gaming, e-commerce, healthcare), task types (routine vs. novel), user expertise (novice vs. expert), and device characteristics (desktop, mobile, wearable). In routine interactions, decisions may be automated and habitual, relying on procedural memory. In contrast, novel tasks demand conscious deliberation, working memory, and problem-solving. The design of interfaces can either support or hinder decision processes. For instance, cluttered interfaces with excessive options can lead to choice overload, increasing cognitive load and potentially causing decision paralysis. Conversely, well-designed interfaces that employ clear visual hierarchies, progressive disclosure, and meaningful defaults can facilitate swift and confident decision-making. Another significant factor influencing decision-making in HCI is the role of affect and emotion. Emotions can affect both the speed and quality of decisions. Positive affect generally enhances creative decision making and fosters exploration, while negative affect may cause users to become risk-averse or hyper-focused. Emotional design, therefore, aims to create experiences that elicit positive emotions, thereby supporting better decision outcomes. For example, recommender systems in e-commerce aid in narrowing down choices, while digital assistants can handle routine decision tasks, freeing cognitive resources for higher-order decisions. However, such systems must be designed with care to maintain user agency and avoid over-reliance on automation, which can erode decision-making skills and user trust. In sum, decision making is a foundational aspect of HCI that intersects with cognitive science, design principles, emotional psychology, and ethical considerations. It underpins every user interaction, shaping usability, user satisfaction, and overall system success. By deeply understanding how users make decisions and thoughtfully designing interactions that support these processes, HCI professionals can create interfaces that not only meet functional requirements but also foster positive, empowering user experiences.

COGNITIVE PROCESSES IN DECISION MAKING: UNDERSTANDING THE MENTAL MECHANISMS

Cognitive processes form the foundation of human decision making, orchestrating the mental activities that enable individuals to interpret information, evaluate alternatives, and select actions. In the realm of Human-Computer Interaction (HCI), understanding these cognitive processes is essential for designing systems that align with human capabilities and limitations (Wien & Peluso, 2021). Decision making is not a singular act but a complex interplay of multiple cognitive components—perception, attention, memory, reasoning, and emotion—all of which interact dynamically within the constraints of time, context, and task demands. This section explores the key cognitive processes involved in decision making, shedding light on their implications for interface design and user experience. Attention is the mechanism by which users allocate their cognitive resources to specific aspects of the interface. In the modern digital environment, where attention is fragmented across multiple streams of information, guiding user attention is critical. Designers use visual cues, motion, color, and layout to direct focus toward important options and away from irrelevant noise. Sustained attention is needed for complex decisions that involve weighing trade-offs or processing unfamiliar information. gnitive bandwidth and facilitate more deliberate decision making (Monsell S., 2021). Long-term memory influences decision making by providing stored knowledge, experiences, and learned associations that inform current choices. Expert users draw upon extensive schemas and mental models that allow them to make quick, efficient decisions in familiar contexts. Novice users, lacking such models, may struggle with interpretation and rely more heavily on interface guidance. Recognition-based interfaces (e.g., icon-based navigation) can ease the cognitive burden compared to recall-based interactions (e.g., command-line inputs). Moreover, repeated interactions lead to habitual decision making, where procedural memory enables users to perform tasks with little conscious effort—an essential consideration for the design of frequently used systems.

Reasoning and problem-solving processes become central in non-routine, complex, or high-stakes decisions. Users engage in either analytical reasoning—a deliberate, systematic evaluation of options—or heuristic reasoning—using mental shortcuts and rules of thumb. Dual-process theories, such as Kahneman's System 1 and System 2 framework, describe these as fast, intuitive vs. slow, deliberative modes of thinking. HCI designers must anticipate which mode users will employ based on task context. For example, in emergency response systems or medical diagnostics, interfaces must support accurate analytical reasoning (Jenke L., et al, 2021). In contrast, for routine shopping or entertainment choices, interfaces can effectively leverage heuristics, such as social proof or default options, to guide decisions. Metacognition—thinking about one's thinking—is another critical cognitive process that influences decision confidence, error detection, and adaptive learning. Interfaces that provide clear feedback, transparency of system state, and opportunities for reflection can enhance metacognitive awareness, enabling users to recognize when they need to seek more information or revise their choices. In contrast, opaque or overly automated systems may erode metacognitive engagement, leading to blind trust or disengagement. Temporal factors also influence cognitive processing in decision making. Time pressure can shift users toward heuristic-driven, fast decisions, potentially increasing error rates. In contrast, interfaces that allow users to pace their decision process enable more thoughtful deliberation. Understanding the temporal dynamics of decision making helps designers create systems that either accommodate natural pacing (as in complex planning tools) or appropriately scaffold rapid decision contexts (as in real-time control systems). In sum, decision making is a multifaceted cognitive endeavor that integrates perception, attention, memory, reasoning, emotion, and social influences (Vuong Q H., 2023). In HCI, understanding these cognitive processes allows designers to craft systems that respect human limitations, leverage strengths, and mitigate vulnerabilities. By aligning interface design with cognitive realities, designers can enhance usability, improve decision quality, and foster positive user experiences. Moreover, as emerging technologies—such as AI-driven decision aids, augmented reality, and brain-computer interfaces—reshape the decision-making landscape, ongoing research into cognitive processes will remain vital to ensuring that human agency, autonomy, and well-being are preserved at the heart of digital interaction.

HEURISTICS AND BIASES: UNDERSTANDING COGNITIVE SHORTCUTS IN DECISION MAKING

Heuristics and biases are central concepts in the study of human decision making, with profound implications for psychology, behavioral economics, and Human-Computer Interaction (HCI). They describe the mental shortcuts and systematic errors that arise from the way our cognitive processes are structured. While heuristics allow us to make quick, efficient judgments under uncertainty or limited cognitive resources, they can also lead to predictable distortions in reasoning, which we call cognitive biases (Mangus & Mahajan, 2022). In the context of HCI, understanding heuristics and biases is essential for designing interfaces and interactions that align with natural human tendencies while minimizing the risk of error or manipulation. Commonly studied heuristics include the availability heuristic, representativeness heuristic, anchoring and adjustment, and affect heuristic. The availability heuristic leads people to judge the likelihood or importance of an event based on how easily examples come to mind. For instance, if a user recently heard about a data breach, they may overestimate the risk of online shopping. The representativeness heuristic causes users to make judgments based on perceived similarity to a prototype, often ignoring statistical probabilities, such as assuming a sleek, high-tech website must also be secure and trustworthy. Anchoring and adjustment describe the tendency to rely too heavily on an initial value (anchor) when making estimates, even if that value is arbitrary. For example, a user might perceive a product as a bargain simply because it is labeled "50% off," regardless of its actual value (Jain J., et al, 2023). The affect heuristic highlights the role of emotions in shaping judgments—positive feelings can bias us toward perceiving options more favorably. Cognitive biases, the systematic errors that arise from the overuse or misapplication of heuristics, can lead to flawed decision-making. Numerous biases have been documented, each influencing user behavior in distinct ways. For example, confirmation bias leads users to seek out information that confirms their existing beliefs, potentially narrowing their exploration of new ideas. Framing effects demonstrate that the same information can lead to different decisions depending on how it is presented—for instance, highlighting "90% success" versus "10% failure" can sway user perception. Loss aversion, a bias where potential losses are weighted more heavily than equivalent gains, can influence how users respond to offers, warnings, or risk-related choices.

To design responsibly in light of heuristics and biases, HCI practitioners must balance several goals: (1) Support decision efficiency by designing interfaces that align with users' natural heuristics; (2) Mitigate bias risks by presenting information transparently and helping users recognize when reflection or deeper analysis is needed; and (3) Protect user autonomy by avoiding manipulative patterns that exploit cognitive vulnerabilities. This requires a nuanced understanding of how different heuristics operate in various contexts and among different user populations (Little & Wu, 2021). Recognizing that biases are not fixed, HCI research also explores ways to foster debiasing—helping users develop greater awareness of their cognitive shortcuts and adjust their behavior. Tools such as just-in-time feedback, visual explanations, and reflective prompts can nudge users toward more deliberate reasoning when needed. For instance, providing clear explanations for automated recommendations can help counter over-reliance on AI systems (automation bias), while alerting users to potentially biased sources of information can promote critical thinking. The influence of cultural and individual differences on heuristics and biases is another key consideration. While certain biases appear universal, their expression can vary across cultures and contexts. Designers must be sensitive to these variations and avoid assuming that a single interface structure will serve all users equally well. In summary, heuristics and biases are intrinsic aspects of human cognition that both enable and constrain decision making (Sethumadhavan et al., 2025). In HCI, they offer both opportunities and challenges: by aligning interfaces with users' natural heuristics, designers can enhance usability and efficiency; by being mindful of biases, they can avoid pitfalls that distort judgment or undermine user trust. The ethical responsibility of HCI practitioners lies in harnessing this knowledge to create interfaces that empower users, foster sound decision making, and respect cognitive diversity. As digital systems grow more pervasive and complex, understanding heuristics and biases will remain critical to designing humane, trustworthy, and effective interactions.

USER-CENTERED DESIGN: EMBRACING THE USER-CENTERED APPROACH

User-Centered Design (UCD) is a foundational philosophy and methodology within Human-Computer Interaction (HCI) that emphasizes designing interactive systems around the needs, abilities, preferences, and contexts of the people who use them. In contrast to technology-driven or business-driven design approaches, which may prioritize technical novelty or organizational goals, the user-centered approach places the user's experience, well-being, and success at the heart of the design process (Chowdhary S., et al, 2025). The guiding principle of UCD is simple yet powerful: understanding users—who they are, what they value, how they think, and what they seek to accomplish—is essential for creating effective, usable, and satisfying digital products and services. Several key principles define the user-centered approach. First and foremost is early and continuous user involvement. By conducting user research—such as interviews, contextual inquiries, surveys, and ethnographic studies—designers build empathy for their target audiences and uncover the nuanced realities of their behaviors, environments, motivations, and pain points. This deep understanding moves beyond superficial demographics to capture users' goals, workflows, mental models, and emotional drivers. Second, UCD emphasizes empirical measurement of usability and user experience. Through usability testing, A/B experiments, eye-tracking, and other techniques, designers gather concrete data on how users interact with prototypes and systems, using these insights to guide iterative refinements (Ahmad & Wu, 2024). Multidisciplinary collaboration is also intrinsic to UCD. Designing with users in mind requires integrating expertise from multiple domains—interaction design, visual design, psychology, cognitive science, anthropology, engineering, business strategy, and more. Cross-functional collaboration ensures that different perspectives inform the design, enriching its relevance and quality. Moreover, UCD fosters an organizational culture where user empathy and advocacy permeate all aspects of product development, from executive decision-making to technical implementation.

One of the most important ways UCD supports decision-making is through transparency and control. Interfaces should provide users with the right amount of information at the right time—neither overwhelming them with unnecessary details nor concealing critical information. Feedback mechanisms—such as confirmations, progress indicators, and error messages—help users understand the consequences of their actions and make informed adjustments. Moreover, UCD champions user autonomy, ensuring that systems do not coerce or manipulate users into decisions but instead empower them with clear, honest choices (Zorzetti M., et al, 2022). The user-centered approach also acknowledges the context of use as a critical factor in design. Users interact with systems in varied environments—while mobile, under time pressure, with different devices, across cultural settings. UCD encourages designers to consider these contexts and design for flexibility, accessibility, and inclusivity. For example, responsive design ensures that interfaces work well across devices and screen sizes; accessible design ensures that people with disabilities can use systems effectively; and culturally sensitive design avoids assumptions that may not hold across different user groups. The rise of AI and intelligent systems presents new challenges and opportunities for UCD (Bin Ahsan W., 2024). As systems become more autonomous and adaptive, maintaining user understanding, trust, and control becomes even more critical. UCD provides principles for designing AI interactions that are explainable, accountable, and aligned with user goals. For example, AI recommendations should be transparent about why certain options are suggested, and users should have the ability to review and override automated decisions. Designing for human-AI collaboration—rather than mere automation—ensures that users remain empowered participants in decision making. In sum, User-Centered Design is more than a set of techniques; it is a humanistic approach that views technology as a means to enhance human capabilities, not an end in itself. By grounding design in empathy, iteration, collaboration, and respect for users' needs and contexts, UCD fosters digital experiences that are effective, ethical, and delightful (Niu Y., et al, 2021). In an increasingly complex digital world—where users must navigate vast amounts of information, interact with intelligent agents, and make decisions with real consequences—the principles of the user-centered approach provide an essential guide for designing systems that truly serve people. Whether in everyday apps or critical decision-support systems, UCD ensures that technology remains human-centered, supporting users not only in what they do but in who they are and who they aspire to be.

INFORMATION PRESENTATION: ENHANCING DECISION MAKING THROUGH EFFECTIVE DISPLAY

Information presentation refers to the methods, formats, and structures used to display data, content, and choices to users within a system or interface. It is a critical aspect of Human-Computer Interaction (HCI), as the way information is conveyed directly influences how users perceive, interpret, and act upon it. Effective information presentation enhances decision making by reducing cognitive load, guiding user attention, supporting comprehension, and enabling efficient navigation of complex content (Awan U., et al, 2021). Poor presentation, by contrast, can lead to confusion, misinformation, decision fatigue, and user frustration. Therefore, understanding the principles of information presentation is vital for designing systems that support accurate, confident, and timely decisions. The modality of presentation also significantly affects usability and decision outcomes. Information can be presented visually, auditorily, or through haptic feedback, with visual formats being the most common. Within the visual modality, designers may use text, images, graphs, charts, tables, icons, and color coding to communicate meaning. Each of these elements serves different purposes: textual explanations are good for detailed information; icons and symbols provide quick cues; graphs and charts are ideal for summarizing trends or comparing quantities. A good interface balances these elements to cater to various cognitive styles, literacy levels, and decision-making contexts. Cognitive load theory plays a central role in guiding how information should be presented. Users have limited working memory, and when interfaces present information in a disorganized, dense, or ambiguous manner, they increase the intrinsic and extraneous load on the user (Sahoo & Goswami, 2023). Effective presentation techniques aim to minimize unnecessary cognitive effort so that users can devote more mental resources to meaningful decision tasks. For example, chunking related items together, avoiding jargon, using familiar terms, and maintaining consistency across the interface can significantly reduce mental strain and improve comprehension. Information timing—when and how information is presented relative to the user's actions or goals—is equally important. In real-time systems or dynamic environments, information must be timely and contextually relevant. Dashboards that update in real time, alerts that notify users of anomalies, and prompts that guide task sequences exemplify context-sensitive information presentation. Delays, lags, or misaligned feedback can cause users to act based on outdated or misunderstood information, leading to errors or inefficiency.

The framing effect—a well-documented cognitive bias—demonstrates how the presentation of equivalent information in different ways can lead to different decisions. For instance, presenting medical information as "90% survival" versus "10% mortality" can significantly influence patients' treatment preferences. Designers must be aware of such framing effects and strive for neutrality and transparency in information presentation, especially in contexts where unbiased decision making is ethically critical. This is especially relevant in news platforms, e-commerce sites, and political applications, where biased framing can manipulate user opinions and behaviors. Another major consideration in information presentation is user diversity. Users differ in age, education, cognitive ability, cultural background, language proficiency, and accessibility needs (Wang & Yin, 2021). Designers must account for these differences to ensure inclusive and equitable access to information. For instance, interfaces should support screen readers, provide alternative text for images, use readable font sizes, and avoid relying solely on color to convey meaning. Culturally inclusive design should also avoid symbols, icons, or idioms that may not translate across cultures. Localization and internationalization efforts are vital for ensuring that information is meaningful and accessible to a global audience. Interactive elements such as tooltips, modals, popovers, and expandable sections also play a vital role in contextual information presentation. They allow users to access help or clarification without leaving the main task flow. However, overuse or poor timing of such elements can disrupt user engagement or create annoyance. Design decisions must carefully consider when and how to introduce supplementary information to support, rather than interrupt, the user experience. In mobile and wearable interfaces, space constraints demand highly efficient information presentation. Minimalist design, icon-based navigation, gesture controls, and contextual feedback are common techniques for conveying information without overloading the screen (Coates J M., et al, 2025). Voice interfaces add another layer, where information must be presented in a linear, auditory format. In such cases, brevity, clarity, and sequential structuring become paramount to ensure comprehension and retention. In sum, information presentation is not a neutral or secondary concern—it is a central driver of how users understand, evaluate, and act on the content they encounter. Whether making everyday decisions or navigating critical tasks, users depend on the structure, clarity, timing, and integrity of presented information. Designers, developers, and decision system architects must therefore approach information presentation as both an art and a science, blending cognitive psychology, visual communication, accessibility standards, and ethical considerations. By doing so, they can create interfaces that empower users, foster trust, and enhance the quality of decision-making in a world increasingly shaped by digital interactions.

FEEDBACK AND ERROR PREVENTION: THE IMPORTANCE OF FEEDBACK IN ENHANCING USER INTERACTION

Feedback and error prevention are fundamental principles in Human-Computer Interaction (HCI), playing a critical role in shaping how users interact with digital systems, how they understand system status, and how they make decisions. In any interactive system, users need to understand whether their actions have been successful, whether the system is responding as expected, and what steps they should take next. Clear, timely, and meaningful feedback ensures that users remain oriented, informed, and in control of their interactions (Nisha Angeline M., et al, 2025). Conversely, poor or missing feedback can leave users confused, frustrated, and prone to errors, ultimately degrading trust and satisfaction. Understanding the importance of feedback and designing effective error prevention mechanisms is, therefore, essential for creating interfaces that support smooth, confident, and error-free decision making. One of the primary roles of feedback is to reduce uncertainty. In complex digital environments, users often perform actions with an incomplete understanding of how the system will respond. For example, when a user submits an online form, they need reassurance that the form was successfully received (Sajadi A., 2024). If no feedback is provided, users may question whether they need to resubmit the form, potentially leading to redundant actions or abandonment. Similarly, when initiating a file download, a visible progress bar informs users that the system is working, reduces impatience, and sets expectations for how long the task will take. Without such feedback, users may become disoriented or believe the system has failed. Delayed feedback can confuse and disrupt the user's mental flow, as users may not be sure whether their action was accepted. However, in some cases, deliberately paced feedback — such as brief animations — can improve perceived performance and usability by making system transitions feel smoother and more natural. Closely tied to feedback is the concept of error prevention. Rather than simply helping users recover from mistakes, well-designed systems strive to minimize the likelihood of errors occurring in the first place. This reflects the proactive principle of "designing for error," recognizing that human errors are inevitable and should be anticipated (Zhang & Hyland, 2023). Effective error prevention involves multiple strategies. Constraints: Limiting the ways a user can interact with the system to prevent invalid inputs. For example, date pickers prevent entering impossible dates, and disabled buttons prevent submission of incomplete forms. Affordances: Designing interface elements that suggest their proper use, reducing the likelihood of user confusion. Real-time validation: Providing immediate feedback when an input error is detected, such as highlighting an invalid email address field. Defaults and smart defaults: Prefilling fields with likely values to guide user input and reduce effort. Confirmation prompts: Asking users to confirm critical or irreversible actions, giving them a chance to reconsider.

In high-stakes environments, such as healthcare, aviation, or financial systems, feedback and error prevention take on even greater importance. Errors in these domains can have serious consequences, so interfaces must be meticulously designed to minimize risks. This includes not only preventing input errors but also providing clear alerts for abnormal conditions, supporting situational awareness, and ensuring that feedback helps users make informed decisions under pressure. With the increasing integration of AI systems and automation, feedback becomes even more critical for fostering trust and transparency (Madaan A., et al, 2023). Users must understand when the system is acting autonomously, what it is doing on their behalf, and why certain recommendations or actions are being presented. Without clear feedback, users may over-rely on automation (automation bias) or under-trust it (algorithm aversion). Explainable AI (XAI) emphasizes feedback mechanisms that help users understand and appropriately calibrate their trust in AI outputs. The design of micro-interactions — small, focused feedback events — has become an area of rich exploration in modern interface design. Examples include animations that indicate progress, subtle transitions when moving between screens, or sound cues that confirm receipt of input. These micro-interactions may seem minor, but they contribute significantly to the perceived responsiveness, polish, and quality of the interface. They also help keep users engaged and provide the sensory confirmation needed for a seamless user experience (Hooda M., et al, 2022). In conclusion, feedback is not merely a technical add-on; it is a core enabler of effective interaction, learning, decision making, and user satisfaction. It shapes how users understand the system, how they build trust, how they avoid errors, and how they recover when things go wrong. Error prevention further enhances user experience by minimizing frustration and guiding users toward successful outcomes. Together, these principles embody a user-centered philosophy of interaction design: respecting the user's time, attention, and capabilities. In an increasingly complex digital world — where systems are adaptive, intelligent, and interconnected — the importance of clear, timely, and empathetic feedback will only grow, ensuring that users remain informed, empowered, and in control of their digital experiences.

DECISION SUPPORT SYSTEMS (DSS): A COMPREHENSIVE OVERVIEW

Decision Support Systems (DSS) are interactive, computer-based systems that support individuals and organizations in making informed, data-driven decisions. These systems combine data, sophisticated analytical models, user-friendly interfaces, and domain-specific knowledge to assist users in solving semi-structured and unstructured problems—those that cannot be addressed by routine procedures alone. DSS aims not to replace human decision-making but to augment and enhance it, especially in complex, dynamic environments where multiple variables, uncertainties, and trade-offs are involved. Originating from the convergence of management science, information systems, and cognitive psychology in the 1960s and 70s, DSS have evolved significantly over time, encompassing a wide variety of tools and technologies that span across sectors such as healthcare, finance, logistics, agriculture, education, and government (Forcina A., et al, 2024). One of the defining characteristics of DSS is their ability to support semi-structured decisions, where part of the decision-making process is known and definable, but human judgment is still essential. For instance, a hospital administrator using a DSS might analyze bed occupancy data, patient admission trends, and staff availability to decide how to allocate resources, balancing quantitative insights with professional expertise. This blend of computation and cognition is what sets DSS apart from traditional information systems that merely report facts without interpretative or predictive capabilities. DSS can be categorized into several types based on their functionality and focus. These include: Data-Driven DSS: These systems emphasize access to and manipulation of large databases of structured data, often through online analytical processing (OLAP) and data visualization tools. Examples include business dashboards, financial analysis platforms, and clinical decision support systems that extract insights from electronic health records. Model-Driven DSS: These systems emphasize access to and manipulation of mathematical models, such as optimization, simulation, or statistical forecasting. They are common in operations management, logistics, and engineering, where scenarios and outcomes must be rigorously analyzed (Mumali F., 2022). Knowledge-Driven DSS (Expert Systems): These utilize knowledge bases and inference engines to provide expert advice or recommendations, often in rule-based formats. Examples include diagnostic systems in medicine or legal advisory tools. Communication-Driven DSS: Focused on collaboration and group decision making, these systems support the sharing of information, voting, brainstorming, and consensus-building among teams. Group Decision Support Systems (GDSS) fall under this category. Document-Driven DSS: These systems help users retrieve and analyze unstructured information stored in documents, reports, and multimedia content. They are particularly useful in legal, academic, and research settings where decision-making depends on qualitative evidence.

Figure 1. Human-Computer Interaction

Human-Computer Interaction (HCI) plays a central role in DSS design and effectiveness. The success of a DSS depends not only on the quality of data or models it uses but also on how intuitively and transparently it communicates insights to users. Good HCI in DSS ensures that users can understand the implications of their choices, explore alternatives, and trust the recommendations provided. Visualizations, dashboards, what-if analysis tools, and interactive scenario simulations all serve to translate complex data into actionable understanding. Poor interface design, by contrast, can lead to misinterpretation of data, over-reliance on automation, or disengagement from the decision-making process (Antomarioni S., et al, 2021). In organizational contexts, DSS contribute to better alignment between strategic goals and operational actions. By providing evidence-based insights and fostering transparency, they help managers justify decisions, manage risks, and anticipate future challenges. DSS are particularly valuable in environments characterized by volatility, uncertainty, complexity, and ambiguity (VUCA), where traditional decision-making approaches may fall short. For example, in crisis management or disaster response, DSS can synthesize information from multiple sources to guide rapid, coordinated action. Despite their many benefits, DSS also face challenges. Data quality and integration are common issues, as effective decision support depends on the availability, accuracy, and consistency of data across sources. User resistance can arise when systems are perceived as difficult to use, threatening to professional autonomy, or opaque in how they generate outputs. Ethical concerns also emerge, especially when DSS influence high-stakes decisions involving health, justice, or public policy (Upadhyaya A N., et al, 2025). In such cases, questions around accountability, fairness, transparency, and bias must be addressed to ensure that DSS serve the public interest. The ethical design of DSS also requires consideration of user diversity and accessibility. Systems should be designed to accommodate a wide range of cognitive styles, literacy levels, and abilities. User-centered design principles ensure that DSS are not only powerful but also inclusive, transparent, and respectful of individual autonomy. In conclusion, Decision Support Systems are indispensable tools for navigating the complexity of modern decision environments. They embody the synergy between data, technology, and human judgment, enabling more accurate, timely, and informed choices across domains. Whether in everyday operations or strategic planning, DSS provides the analytical backbone for evidence-based decision making. Their effectiveness depends on the thoughtful integration of robust models, reliable data, intuitive interfaces, and ethical design principles. As technology continues to evolve, DSS will play an increasingly vital role in shaping intelligent, responsive, and responsible decision-making ecosystems across society.

INTERACTIVE SYSTEMS: FOUNDATIONS, FUNCTIONS, AND IMPORTANCE IN HUMAN-COMPUTER INTERACTION

Interactive systems are computer-based systems designed to support a dynamic, continuous exchange of information and actions between users and technology. Unlike passive systems, which simply perform tasks or provide output upon request, interactive systems are characterized by their responsiveness, adaptability, and engagement with users in real time. They serve as the core interface between human cognitive processes and computational power, allowing users to input data, make choices, receive feedback, and refine their actions based on system responses. In Human-Computer Interaction (HCI), interactive systems are not merely tools; they are co-participants in a human-centered dialogue, shaping the way decisions are made, tasks are performed, and experiences are created (Thirunagalingam & Selvakumar, 2025). The degree of interactivity in a system can vary widely, depending on the context and the complexity of user tasks. Some systems offer basic forms of interactivity, such as click-based navigation or form submission. Others support rich interactivity, where users engage in multiple, simultaneous inputs, manipulate data in real time, or collaborate with other users through shared interfaces. Richly interactive systems include simulation environments, collaborative design platforms, immersive learning applications, and intelligent virtual assistants. The interface serves as the point of contact between user intent and system functionality. It must be visually coherent, contextually aware, and accessible to diverse users. Principles such as affordance (visual cues that suggest how an element is used), feedback (immediate response to user actions), visibility (clear availability of options), and consistency (uniform behaviors across similar elements) guide the design of effective interfaces (Bala & Damla, 2021). For example, an interactive data visualization system that allows users to explore trends by dragging sliders or clicking on graph elements must indicate what interactions are possible and provide real-time feedback to reinforce understanding. Modern interactive systems also emphasize adaptability—the ability to tailor responses based on user preferences, context, and behavior. Adaptive systems can learn from past interactions to provide more personalized experiences, such as recommending content, auto-completing input, or adjusting interface complexity based on user proficiency. This adaptability can enhance efficiency and satisfaction, but also raises concerns about transparency and user control. Users should be informed when systems adapt and can override or adjust these behaviors as needed.

Collaborative interactive systems are designed to support multiple users working together—either synchronously or asynchronously—on shared tasks. These include systems for co-authoring documents, designing products, or managing workflows. Collaboration in such systems requires careful coordination of inputs, awareness of others' actions, and mechanisms for communication and conflict resolution (Alkatheiri M S., 2022). Interactive systems that support collaboration must balance individual control with group coherence, often through shared visualizations, real-time updates, and version tracking tools. As interactive systems grow more intelligent, the line between passive and active engagement is blurring. Conversational agents and chatbots exemplify this evolution, using natural language processing to simulate human-like interactions. These systems often serve as intermediaries between users and more complex back-end processes, helping schedule meetings, answer questions, or guide users through transactions. The key to successful conversational interaction is maintaining context, managing ambiguity, and providing fallback options when misunderstandings occur (Augstein M., et al, 2023). Interactive systems are also vital in learning environments, where educational platforms leverage interactivity to improve engagement, retention, and motivation. Interactive simulations, quizzes, virtual labs, and adaptive learning paths empower learners to explore content at their own pace and receive immediate feedback. Gamification techniques—such as points, badges, and challenges—further increase motivation and deepen engagement. These educational interactive systems personalize learning and support continuous improvement, both for learners and educators. In conclusion, interactive systems are the dynamic interface through which humans and technology collaborate. They are indispensable in modern digital life, enabling users to control, explore, create, and decide with the support of responsive, adaptable, and user-centered technology (Blackley C., et al, 2021). Whether in everyday applications or high-stakes professional environments, interactive systems transform passive information processing into active problem-solving. As these systems continue to evolve—becoming more multimodal, intelligent, and embedded in our environments—their role in shaping human cognition, productivity, and decision making will only grow more significant. The future of HCI depends not just on more powerful technologies but on more thoughtful, humane, and responsive systems that honor the full spectrum of human needs and abilities.

COGNITIVE LOAD AND DECISION FATIGUE: THE IMPACT OF COGNITIVE LOAD ON DECISION MAKING

Cognitive load refers to the amount of mental effort used in the working memory at any given time. In the context of Human-Computer Interaction (HCI), cognitive load plays a pivotal role in determining how effectively users can process information, make decisions, and interact with digital systems. The term encompasses different types of mental demands—such as remembering, reasoning, problem solving, and paying attention—each drawing from a limited cognitive resource pool. When cognitive load becomes excessive, it can overwhelm users, degrade decision quality, and lead to decision fatigue—a psychological state in which prolonged or intense decision-making effort results in mental exhaustion and reduced capacity for further decisions (Howie E E., et al, 2023). Understanding how cognitive load impacts decision making is critical for designing user interfaces and interactive systems that promote efficiency, accuracy, and user well-being. Cognitive load is generally categorized into three types: intrinsic, extraneous, and germane. Intrinsic load is inherent to the complexity of the task itself; it reflects how difficult the material or decision problem is by nature. For example, diagnosing a rare medical condition or analyzing a complex investment portfolio naturally imposes high intrinsic cognitive demands. Extraneous load results from how information is presented and how the task environment is designed; poor interface design, confusing navigation, cluttered layouts, or ambiguous language can all unnecessarily increase this type of load. Germane load is the beneficial mental effort required for learning and understanding; it supports schema development and long-term knowledge construction. The designer's goal is to manage and optimize these loads—minimizing extraneous load, supporting germane load, and aligning intrinsic load with the user's capabilities (Han

Y., et al, 2021). Research by psychologists such as Roy Baumeister and Kathleen Vohs has demonstrated that decision fatigue is a real and measurable phenomenon. Their studies show that as cognitive resources are depleted, self-control diminishes, attention wanes, and risk assessment becomes less reliable. This has important implications for system design: interactive systems should not merely assume unlimited user attention and effort. Instead, they should actively support users in managing their cognitive load by structuring workflows, offering clear guidance, and reducing unnecessary decision points.

Information chunking—grouping related information into manageable units—is another widely used method for reducing cognitive load. Instead of overwhelming users with raw data, well-designed systems present information in digestible "chunks" that align with the brain's natural processing capabilities. For example, breaking a complex form into several simple steps or using visual groupings to segment related content can dramatically improve user comprehension and decision quality. Cognitive load also varies depending on user expertise. Novices experience higher intrinsic load when facing unfamiliar tasks, requiring more guidance, scaffolding, and simplified interfaces. Experts, on the other hand, can handle greater information density and may even be hindered by overly simplistic designs (Souchet A D., et al, 2022). Adaptive systems that tailor interface complexity to user proficiency can optimize cognitive load dynamically, supporting both learning and performance across diverse user populations. In collaborative decision-making environments, cognitive load is distributed across team members, but this introduces its challenges. Coordination overhead, managing multiple perspectives, and maintaining shared situational awareness can all increase cognitive demands. Professionals in these fields often operate under time pressure, ambiguity, and emotional stress—all of which exacerbate cognitive demands. Interactive systems must be rigorously designed to present only the most relevant information at each decision point, provide clear and timely feedback, and support error prevention. Failure to manage cognitive load effectively in such environments can have catastrophic consequences. In summary, cognitive load profoundly influences the quality of user decision making, the efficiency of interaction, and the overall user experience (Hubballi R B., et al, 2025). Excessive cognitive load leads to errors, fatigue, disengagement, and poor outcomes. Conversely, well-designed systems that manage cognitive demands intelligently promote learning, confidence, and sustained performance. As digital environments grow increasingly complex, the importance of understanding and designing for cognitive load will only intensify. Future interactive systems must prioritize not only what users can do, but what they can do well, comfortably, and sustainably, recognizing that human cognitive resources are both precious and limited.

EMERGING TRENDS IN HCI AND DECISION MAKING: SHAPING THE FUTURE OF DIGITAL INTERACTION AND COGNITIVE SUPPORT

The landscape of Human-Computer Interaction (HCI) is undergoing a profound transformation, particularly in how digital systems support and shape human decision-making. As the volume, complexity, and velocity of information continue to grow, traditional models of interaction are giving way to more adaptive, intelligent, and human-centered approaches. The future of HCI in decision making will be defined by emerging trends in artificial intelligence (AI), multimodal interaction, adaptive user interfaces, ethical and responsible design, cognitive augmentation, and collaborative intelligence (Maheswaran K., et al, 2025). Together, these trends promise to reshape how humans interact with technology, empowering individuals and groups to make better, faster, and more informed decisions, while also raising new challenges for designers, technologists, and society at large. One of the most significant trends is the deepening integration of AI and machine learning into decision support systems. AI-driven interfaces are no longer limited to automating repetitive tasks; they are increasingly taking on roles as active collaborators in complex decision-making. From clinical diagnostics and legal reasoning to financial planning and strategic management, AI systems are being trained to analyze large datasets, detect patterns, and generate insights that would be difficult or impossible for humans to discover alone (Vetrivel S C., et al, 2024). These AI-enhanced systems offer recommendations, highlight anomalies, simulate outcomes, and even co-create solutions with users. As generative AI and large language models (LLMs) advance, conversational agents are also emerging as decision-making partners, providing natural language explanations, answering queries, and engaging users in dialogue to clarify goals and preferences. The rise of AI as a decision collaborator shifts the role of HCI from simply facilitating interaction to orchestrating an effective human-AI partnership, where trust, transparency, and mutual understanding are key. Multimodal interaction is another emerging trend that will reshape how users engage in decision-making. Beyond traditional graphical user interfaces (GUIs), users are now interacting with systems through voice, gesture, gaze, haptics, augmented reality (AR), and even brain-computer interfaces (BCIs) (Fu Z., et al, 2025). These modalities can be combined to create more natural, immersive, and accessible decision environments. For example, an AR-enhanced industrial system might allow an engineer to visualize machine data in situ and adjust settings through hand gestures, while an executive might interact with a strategic simulation through voice and touchscreen simultaneously. Multimodal systems also broaden access for users with disabilities, enabling more inclusive participation in digital decision spaces. As sensors and processing technologies improve, multimodal HCI will allow for richer and more intuitive forms of cognitive engagement with complex information.

Explainability and responsible AI will become increasingly central to the design of decision support systems. As AI takes on greater roles in shaping decisions, users must be able to understand how conclusions are reached. Emerging HCI trends emphasize the importance of explainable AI (XAI)—interfaces that reveal the logic, evidence, and confidence behind AI recommendations. Transparent explanations not only build user trust but also empower users to make informed judgments about when to accept, reject, or further investigate AI-generated insights. Responsible AI design also addresses concerns about bias, fairness, privacy, and accountability—critical factors when AI is used to influence decisions in sensitive domains such as hiring, lending, or criminal justice. Designers must balance performance with ethics, ensuring that systems support human autonomy and align with societal values. Emotion-aware interfaces are also emerging as a frontier in HCI (Tang W., et al, 2025). Advances in affective computing allow systems to detect and respond to users' emotional states through voice tone, facial expression, physiological signals, or behavioral patterns. Emotion-aware systems can adjust content, interaction pacing, or level of support based on the user's stress level, frustration, or engagement. In decision support contexts, this can help mitigate decision fatigue, reduce errors caused by emotional overload, and foster more positive experiences. For example, a tutoring system might encourage a student when they are struggling or suggest a break when signs of fatigue are detected. Emotion-aware HCI introduces new opportunities for empathy and personalization in decision environments, while also raising important questions about privacy and consent. Cross-device and cross-platform decision support is becoming essential in a world where users engage across multiple devices—smartphones, tablets, desktops, wearables, and more. Future HCI systems must provide seamless continuity, allowing users to pick up a decision process where they left off, move fluidly between contexts, and integrate inputs from different sources (Ding Z., et al, 2024). For example, a traveler planning a trip might research destinations on a tablet, book flights on a desktop, and check itineraries on a smartphone, expecting a coherent and consistent decision experience across platforms. Designing for continuity and coherence in multi-device ecosystems is key to reducing cognitive fragmentation and supporting effective decision workflows. In conclusion, the future of HCI and decision-making is being shaped by a confluence of powerful trends: AI integration, adaptive interfaces, multimodal interaction, collaborative intelligence, explainable AI, cognitive augmentation, emotion-aware systems, real-time decision environments, cross-device continuity, and ethical, human-centered design. Together, these trends promise to create richer, more intuitive, and more empowering decision support experiences—while also raising new questions about trust, responsibility, and the nature of human-computer collaboration. As designers, technologists, and society confront these opportunities and challenges, the guiding vision must be clear: to build systems that respect human agency, enhance cognitive potential, and support decisions that are not only smart but also wise and humane.

CONCLUSION

Artificial Intelligence (AI) and Machine Learning (ML) are fundamentally transforming the landscape of decision support systems (DSS) across virtually every domain—healthcare, finance, education, logistics, public policy, and beyond. Traditionally, decision support in Human-Computer Interaction (HCI) focused on organizing and presenting data in a way that allowed human users to analyze, interpret, and choose a course of action. Today, with the emergence of sophisticated AI algorithms and large-scale machine learning models, decision support systems are evolving from passive information providers into active, intelligent collaborators. AI now plays a critical role in enhancing decision quality, improving efficiency, and enabling decisions at scales and complexities that would overwhelm unaided human cognition. Beyond raw data processing, AI enhances decision-making by automating complex analytical tasks, reducing the cognitive load on human users. Techniques such as natural language processing (NLP), computer vision, and predictive modeling allow AI to transform unstructured data—text, images, videos—into actionable insights. For example, legal professionals can use AI-powered document review tools to quickly extract key clauses and precedents from large volumes of legal texts. In journalism, AI can analyze social media sentiment, news trends, and public discourse to inform editorial decisions. In these contexts, AI serves as a cognitive amplifier, allowing human decision makers to focus on higher-level reasoning and judgment rather than routine analysis. An emerging frontier of AI in decision support is its integration into conversational interfaces. Powered by large language models (LLMs) and advanced NLP, conversational agents can now engage users in natural language dialogues to support decision making. These agents can explain options, clarify user goals, provide contextual knowledge, and guide users through decision processes in an interactive, human-like manner. For example, in healthcare, AI-driven virtual assistants can answer patient questions, remind users about medications, and support lifestyle decisions. In enterprise settings, AI chatbots can assist employees in navigating complex internal knowledge bases to inform business decisions. The conversational modality lowers barriers to entry, supports more inclusive interaction, and makes decision support accessible to users with varying levels of expertise. However, as AI takes on greater roles in decision-making, trust and explainability become paramount. Users must understand how AI systems generate their outputs to make informed judgments about when to rely on them. Explainable AI (XAI) techniques—such as feature attribution, counterfactual examples, and visual explanations—are being developed to open the "black box" of AI reasoning. Moreover, AI-driven decision systems must respect user autonomy, providing clear choices rather than manipulative nudges or default paths that overly steer user behavior. Particularly in sensitive domains—healthcare, justice, employment—AI-supported decisions must remain transparent, contestable, and subject to human oversight. AI's expanding role in decision support also intersects with emerging interface paradigms. Multimodal interaction—combining voice, touch, gesture, gaze, and haptics—enables more natural and intuitive engagement with AI-powered systems. Augmented reality (AR) and virtual reality (VR) environments are being used to immerse decision makers in rich data contexts, enhancing spatial understanding and collaborative sensemaking. Brain-computer interfaces (BCIs), though still experimental, hold potential for even deeper integration between AI and human cognition. These technologies are extending the boundaries of how AI can support, amplify, and integrate with human decision-making. In conclusion, AI and machine learning are revolutionizing decision support by enabling systems to process vast data, generate actionable insights, personalize experiences, simulate scenarios, and engage users in naturalistic dialogue. The role of AI is evolving from tool to collaborator, augmenting human cognition and enabling decision-making at new levels of complexity and speed. At the same time, challenges of transparency, bias, trust, and ethics must be addressed to ensure that AI-supported decisions remain aligned with human values and social good. The future of HCI and decision support will be defined by how effectively we integrate AI's capabilities with human judgment, creativity, and ethical reasoning, creating systems that not only enhance decisions but also enrich the human experience of making them.

REFERENCES

Ahmad, M., & Wu, Q. (2024). Heuristic-driven biases as mental shortcuts in investment management activities: A qualitative study. *Qualitative Research in Financial Markets*, *16*(2), 291–309. DOI: 10.1108/QRFM-10-2022-0167

Alkatheiri, M. S. (2022). Artificial Intelligence Assisted Improved Human-Computer Interactions for Computer Systems. *Computers & Electrical Engineering*, *101*, 108024. Advance online publication. DOI: 10.1016/j.compeleceng.2022.107950

Antomarioni, S., Lucantoni, L., Ciarapica, F. E., & Bevilacqua, M. (2021). Data-Driven Decision Support System for Managing Item Allocation in an ASRS: A Framework Development and a Case Study. *Expert Systems with Applications*, *185*, 115622. Advance online publication. DOI: 10.1016/j.eswa.2021.115622

Augstein, M., Herder, E., & Wörndl, W. (2023). *Personalized Human-Computer Interaction*. De Gruyter., DOI: 10.1515/9783110988567

Awan, U., Shamim, S., Khan, Z., Zia, N. U., Shariq, S. M., & Khan, M. N. (2021). Big data analytics capability and decision-making: The role of data-driven insight on circular economy performance. *Technological Forecasting and Social Change*, *168*, 120766. Advance online publication. DOI: 10.1016/j.techfore.2021.120766

Bala, M. Y., & Damla, K. (2021). "A Review of Human-Computer Interaction Design Approaches Towards Information Systems Development." BRAIN. *Broad Research in Artificial Intelligence and Neuroscience*, *12*(1), 229–250. DOI: 10.70594/brain/12.1/180

Bin Ahsan, W. (2024). *The EDIT UX Framework: A User-Centered Approach to Effective Product Redesign*. Userhub., DOI: 10.58947/PBSY-TGJR

Blackley, C., Redmond, P., & Peel, K. (2021). Teacher Decision-Making in the Classroom: The Influence of Cognitive Load and Teacher Affect. *Journal of Education for Teaching*, *47*(4), 548–561. DOI: 10.1080/02607476.2021.1902748

Chowdhary, S., Selvakumar, P., Sangeetha, B., Sarvaani, V. V., Anute, N., & Sharma, M. (2025). The Role of Social Capital in Entrepreneurial Development. In Martínez-Falcó, J., Sánchez-García, E., & Marco-Lajara, B. (Eds.), *Support Networks for Entrepreneurs: Social Capital and Strategic Collaborations* (pp. 311–338). IGI Global Scientific Publishing., DOI: 10.4018/979-8-3693-9836-4.ch013

Coates, J. M., Santhanalakshmi, K., Dinesh, N., Gandhimathi, S., & Chandel, P. S. Mohit & Selvakumar P, (2025). "Assessment and Evaluation in Transnational Staff Development." In G Morris & J Ye (Eds.), Training and Development in Transnational Higher Education (pp. 387–406). IGI Global Scientific Publishing. https://doi.org/DOI: 10.4018/979-8-3373-1897-4.ch019

Ding, Z., Ji, Y., Gan, Y., Wang, Y., & Xia, Y. (2024). Current Status and Trends of Technology, Methods, and Applications of Human–Computer Intelligent Interaction (HCII): A Bibliometric Research. *Multimedia Tools and Applications*, *83*(27), 69111–69144. DOI: 10.1007/s11042-023-18096-6

Forcina, A., Silvestri, L., De Felice, F., & Falcone, D. (2024). Exploring Industry 4.0 Technologies to Improve Manufacturing Enterprise Safety Management: A TOPSIS-Based Decision Support System and Real Case Study. *Safety Science*, *169*, 106351. Advance online publication. DOI: 10.1016/j.ssci.2023.106351

Fu, Z., Barbara, A., & Scupelli, P. (2025). *Digital Futures in Human-Computer Interaction: Design Thinking for Digital Transformation.* CRC Press., DOI: 10.1201/9781032693606

Han, Y., Diao, Y., Yin, Z., Jin, R., Kangwa, J., & Ebohon, O. J. (2021). Immersive Technology-Driven Investigations on Influence Factors of Cognitive Load Incurred in Construction Site Hazard Recognition, Analysis and Decision Making. *Advanced Engineering Informatics*, *48*, 101298. Advance online publication. DOI: 10.1016/j.aei.2021.101298

Hooda, M., Rana, C., Dahiya, O., Rizwan, A., & Hossain, M. S. (2022). Artificial Intelligence for Assessment and Feedback to Enhance Student Success in Higher Education. *Mathematical Problems in Engineering*, *2022*, 5215722. Advance online publication. DOI: 10.1155/2022/5215722

Howie, E. E., Dharanikota, H., Gunn, E., Ambler, O., Dias, R., Wigmore, S. J., Skipworth, R. J. E., & Yule, S. (2023). Cognitive Load Management: An Invaluable Tool for Safe and Effective Surgical Training. *Journal of Surgical Education*, *80*(3), 311–322. DOI: 10.1016/j.jsurg.2022.12.010 PMID: 36669990

Hubballi, R. B., Selvakumar, P., Seenivasan, R., Basava Aradhya, S. G., Dinesh, N., & Seelam, P. K. (2025). Overview of Current AI Technologies in Education. In Kyei-Blankson, L., & Ntuli, E. (Eds.), *Transformative AI Practices for Personalized Learning Strategies* (pp. 1–26). IGI Global Scientific Publishing., DOI: 10.4018/979-8-3693-8744-3.ch001

Jain, J., Walia, N., Singla, H., Singh, S., Sood, K., & Grima, S. (2023). Heuristic biases as mental shortcuts to investment decision-making: A mediation analysis of risk perception. *Risks*, *11*(4), 72. Advance online publication. DOI: 10.3390/risks11040072

Jenke, L., Bansak, K., Hainmueller, J., & Hangartner, D. (2021). Using eye-tracking to understand decision-making in conjoint experiments. *Political Analysis*, *29*(1), 75–101. DOI: 10.1017/pan.2020.11

Little, A. S., & Wu, S. J. (2021). Cognitive bias and neurosurgical decision making. *Journal of Neurosurgery*, *137*(1), 307–312. DOI: 10.3171/2021.9.JNS212058 PMID: 34972092

Madaan, A., Tandon, N., & Gupta, P.. (2023). Self-refine: Iterative Refinement with Self-Feedback. *Advances in Neural Information Processing Systems*, *36*, 88–102. DOI: 10.48550/arXiv.2303.17651

Maheswaran, K., Selvakumar, P., Shaikh, M., Shashank, G., Gupta, S., & Manjunath, T. C. (2025). The Evolution of Smart Urban Mobility With Emerging Technologies. In Ponnusamy, S., Assaf, M., Youssef, B., Jeon, G., & Kalyanaraman, S. (Eds.), *Leveraging VANETs and Blockchain Technology for Urban Mobility* (pp. 1–28). IGI Global Scientific Publishing., DOI: 10.4018/979-8-3373-0265-2.ch001

Mangus, C. W., & Mahajan, P. (2022). Decision making: Healthy heuristics and betraying biases. *Critical Care Clinics*, *38*(1), 37–49. DOI: 10.1016/j.ccc.2021.07.002 PMID: 34794630

Monsell, S. (2021). "Control of mental processes." In Unsolved Mysteries of the Mind, pp. 93–148. DOI: DOI: 10.4324/9781315784960-4

Mumali, F. (2022). Artificial Neural Network-Based Decision Support Systems in Manufacturing Processes: A Systematic Literature Review. *Computers & Industrial Engineering*, *165*, 107964. Advance online publication. DOI: 10.1016/j.cie.2022.107964

Nisha Angeline, M., Selvakumar, P., Gupta, M. K., Nazar, N., Elumalai, D., & Bhattacharya, S. (2025). Emerging Materials for Energy Storage and Environmental Applications. In Muduli, K., Moharana, B., Ales, S., & Biswal, D. (Eds.), *Using Computational Intelligence for Sustainable Manufacturing of Advanced Materials* (pp. 53–72). IGI Global Scientific Publishing., DOI: 10.4018/979-8-3693-7974-5.ch003

Niu, Y., Ying, L., Yang, J., Bao, M., & Sivaparthipan, C. B. (2021). Organizational business intelligence and decision making using big data analytics. *Information Processing & Management*, *58*(6), 102725. Advance online publication. DOI: 10.1016/j.ipm.2021.102725

Sahoo, S. K., & Goswami, S. S. (2023). A comprehensive review of multiple criteria decision-making (MCDM) Methods: Advancements, applications, and future directions. *Decision Making Advances*, *1*(1), 25–48. DOI: 10.31181/dma1120237

Sajadi, A. (2024). Optimizing Data Input Interfaces Through User Error Analysis: A Statistical and HCI-Driven Approach. *International Journal of Advanced Human Computer Interaction*, *2*(2), 67–77. DOI: 10.5281/zenodo.10543292

Sethumadhavan, R. P. P. S., Selvakumar, P., Kadu, S. G., Chandel, P. S., & Gupta, S. (2025). Environmental Sustainability and Employee Participation. In Şeker, C., Örücü, E., & Kaya, A. (Eds.), *Green Leadership Strategies for Increased Employee Engagement and Performance* (pp. 275–298). IGI Global Scientific Publishing., DOI: 10.4018/979-8-3373-2106-6.ch009

Souchet, A. D., Philippe, S., Lourdeaux, D., & Leroy, L. (2022). Measuring Visual Fatigue and Cognitive Load via Eye Tracking While Learning with Virtual Reality Head-Mounted Displays: A Review. *International Journal of Human-Computer Interaction*, *38*(9), 801–824. DOI: 10.1080/10447318.2021.1976509

Tang, W., Chen, S., Shao, J., Du, X., Wu, J., Zhang, X., Niu, Y., & Xue, C. (2025). Emerging Themes and Future Directions in Neurodesign and Human-Computer Interaction: A Systematic Review. *International Journal of Human-Computer Interaction*, *41*(2), 1652–1673. DOI: 10.1080/10447318.2024.2362486

Thirunagalingam, A., & Selvakumar, P. (2025). Application of Automation and Robotics in the Oil and Gas Industry. In Abdullayev, V., & Khang, A. (Eds.), *Revolutionizing AI and Robotics in the Oil and Gas Industry* (pp. 229–244). IGI Global Scientific Publishing., DOI: 10.4018/979-8-3693-8156-4.ch011

Upadhyaya, A. N., Debnath, P., Rani, P., Selvakumar, P., Manjunath, T. C., & Bhattacharya, S. (2025). AI-Based Disaster Prediction and Early Warning Systems. In Mobo, F. (Ed.), *AI and Emerging Technologies for Emergency Response and Smart Cities* (pp. 55–74). IGI Global Scientific Publishing., DOI: 10.4018/979-8-3693-9770-1.ch003

Vetrivel, S. C., Sowmiya, K. C., Sabareeshwari, V., & Arun, V. P. (2024). Navigating the Digital Economy: The Crucial Role of Human-Computer Interaction. In *Social Reflections of Human-Computer Interaction in Education, Management, and Economics* (pp. 88–102). IGI Global., DOI: 10.4018/979-8-3693-3033-3.ch010

Vuong, Q. H. (2023). *Mindsponge theory*. Walter de Gruyter., DOI: 10.2478/9788367405157

Wang, X., & Yin, M. (2021). "Are explanations helpful? A comparative study of the effects of explanations in AI-assisted decision-making." *Proceedings of the 26th International Conference on Intelligent User Interfaces (IUI '21)*, pp. 26–36. DOI: DOI: 10.1145/3397481.3450650

Wien, A. H., & Peluso, A. M. (2021). Influence of human versus AI recommenders: The roles of product type and cognitive processes. *Journal of Business Research, 137*, 13–27. DOI: 10.1016/j.jbusres.2021.08.016

Zhang, Z., & Hyland, K. (2023). The Role of Digital Literacy in Student Engagement with Automated Writing Evaluation (AWE) Feedback on Second Language Writing. *Computer Assisted Language Learning*. Advance online publication. DOI: 10.1080/09588221.2023.2256815

Zorzetti, M., Signoretti, I., Salerno, L., Marczak, S., & Bastos, R. (2022). Improving agile software development using user-centered design and lean startup. *Information and Software Technology, 141*, 106718. Advance online publication. DOI: 10.1016/j.infsof.2021.106718

Chapter 4
Explaining Hallucination of Artificial Intelligence in Data-Driven Decision-Making

Zian Shah Kabir
https://orcid.org/0000-0001-8710-5283
University of Technology Sydney, Australia

ABSTRACT

Artificial intelligence transforms a paradigm shift in delivering data-driven innovative knowledge, where hallucinated information can influence users' decision-making. It demands exploring the effects of hallucinations towards the trustworthiness and ethics of AI in decision-making. Thus, this chapter aims to explain the implications of hallucinations across data-driven decision-making. The chapter conceptualizes a framework for explaining a process flow of mitigating AI-driven hallucinations that engages critical flaws in promoting innovative and strategic decision making for professional practices. Theoretically, the framework explains the effects of hallucinations and how it impacts decision making, following multiple theories and models. We present a metatheoretical framework to study the hallucinations of LLMs-based AI, which will allow researchers and practitioners to capture a multidisciplinary theoretical lens to investigate the complexities of AI-driven hallucinations in the data-driven innovative decision-making process.

DOI: 10.4018/979-8-3373-5702-7.ch004

INTRODUCTION

The 21st century stands as a period of unprecedented technological advancement, largely driven by the rapid evolution of Artificial Intelligence (AI). From its nascent development rooted in early computational theories and symbolic reasoning, AI has advanced through distinct periods of fervent optimism and challenging setbacks, often referred to as "AI Springs" and "AI Winters" (Haenlein & Kaplan, 2019). The AI era represents a new "Spring," fueled by breakthroughs in machine learning, vast datasets, and increased computational power, leading to incorporate AI systems capabilities of performing tasks once considered exclusively within the realm of human intellect. This accelerating progress is not only introducing new tools for reshaping industries, economies, and the new fabric for society, but also presenting critical challenges in disseminating false realities and misinformation through hallucinations (Rawte et al., 2023). However, detection capabilities of those hallucinations can determine a data-driven decision-making process through change management, pre-learning, reconfiguration, and governance of AI (Guida et al., 2023; Williamson & Prybutok, 2024). The core research problem addressed in this chapter is to understand and prepare the adaptations of mitigating hallucinations toward data-driven decision-making in complex situations, particularly rapidly approaching AI-dominated industries.

The integration of AI into organizational decision-making, especially in critical functions such as autonomous driving, remote surgical operation, smart logistics, recruitment, vendor selection, and performance evaluations, is rapidly increasing across various sectors. AI-driven tools promise efficiency, consistency, and objectivity in delivering tasks. However, their deployment has raised significant ethical concerns around algorithmic bias and hallucinations, transparency, trustworthiness, and fairness (Hunkenschroer & Kriebitz, 2022; Tabassam et al., 2023). The well-known case of Amazon's biased hiring algorithm in 2018 exemplifies the potential ethical pitfalls of poorly governed AI systems, illustrating severe implications for fairness and equality (Hunkenschroer & Kriebitz, 2022). Despite acknowledging these ethical and AI hallucinated risks, many organizations lack robust governance frameworks capable of effectively aligning ethical principles with technical AI development. However, addressing those AI hallucinations in data-driven decision-making is still unexplored. A key challenge is the translation of those complex ethical and hallucination concepts that often lead to falsifications or misinterpretations in practical implementation (Hartmann et al., 2022).

The growing adoption of AI-powered autonomous systems in business operations has brought critical needs to alleviate hallucinations for AI in delivering the data-driven decision-making. These hallucinations may inadvertently be reinforced in decision settings with AI assistance. Humans might have too much confidence in

AI-generated results and take advice without question, which could lead to less-than-ideal or even unexpected decision outcomes (Ahmed, 2025). Addressing the risk of hallucination and bias undermining judgment is becoming increasingly crucial as AI tools become more integrated into the data-driven decision-making procedures. Although AI may improve precision in taking decisions, its proper application requires practitioners to retain a critical mindset and acknowledge the limitations of automated systems. The main goal of this chapter is to find mitigation strategies to lessen hallucinations and improve data-driven decision-making processes.

Recently, the rapid adoption of generative AI highlights the enormous potential of large language models (LLMs) in solving complex decision making (Zhao et al., 2023). These models excel in natural language processing (NLP) tasks such as context awareness and question-answering. With billions of parameters, LLMs have approached or surpassed human-level performance in language comprehension and knowledge synthesis. AI has achieved remarkable success in various tasks; however, it occasionally can deliver information that is factually incorrect or inconsistent with the expected output, a phenomenon known as hallucination (Li et al., 2024; Guerreiro et al., 2022). Furthermore, these models can generate misinformation that appears credible but is inaccurate (Huang et al., 2024). This chapter addresses how the hallucination issue significantly limits the deployment of AI models in critical service delivery processes through a data-driven decision-making framework, where accuracy is paramount (Li et al., 2024).

As AI plays an increasingly central role in shaping the digital economy, their ethical and trust implications, particularly regarding accountability, fairness, and transparency, have emerged as critical concerns. Mitigating hallucinations and misinformation produced by LLMs is a relatively recent area of research that has gained increasing significance to develop decision-making framework that can enhance the quality of advanced LLM-generated outcomes (Liu et al., 2024). This chapter uniquely presents a framework that identifies the process mechanisms of managing the hallucination of LLMs-based AI towards data-driven decision-making. Against this backdrop, the chapter aims to explicate the prevention, detection, and mitigation of hallucination of LLMs. As LLMs are a multifaced and multidimensional idea, this chapter, therefore, adopts a multi-theoretical approach to ground up the discussion of mitigation for hallucination in LLMs to present an expectation-execution-mitigation framework in decision-making. The framework is underpinned by cognitive system theory (CST), dynamic capability view (DCV), and social license to operate (SLO) that encompass cognition capability, and community within an AI-driven organizational context.

With the advent of AI's speed and ease of use, establishing national ethical standards and ensuring credible information have become imperative for society. AI has reached its forward steps in broad use for clinical applications, supply chain, agricultural activities, industrial operations, smart city applications, and autonomous driving. However, developers, professionals, governments, and regulators can follow AI ethics that ensure quality deliberations of performance by alleviating hallucinations. The chapter explores the issues by examining hallucinations and offering solutions to mitigate these hallucinations to maintain the sanctity of data-driven decision-making.

The chapter begins with an introduction to the significance of research, followed by the aims and objectives, which outlines the intent of the proposed decision-making framework. The literature review section then evaluates existing research and identifies existing gaps, leading to a discussion with the significance of the chapter to justify its contribution. Then, the chapter describes future directions, along with a conclusion that summarizes the key components of the chapter.

LITERATURE REVIEW

Artificial Intelligence Models

Artificial Intelligence is an advanced and interactive technology where a system or machine performs tasks using human intelligence through a combination of algorithms, iterative improvements, access, and data processing (Haenlein & Kaplan, 2019). In this context, machine learning is a subset of AI which is an algorithm that makes a system absorb data and act gradually without being explicitly programmed.

AI enhances analytical capabilities through different types of models that can allow a system to execute complex tasks autonomously and move toward a new paradigm shift in data analytics using contextual data sources. Recent studies discussed how large language models integrate AI systems with social actors, users, and contextual training data, where the roles of different agents can influence autonomous AI actions (Umbrello & Natale, 2024). However, there might be risks and data inconsistencies due to inadequate data sources for initial trained data. The process flow of misinformation with risk profiles in AI is still unrealized, particularly in ensuring information authenticity and reliability through data-driven decision-making (Aquino et al., 2022).

AI models denote an autonomous system that enhances functional capabilities for decision-making based on pre-trained data, as well as approaching data related to user preferences and target-oriented behavior. Further, previous studies define AI models like LLM as an information system that can act autonomously and acclimate to recommendations and opinions that subsequently influence data-driven decision-making (Moniz, 2021). This is a nascent field of study that can address the adverse effects of AI and how to mitigate the problems to advance the capabilities of existing AI systems that can support real-world applications. The AI engages in advanced analytical options to pursue purposes, and contexts, and acclimatize dynamic situations. However, the effects of AI on data-driven innovation can be investigated, considering hallucinations.

Hallucination In Artificial Intelligence

Before defining hallucination, it is crucial to understand the concept of LLM and the training stages of LLM. Typically, LLMs refer to a series of general-purpose models that leverage the transformer-based language model architecture and undergo extensive training. Previous studies explain three stages: (i) the foundational stage as pre-training to gather knowledge and capacities; (ii) supervised fine-tuning; and (iii) reinforcement learning (Umbrello & Natale, 2024).

LLMs are known for their extensive world knowledge, which enables them to handle a variety of tasks. However, they are particularly prone to hallucination, especially in open-domain scenarios (Huang et al., 2024). As use of AI models expands, critical concerns about misinformation have intensified (Liu et al., 2024). The generation of biased or misleading information exhibits enormous challenges for real-world applications (Lin et al., 2024) and undermines the credibility of LLMs, making them unsuitable for data-driven decision-making (Rawte et al., 2023) as well as for generating sensitive outputs. Misinformation produced by LLMs can range from unintentional factual errors to deliberately deceptive content, leading to severe consequences in complex adaptive environments (Liu et al., 2024).

Thus, hallucination can be defined as the tendency to produce incorrect information, poses a major challenge to ensuring the reliability and trustworthiness of LLMs in practical applications (Li et al., 2024). In NLP, this term is referred to as an instance, where the AI model generates nonsensical or off-topic content that deviates from the source material (Ji et al., 2023), drawing a parallel phenomenon to hallucinations as occurred in human psychology.

The recent advancements in LLMs have engaged dependencies in AI-driven analytics. However, hallucinations create an obstacle to disseminating true insights from AI analytics, which demands extending the research to develop a comprehensive framework that can explain mitigating hallucination in LLMs, especially for human-AI interactions (Rawte et al., 2024; Umbrello & Natale, 2024). Recently, the research phenomenon of addressing hallucination in agentic AI through LLMs has been studied but not emphasized on explaining causes of hallucination and the mitigation process, and there is a scope to develop a framework considering relevant theories.

Umbrello and Natale (2024) described that misinformation and inconsistencies, existing knowledge boundaries, risk alignment, decoding strategies, over-confidence, and reasoning failure would be significant factors in defining causes of hallucinations in LLM from training data. At the same time, the detection of hallucination can provide a benchmark and faithfulness with pre-trained data (Rawte et al., 2023). Further, mitigating data-related hallucinations can be investigated further to describe data-driven decision-making for confirming data evaluation through benchmark data filtering and editing the retrieval-augmented generation process.

Data-Driven Decision Making

In a heterogenous environment, there are diverse types of uncertainties, where AI can employ data-driven decision-making. Decision-making was initially defined as a mental state in performing a course of action after gathering observations from the respective environment. In system engineering, the process is explained through sequential steps: determining the problem, analyzing the situation, exploring and assessing options, and making decisions. With the evolution of advanced technologies, the decision-making process is not only related to human intelligence, but also it engages machine intelligence. Therefore, decision making is a fundamental process that involves human-machine interaction, socio-technological impacts, uncertainties, ethics, and moral judgement.

The monetization of decision-making through data-driven models has the potential to transform business performance, particularly for organizations with extensive data resources. Data-driven decision-making refers to the process of choosing an option by justifying valuable insights and maximum benefits from the available data and confirming it after the fulfilment of perceived experiences (Moniz, 2021). By leveraging data analytics of AI, organizations can develop innovative solutions that can provide strategic advantages, improving both innovative decision-making and overall performance (Aquino et al., 2022). Many companies are already utilizing advanced AI analytics within data-rich ecosystems to gain a competitive edge in data-driven decision-making. However, hallucinations create an obstacle to disseminating true insights from AI analytics, which demands extending the research to develop a comprehensive framework that can explain mitigating hallucination in LLMs, especially for human-AI interactions (Rawte et al., 2024). Considering this, this chapter can advocate for the adoption of data-driven decision-making to enhance internal operations, refine decision-making, and drive innovation by mitigating hallucinations in the context of AI (Wixom & Ross, 2017).

Hallucination In The Data-Driven Decision-Making

In recent years, industries have adapted different types of AI models, particularly LLMs for various data-driven innovations. For example, Amazon's autonomous recommendation engine utilizes predictive analytical models in data products (Jagannathan & Udaykumar, 2020). Furthermore, Google allows users to interact with visual analytics-based platforms that disseminate automated recommendations to influence users' data-driven decision-making by exploiting users' preference through every possible touchpoint (Novet, 2017). Leading platforms for different industries have established data-driven innovative features through advanced analytical tools, using advanced AI models (Umbrello & Natale, 2024; Jakubik et al., 2024). However, hallucinations of AI may pose critical challenges with unpredictability and misinformation, which can be addressed to detect and mitigate the associated risks (Williamson & Prybutok, 2024; Rawte et al., 2023). There is limited knowledge of explaining the process flow to understand hallucinations in AI platforms, considering data-driven decision-making (Moniz, 2021).

Previous studies classify hallucination in LLM-based AI through factuality: LLM's output contains facts that can be grounded in real-world information but present contradictions; and faithfulness that the use of LLMs can shift towards more user-centric applications, demanding instruction, context, and logical inconsistency (Umbrello & Natale, 2024; Rawte et al., 2023). However, those grounded classifications of hallucinations of LLMs can be investigated to develop a framework for data-driven decision-making that can exhibit unique features such as continuity, interactivity, and parallelism within the company's platform or digital ecosystem.

Adopting AI-based applications allows access to interact with training data to optimize user feedback. At present, with advances in LLMs, current AI enable digital applications supported by machine learning to offer values through recommendations by addressing user preferences, contextual information, and coordinated efforts (Rawte et al., 2024). The emergence of AI can expose hallucination in delivering data-driven decision-making, which can be explored through a comprehensive framework. At different stages of decision-making, AI hallucination can confuse users through unwanted messages that provoke adverse effects (Williamson & Prybutok, 2024). For example, misinformation, deception, and user beliefs can negatively impact decision-making, and hallucination can create an ambiguous situation. Further, the data acquisition and refinement stage raises training data issues that can cause biases and inconsistencies, leading to hallucinations in AI (Clatterbuck et al., 2024). Finally, the feedback loop, recommendations, and multi-modal interactions with platforms can cause unwanted and adversarial outcomes through AI platforms that may hamper data-driven decision-making.

AI deception through misinformation can result in delivering unwanted messages, discriminatory information, restricting access to information for users, predatory marketing, and defamation (Umbrello & Natale, 2024) that may result in a hallucination of LLM-based AI. Therefore, protecting users from unfavorable impacts within data-driven decision-making through AI models is an exciting research avenue.

Main Focus of the Chapter

The LLMs-based AI system can improve decision-making through real-time processing and pre-training data simulations that act as an advisory agent. The system can influence humans and machines in their decision-making for complex and uncertain situations (Jarrahi, 2018). However, AI complement human decision-making through faster responses but may occur in creating hallucinations, misinformation, algorithmic bias, data privacy violations, information accuracy, and ethical violations, which can build mistrust towards AI-driven machine intelligence (Davenport et al., 2019). If professionals are over-reliant on AI technology and have a lack of knowledge and expertise in assessing suggested alternatives, it can create long-term risks for any organization.

The attributes and behaviors of LLMs are deeply influenced by their training stages, and LLMs typically undergo three main phases: pre-training, supervised fine-tuning (SFT), and reinforcement learning from human feedback (RLHF) (Huang et al., 2024). Understanding these stages is crucial for analyzing the origins of hallucinations in LLMs, as this chapter addresses each phase that can create hallucinations in distinct characteristics. Pre-training is widely regarded as a foundational stage for LLMs to confirm innovative decision-making that supports AI capabilities (Zhou et al., 2024). During this phase, LLMs perform autoregressive prediction of subsequent tokens in sequences (Huang et al., 2024). Through self-supervised training on large-scale textual corpora, LLMs gain an understanding of language syntax, world knowledge, and reasoning abilities, thereby establishing a solid foundation for further fine-tuning. Moreover, Previous study suggests that predicting subsequent words in this context is like compressing important information without loss that can engage in data-driven decision-making (Delétang et al., 2023). At its core, there is a scope to extend the model's task to predict the probability distribution of upcoming words, with accurate predictions reflecting a deep grasp of knowledge and a sophisticated understanding of detecting and mitigating hallucinations before taking data-driven decision-making (Huang et al., 2024).

This chapter further identifies a few critical factors that can engage in hallucinations in AI-driven systems. Firstly, biased and inappropriate data can mislead AI systems to generate falsified information that impacts taking inaccurate decisions. Secondly, complex situations with a lack of human interactive and contextual data can deliver fabricated outputs. Thirdly, it can breach ethical standards if the AI system is not properly trained through data privacy, accountability, honesty, and norms, following professional code of ethics. Finally, inappropriate interpretations and explanations can affect user trust in decision-making.

Hallucination remains a significant challenge in AI, leading to extensive research on the emerging topic (Ji et al., 2023; Rawte et al., 2024). While prior studies have examined LLM hallucinations from different perspectives, Huang et al. (2024) provide a more targeted analysis of existing contributions in decision-making. Further, Ji et al. (2023) focused on hallucinations in pre-trained models for Natural Language Generation (NLG) but did not specifically analyze the effects of hallucinations for LLMs in decision-making. However, Tonmoy et al. (2024) explored mitigation techniques for LLM hallucinations, while Liu et al. (2023) investigated LLM trustworthiness broadly without emphasizing hallucinations. In another context, Moniz conducted research to explain data-driven decision-making (Moniz, 2021), particularly to ensure factuality and faithfulness by alleviating hallucinations in AI. However, this chapter can conceptualize a detailed data-driven decision-making framework by addressing a distinctive taxonomy and structure, defining detection, execution, and mitigation of hallucinations and directly linking mitigation strategies in taking data-driven decision-making (Huang et al., 2024).

Detection and mitigation strategies through real-time experience and decision-making tools can balance AI performance with external human factors to create robust and data-driven decision-making processes (Tonmoy et al., 2024). The balance between AI and humans supports judgmental decision making by preventing hallucinations that can build user trust in complex and critical situations. However, a comprehensive decision-making framework can be developed for enhancing accuracy, faster responses, real-time insights, strategic decisions, risk management, and data-driven innovations to explore predictive and responsive decision-making.

Expectation-Execution-Mitigation Framework

This chapter investigated different theories, such as – information processing theory, dynamic managerial capability, Decision Field Theory (DFT), and social license to operate (SLO), that were applied to understand the hallucination of AI in data-driven decision-making through functionalities, capabilities, ethical values, and technological implications.

Different theories and models have been applied to assess the hallucination of AI in decision-making, particularly with the advancement of AI (Umbrello & Natale, 2024). Various frameworks exist in the context of AI in decision-making, including theories and models such as the information processing theory (IPT), dynamic managerial capability (DMC), Decision Field Theory (DFT), and social license to operate (SLO) (Newell, 1992; Helfat & Peteraf, 2003; Busemeyer & Townsend, 1993; Moffat et al., 2016). However, a research framework can be developed following theories, considering inconsistencies through hallucinations in AI to explain data-driven decision-making.

The information processing theory (IPT) can be applied to explain the information processing of AI-driven platforms through data sources, retrieval, and reinforcement as explained by the human mind process like cognitive psychology (Newell, 1992), which usually can postulate a framework for explaining the information filtering, data architecture, and decision-making for machine intelligence (Guida et al., 2023). Further, the IPT model can explain how AI processes information that considers hallucination to explain a framework with neural information processing toward data-driven decision-making (Moniz, 2021). The strengths of the IPT model can be associated with AI in three stages: how information is gathered and processed, and how it provides intelligent decisions through pre-training data. Mainly, assessing hallucination in a dynamic AI system is crucial because the system delivers the decision autonomously, based on the processed information through training data that may provide inconsistencies or unwanted information to users, and subsequently can detect hallucination leaned from outcomes and adjust decision-making processes (Williamson & Prybutok, 2024). However, the theory can be applied to develop a framework that explains hallucination through AI models in the context of data-driven decision-making.

In another context, dynamic managerial capability (DMC) defines the organizational capacity to explain adaptability, change effect, learning, reconfiguration, and system governance with external sources of information. However, Helfat and Peteraf (2003) apply DMC to explain systems capabilities through continuous adaptations and manage to maintain qualities in dynamic environments, including regulatory shifts, technological change, ethical code of practices, and market dynamics, where the theory can be used for AI as a technological system to evolve AI's capabilities through reinforced learning, improvements of algorithms, and re-adjustments based on human-AI interactions. The dynamic nature of AI means that its creators and managers must be able to manage these constant changes in operational environments through proper execution. Effective management of AI requires leaders to have dynamic managerial capability by enhancing decision making to oversee its learning, and integration into existing systems. This is crucial for aligning AI with evolving business or societal goals in complex situations.

The Decision Field Theory (DFT) shows how decision-making can evolve through human-machine interactions over time, considering the critical level of uncertainties (Busemeyer & Townsend, 1993). The DFT theory explains the process flow of getting valuable insights through data-driven decision-making, considering hallucinations. A recent study followed by the DFT to examine the decision-making process, and this develops a framework to adopt the functionalities of AI on data-driven decision-making (Hancock et al., 2024). The chapter applies to the DFT in explaining the mitigation process for the AI-driven decision-making framework.

The Social License to Operate (SLO) was initially rooted in the mining industry and expanded to other sectors, including oil, gas, dam building, and supply chain management (Lee et al., 2019; Demajorovic et al., 2019). The SLO process includes promoting, gaining, and maintaining community support, which requires companies to align their activities with stakeholders' expectations. The three levels of SLO—legitimacy, credibility, and trust—are critical for maintaining continuous community approval (Thomson & Boutilier, 2011), and companies must focus on building trust and addressing community concerns to ensure long-term approval. However, the SLO can be applied to explain the dependencies of AI with the final level of information dissemination without hallucinations in AI platforms (Thomson & Boutilier, 2011). This chapter addresses the SLO to understand critical factors influencing community engagement, trust, fairness, sustainability concerns, and local benefits of using AI by alleviating hallucination (Lee et al., 2019).

This chapter develops a standardized framework to address hallucination, where the process flows of hallucinations in AI that can involve unwanted recommendations to users, provoking data-driven decision-making. The proposed framework (as shown in Figure 1) explains the process flow of AI in delivering data-driven decision-making with the shadow side of hallucinations. Firstly, the framework includes expectation for hallucinations of AI in the process of data-driven decision-making that can explain the information filtering process, training data accommodation, interactions capability, legitimacy, cultural values, and ethical standards.

Then, the execution phase outlines identifications of processing information as outcomes of AI toward data-driven decision-making to address inspection mechanisms, misinformation, evaluative techniques, decision judgment, and stakeholder engagement. This stage is crucial for data analytics to identify hallucinations in the AI platform to ensure data-driven innovation that ultimately influences data-driven decision-making.

Finally, the mitigation phase describes the confirmation stage to achieve the data-driven outcomes from the AI by minimizing autonomous & dynamic flow, inconsistencies, unwanted decision-making, reasoning failure, inferior alignment, over-confidence, and inadequate information retrieval process. This stage is critical for developers to make strategies to get optimum values from AI platforms.

The expectation-execution-mitigation framework includes data analytics, intelligent sensing, causal understanding, ethical and cultural values, and human-AI collaboration. It moves beyond static decision trees or dashboards to a complex adaptive system that learns, reasons, and supports critical decision-making in connected healthcare.

Figure 1. Expectation-Execution-Mitigation Framework in decision making

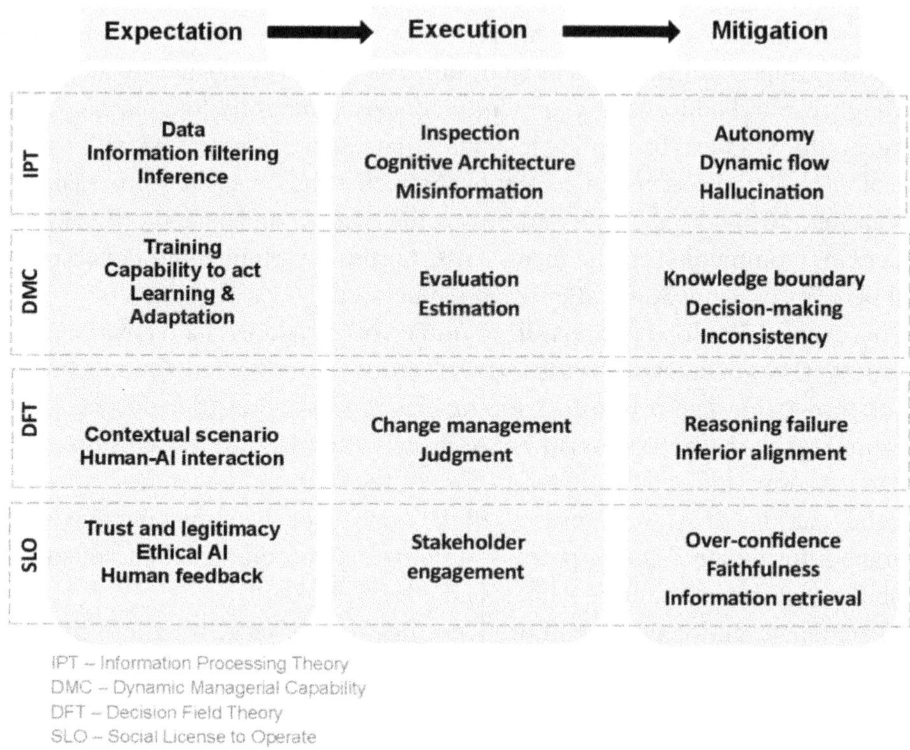

The framework included expectation, execution, and mitigation processes to address hallucinations in LLMs-based AI. Initially, data validation, inference, training, capability to act, human-AI interaction, ethical AI, cultural values, and human feedback can explain the expectation mechanism to maintain authentic information for users in different contexts. Then, the framework includes inspection, cognitive architecture, estimation, change management, judgement, and stakeholder engagement to define the execution process. Finally, we explain autonomy, dynamic flow, knowledge boundary, reasoning failure, over-confidence, and information retrieval to define the mitigation process of hallucinations in LLMs-based AI. To explain the impact of hallucinations in AI, the framework includes trust erosion, inaccurate outcomes, dissemination of misinformation, and dissemination of erroneous content. In the context of data-driven innovation in LLMs-based AI, we explain data-driven decision-making conceptualization towards data acquisition, data refinement, and data storage retrieval that ultimately delivers data-driven decision-making.

Use Case Scenario 1: Decision Making on AI Deepfake

AI uses a technology called Deepfake to create fictitious situations by substituting a person's face for different people in pictures, videos, and audio. The rapid development and widespread availability of AI-generated deepfake technology have changed the digital media platform and has a significant impact on privacy, security frameworks, and social trust. Advanced Generative Adversarial Networks (GANs) like StyleGAN and StyleGAN2, which allow users to create realistic synthetic photographs, videos, and audio content, are used in Deepfake technology to make it harder to distinguish between real and fake content.

Among the readily available deepfake production tools that raise the possibility of misuse and the rapid spread of misleading information on social media are FakeApp, DeepNude, DeepFaceLab, and FaceSwap. Because it becomes more difficult to discern between original and modified content, these shifts pose significant risks that could arise challenges in decision-making. According to Qadir et al. (2024), the public's faith in both traditional and digital media ecosystems is severely impacted by this technical innovation and AI-driven hallucinations, which creates uncertainties in critical situations.

Complex issues of ethics are also emerging around the fast-emerging AI-generated deep-fake technology in various areas. Deep fakes, which are synthetic multimedia that distort audio, video or image are associated with data privacy, user trust, consent matter, and public debates. With the advent of easier access and reality coupled to the legitimacy of the content and biometric information, differentiating between the real and the fake becomes complicated, and this can break norms in society pertaining to authenticity and truth. This can raise biases in people's minds, leading to reluctance in understanding and using AI deepfakes that can misguide in the decision-making process.

Use Case Scenario 2: Ethical Challenges for Decision-Making in Organization

As AI tools increasingly influence high-stakes decisions, such as hiring, promotion, vendor selection, and performance evaluations, they present significant risks relating to algorithmic biases and hallucinations, lack of transparency, and diminished accountability (Ghasemaghaei & Kordzadeh, 2024). Algorithmic bias, resulting from inherent biases in data or flawed algorithmic designs, poses significant risks, often perpetuating discrimination against marginalized groups. A widely discussed example includes Amazon's AI recruitment tool, which inadvertently discriminated against female candidates due to gender-biased training data. These cases underscore the urgent need for robust ethical governance mechanisms that effectively align AI systems with ethical norms and societal expectations (Hartmann et al., 2022).

Leadership accountability is further crucial in mitigating the ethical risks associated with AI deployments. Nevertheless, as highlighted by Ghasemaghaei and Kordzadeh (2024), existing organizational frameworks often fail to explicitly integrate leadership accountability with technical AI governance practices, demonstrating a notable shortcoming in current literature and practice. Specifically, it focuses on the persistent gap between technical AI development and organizational governance frameworks, a gap that can often leads to ethical failures such as algorithmic bias, opacity, and a lack of accountability in AI-driven decision-making tools.

Use Case Scenario 3: Decision-Making in AI-driven Healthcare

In recent years, there have been many groundbreaking applications of LLMs showing a great deal of potential in both clinical and patient-centric healthcare settings. Due to LLMs' ability to process large quantities of medical data, such as clinical notes and diagnostic imaging reports, they are incredibly useful for improving patient outcomes by improving diagnostic accuracy and decision making for clinicians and health professionals (Chkirbene Z. et al, 2024). However, despite the benefits LLMs provide in healthcare, their adoption in clinical practice may face several challenges and drawbacks that can impede the decision-making process in the healthcare industry.

All LLMs are prone to hallucinations where they fabricate information as if it were real, despite it being inaccurate (Liu et al., 2024). In an industry like healthcare where precision and patient safety are more important than anything else, inaccuracies in generated content due to hallucinations can be extremely detrimental as they can lead to diagnostic inefficiencies, jeopardizing patient outcomes (Chkirbene et al, 2024). Although using domain-specific LLMs such as Google's Med-PaLM, which is specifically designed for the medical domain, can reduce the instances of generating inaccurate information, they are still susceptible to hallucinations just like general-purpose LLMs. Another issue with LLMs is the lack of bias from the training data they were trained on, such as historical biases in treatment recommendation across different demographics. Privacy and security are also potential concerns in the context of LLMs as they may be trained on sensitive patient data.

The use of LLMs in clinical practices such as diagnostic decision making is still in its infancy and is approached with extreme caution in the real-world healthcare industry due to possibilities of hallucinations. However, LLMs are still being used in the industry for tasks like drug discovery, as well as administrative and documentation tasks such as automating clinical note taking, summarizing patient records, and streamlining billing processes in decision-making to improve operational efficiency (Nazi & Peng, 2024).

Use Case Scenario 4: Decision-Making in Log Analysis and Anomaly Detection

The emergence of LLMs presents new opportunities for log analysis and anomaly detection. LLMs like Bidirectional Encoder Representations from Transformers (BERT), Generative Pre-trained Transformer (GPT), Large Language Model Meta AI (LLaMA), and Mistral have shown remarkable capabilities in understanding and generating human language, which can enhance decision-making by extending processing and analyzing semi-structured log data. Recent research has investigated using LLMs for a variety of log-related tasks, such as parsing, anomaly detection, and root cause analysis (Pang et al., 2024) for critical industries. Real-time log monitoring involves continuously gathering, analyzing, and displaying log data as it is generated by applications, systems, or infrastructure. This proactive approach can help organizations swiftly by identifying anomalies, security threats, or operational issues, allowing for rapid detection and resolution. The integration of fine-tuned LLMs with real-time log monitoring tools represents a practical extension of theoretical advances into operational environments. However, as systems keep becoming more complex and large-scale, the techniques may develop hallucinations, and it can create critical factors in taking decision-making for preserving security and operational integrity in contemporary digital infrastructures.

Use Case Scenario 5: Decision-making in Autonomous Driving

Ethical and cultural values play critical roles in decision-making processes for autonomous driving. People from different cultural backgrounds have their own levels of perceptions, and social norms in adopting information systems like autonomous driving vehicles (Hofstede, 2011). Further, people inherit their virtues, moral values, and norms from their families, peer groups, and organizations. Moreover, professional code of ethics can pursue their obligations to take decisions before performing their activities. All those critical factors can influence people's decision-making that involves their choice of architecture. For instance, there would be critical roles for IT professionals and developers to confirm decision-making logics considering cultural values, ethics, safety, principles of reducing harms, and maximizing benefits to make autonomous driving vehicle as a moral machine

DISCUSSION

This research presents a framework to manage the hallucination of LLMs-based AI models within the context of data-driven decision-making. Through the illustration of findings grounded on the literature review, a range of underlying processes have been expressed that impose the importance of building relevant skill sets at the organizational level to successfully address the critical complications of hallucinations in AI. Following the data-driven view, we argue that mitigating hallucination is a dynamic mechanism to enhance data-driven decision-making. This chapter extends the research to explain the data-driven decision-making process by alleviating hallucinations in AI that is in line with existing literature (Umbrello & Natale, 2024; Jakubik et al., 2024). The term data-driven decision-making shall be considered as synchronous emergence, therefore in this chapter, we argue that our identified hallucination mitigation process is a synchronous emergent dynamic mechanism that can capture the essence of interactions between human and AI to deliver credible and authentic information.

As hallucination continues to be a significant challenge in AI, this chapter presented a comprehensive framework to outline the processes for expecting, executing, and mitigating hallucinations to promote data-driven decision-making. Consistent with previous research (Huang et al., 2024), the proposed framework centers on execution and mitigation strategies for hallucinations in large language model (LLM)-based AI within the context of data-driven analytics. This chapter specifically explores how hallucinations can be mitigated by defining key terms and emphasizing the data-driven decision-making process that improves theoretical knowledge. Rawte et al. (2023) explains related to this chapter, focusing on the definition of LLM hallucinations, as well as techniques for mitigation. However, this chapter proposes a framework based on the information processing theory, the dynamic managerial capability theory, the decision field theory, and the social license to operate theory to explain the critical process of identifying and mitigating the hallucinations for data-driven decision-making. The proposed framework differentiates itself by offering unique constructs and stands out as an organized process flow for mitigating hallucination in AI, aimed at delivering data-driven, strategic, and innovative decision-making. We provide a comprehensive explanation of how to expect, execute, and mitigate hallucinations in AI based on large language models. Additionally, we explore a critical concept of data-driven decision-making, linking our proposed stages of expectation, execution, and mitigation with the impacts of hallucinations in AI.

The study clearly outlined the consequences of LLM-based AI, specifically addressing hallucinations and their impacts within key phases of data-driven decision-making. The framework remains consistent with the proper validation process through pre-training data, post-training data, human factors, and contextual human-AI interactions-based information. Furthermore, the proposed framework incorporates validation, intervention, activation, inspection, estimation, evaluation, verification, ethical judgment, data privacy, methods, training, governance, and information retrieval as essential indicators for defining the expectation, execution, and mitigation of hallucinations in the context of LLMs-based AI, aiming to achieve data-driven decision-making.

The findings confirmed that the pre-training and post-training data in AI are based on factual knowledge obtained from human-AI interactions ethical, and cultural values. These interactions enable an autonomous system to expect, execute, and mitigate hallucinations by utilizing various contextual information. This mechanism contributes to effective hallucination management and supports data-driven decision-making. The study explored the impacts of hallucinations through few stages: (i) trust erosion, (ii) inaccurate outcomes, (iii) dissemination of misinformation and false realities, (iv) dissemination of erroneous content, and (v) ethical and cultural values. AI based on LLMs can analyze pre-training data, implement governance methods, and utilize R&D mechanisms with contextual user information to prevent hallucinations and foster data-driven decision-making. This movement is essential for identifying hallucinations and requires executing the root causes to mitigate inconsistencies and avoid unwanted outcomes in complex situations. In a different context, this chapter includes mechanisms for validation, verification, intervention, and activation to mitigate the spread of misinformation and unwanted content. This delivers credible decision-making through data-driven innovation and effective execution and mitigation strategies. For instance, hallucinations from LLM-based AI may influence digital platform outputs, such as reinforcing gender stereotypes and ethnic biases. When these hallucinations involve misinformation, like fake news or rumors, these can mislead users into making decisions based on faulty recommendations. This chapter highlights the increasing need for alleviating effective hallucination through expectation, execution, and mitigation in LLM-based AI, particularly in the realm of data-driven decision-making.

The dispersal of knowledge in the execution and mitigation of hallucinations is insufficient, indicating that the knowledge on handling hallucinations is limited to addressing data-driven decision-making. This chapter outlines the process flow of the hallucination management framework, which explains the execution and mitigation techniques in LLM-based AI for data-driven decision-making. We addressed the issues of hallucinations in AI by outlining a comprehensive strategy for expectation, execution, mitigation, and understanding of the effects of hallucinations on data-driven decision-making, which led to the development of a framework for managing hallucination challenges. Addressing long-tail knowledge through extended temporal boundaries, the framework suggests attributes to understanding the expectation, execution, and mitigation processes of hallucinations in LLM-based AI, which subsequently influences system performance through data-driven decision-making.

As a result, AI based on LLMs may produce misinformation or misappropriate content, particularly given existing knowledge limitations and restricted training data. This chapter offers valuable insights into various and extensive mitigation techniques while avoiding reliance on outdated information. Current LLM-based AI is legally restricted to training data derived from existing information, which limits its ability to address new contextual cases and user-centric data in critical infrastructures. This chapter uniquely addressed the practical knowledge needed to execute and mitigate hallucinations in LLM-based AI, which may enhance data-driven decision-making in critical systems. The chapter explored the knowledge gap related to hallucinations in LLMs-based AI for data-driven decision-making, highlighting the insufficient understanding that leads to biased outputs and hallucinations.

The managerial implications include emphasizing the serious importance of mitigating hallucination in the current decision-making environment that will assist managers to enhance the expectation, execution, and mitigation of hallucinations in LLMs based AI models across the organizational boundary in a sustainable manner. Managers and IT professionals can apply the proposed framework to design and develop their AI-driven systems to prevent and mitigate hallucinations that can enhance the data-driven decision-making process.

LIMITATIONS AND RECOMMENDATIONS

Aside from technical and validation issues, the ethical, cultural, regulatory, and integration of AI into personalized recommendations raise substantial barriers. Ethical aspects range from avoiding hallucination that may result in discriminatory decisions, misguiding informed consent for the use of AI for service delivery, and accounting for responsibility when AI-informed decisions cause adverse effects. Regulation of AI as a recommendation device is still in flux, with a lack of practical data to formulate policies. However, detailed guidelines for AI systems are required to support the potential for AI with the resultant safety concerns, as well as to secure the acceleration of widespread usage by minimizing hallucinations.

Decision-making uptake and use of AI tools are also problematic, usually being deterred by limited evidence on input data and acceptance of fake information, with potential of disseminating misinformation. Organizational issues, such as the absence of infrastructure and adequate data management, may also hinder effective implementation. Exploring these interrelated ethical, social, cultural, regulatory, and data integration aspects is essential in order to suggest feasible guidelines and policies on ensuring safe and effective AI implementation in delivering data-driven decision-making.

In short, AI is of tremendous potential in revolutionizing individualized service delivery mechanisms by enhancing information precision and personalizing recommendations, and its effective and responsible application is contingent on tackling serious issues of data. The AI platforms enhance predictive performance, maintaining strict reliability and safety, and dealing with the intricate ethical, cultural, regulatory, and data environment. That background forms the logical reasoning for this chapter initiative, with the objective of examining those specific domains that can support implementing a secure and trustworthy AI in the decision-making field.

This chapter intends to suggest broad recommendations that can help organizations, professionals, technologists, regulators, and policymakers explore potentials of AI while mitigating connected challenges. The proposed framework can guide all stakeholders to adapt AI in data-driven decision-making.

1. Centralized monitoring system for AI Performance: Organizations shall establish a centralized monitoring system to ensure regular checking mechanisms for assessing AI system performance that can enhance data-driven decision making. This requires system level monitoring with the measurement of information accuracy, system reliability, and user feedback while recognizing AI capabilities.
2. Data Privacy and Security: The AI system can ensure user consent for data usage that is obtained by the original authors, and all data have been fully de-identified. For this research, participants' responses can be stored locally on secure and access-controlled devices. Prior to participation, all individuals will be required to provide informed consent for the use of their data inputs to secure the credibility and authenticity of data. No Personally Identifiable Information (PII) will be collected, recorded, or disclosed at any stage of the data collection phase.
3. Ethical Considerations: This AI system will not involve the use of actual AI-generated suggestions, nor will any participant data be submitted to an AI system. By using manually crafted predictions as the ground truth, the system shall minimize ethical concerns commonly associated with the use of AI in taking data-driven decision-making.
4. Integration of AI Models: Before executing AI systems, organizations shall conduct sensible assessment of several AI models approaches. This can engage evaluating the efficacy of various models, measuring their impact on user trust, and accepting their integration conditions. Organizations can select applicable and reasonable methods to gauge the system requirements that can improve the data-driven decision-making.

5. Evaluation Metrics: An integrated AI system can develop evaluative metrics to assess the outcomes of the data execution through a data-driven decision-making. The system can accommodate features to ensure accuracy in detecting hallucinations, hallucination Reduction Rate, transparency improvement score, and stakeholder satisfaction. The feature on confirming accuracy in detecting hallucinations can contribute with demographic group outcomes to assessing precise detection. Further, the bias reduction rate can compute the average change in group-based selection disparities before and after hypothetical de-biasing interventions. Also, the transparency improvement score can measure through the development and review of transparency reports, focusing on frequency, clarity, and stakeholder accessibility. Moreover, the stakeholder satisfaction can assess the perceived effectiveness of fairness measures using Likert-scale survey results, supported by qualitative feedback.
6. Transparency in Decision-Making: The hallucination and derivative issues of AI systems is often hidden in "black box" decision-making, so improving the interpretability, explainability, and transparency of AI algorithms is critical. The counterfactual fairness can provide a way to effectively assess whether an algorithm is fair or not. A comprehensive method can be executed to test whether the algorithmic results of an AI system change if the sensitive attributes of outcomes change. In addition, providing counterfactual reasoning helps not only to systems, but also to organizations to understand their weaknesses or areas of mismatch with the desired outcomes, which can be used to improve performance in the future decision-making process. We can introduce counterfactual simulations with a fairness delivery mechanism, and a potential discriminatory logic can be identified and corrected at the model development stage.
7. Internal Audit Trails: Businesses and organizations can arrange continuous internal audit trails in place, as frequent audits can reveal systemic problems that need to be corrected. The regular internal audits allow companies to monitor, identify, and correct biased algorithms. Audit systems can help companies track differences between groups in model outputs and identify structural biases in gender, age, ethnicity, cultural background, etc. that may be implicit in the data promptly. To ensure that audits are scientific and sustainable, future internal audit mechanisms need to focus not only on AI model accuracy but also introduce group fairness indicators. It focuses on whether the system treats individuals belonging to different social groups in a similar manner. Regular audits with model training or deployment can help detect and correct hallucination.
8. Human-AI Collaboration: Replacing humans is not the goal of AI recruitment systems. To minimize the issues of hallucinations and derivative issues, the key to solving the problem is human-machine collaboration. We can't let AI be the sole decision maker, but rather a supporting decision maker. AI can help with

rapid processing and screening of tasks, but truly complex judgements - such as a task's potential, diversity, and cultural fit - still require human intelligence and ethical judgement. Humans still have irreplaceable strengths in interview judgement, cultural analysis and unstructured information interpretation. Human-machine collaboration not only helps to enhance decision-making plurality but also serves as a monitoring mechanism that effectively reduces the risk of hallucinations brought about by AI systems dictating at key nodes. Therefore, the integration of humanity and technology is critical, and it is a necessary path to alleviate hallucinations and other issues.

Technology is not neutral. We have a responsibility to ensure that every judgement made by AI recruitment systems is consistent with human dignity and equal rights. This is not just a technical challenge, but an ethical commitment. In modern society, the implementation of AI technology in job accomplishments is an irreversible trend. AI technology can only minimize hallucinations if it improves the transparency of data-driven decision-making, if organizations introduce accountability mechanisms, and if collaboration between humans and AI is sensibly developed and used. However, even efforts to promote fairness in machine learning may end up abstracting important contextual information and imposing artificial fixity on groups.

CONCLUSION

This chapter developed a framework for managing hallucinations in LLMs-based AI, focusing on the expectation, execution, and mitigation processes involved in delivering data-driven decision-making. The proposed framework consists of three stages that explain every aspect of expectation, execution, and mitigation concerning the impacts of hallucinations. These stages aim to deliver data-driven decision-making through sequential phases. This will enhance the intelligent recommendation process, improve predictive analytic platforms, and strengthen digital networks to expect, execute, and mitigate issues related to AI, ensuring ongoing data-driven decision-making. The framework explains the unique nature of identifying process flow for hallucination management capability in LLMs-based AI toward data-driven decision-making with the emergence of hallucinations.

As this study on hallucinations in AI expands within a mitigation framework, future research could explore different industries such as health, education, retail, agriculture, and tourism. This includes examining the impact of hallucinations in AI on data-driven decision-making in Large Language Models (LLMs), knowledge boundaries, and other predictive analytical models. Future studies can expand hallucinations for a predictive analytical platform for AR/VR or human-computer interfaces. The framework for managing hallucinations in AI can explore the reinforcement process to prevent unwanted content generation when applied in various real-world contexts. Thus, there is a critical need for a comprehensive strategy to reduce hallucinations in large-scale human-AI interaction platforms and ensure accurate information. Future studies may investigate qualitative and quantitative methods to identify significant factors and their relationships involving immersive technologies such as AR/VR, robotic arms, the Internet of Things (IoT), autonomous vehicles, cloud computing, drone applications, and brain-computer interfaces.

REFERENCES

Ahmed, W. (2025). The Role of Machine Learning and Deep Learning in Revolutionizing Healthcare: A Review. *Premier Journal of Artificial Intelligence, 2025*(3), 100010. DOI: 10.70389/pjai.100010

Aquino, G. J., Chamberlin, J., Yacoub, B., Kocher, M. R., Kabakus, I., Akkaya, S., Mercer, M., Waltz, J., Fiegel, M., Leaphart, N., Jacob, A., Gulsun, M. A., Gilkes, J., Stephenson, J., Sharma, P., Sahbaee, P., Schoepf, J., Zimmerman, S., Field, M. E., & Burt, J. R. (2022). Diagnostic accuracy and performance of artificial intelligence in measuring left atrial volumes and function on multiphasic CT in patients with atrial fibrillation. *European Radiology, 32*(8), 5256–5264. DOI: 10.1007/s00330-022-08657-y

Busemeyer, J. R., & Townsend, J. T. (1993). Decision field theory: A dynamic-cognitive approach to decision making in an uncertain environment. *Psychological Review, 100*(3), 432.

Chkirbene, Z., Hamila, R., Gouissem, A., & Devrim, U. (2024). *Large Language Models (LLM) in Industry: A Survey of Applications, Challenges, and Trends. In 2024, IEEE 21st International Conference on Smart Communities: Improving Quality of Life using AI, Robotics and IoT (HONET)*. IEEE., https://ieeexplore.ieee.org/abstract/document/10822885

Clatterbuck, H., Castro, C., & Morán, A. M. (2024). Risk Alignment in Agentic AI Systems. https://doi.org//arxiv.2410.01927.DOI: 10.48550

Davenport, T. H., Brynjolfsson, E., McAfee, A., & Wilson, H. J. (2019). *Artificial intelligence: The insights you need from Harvard Business Review*. Harvard Business Press.

Delétang, G., Ruoss, A., Duquenne, P. A., Catt, E., Genewein, T., Mattern, C., & Veness, J. (2023). Language modeling is compression. arXiv preprint arXiv:2309.10668.

Demajorovic, J., Lopes, J. C., & Santiago, A. L. F. (2019). The Samarco dam disaster: A grave challenge to social license to operate discourse. *Resources Policy, 61*, 273–282.

Ghasemaghaei, M., & Kordzadeh, N. (2024). Understanding how algorithmic injustice leads to making discriminatory decisions: An obedience to authority perspective. *Information & Management, 61*(2), 103921. Advance online publication. DOI: 10.1016/j.im.2024.103921

Guerreiro, N. M., Voita, E., & Martins, A. F. (2022). A Comprehensive Study of Hallucinations in Neural Machine Translation. In *Proceedings of the 17th Conference of the European Chapter of the Association for Computational Linguistics* (pp. 1059–1075). Association for Computational Linguistics., https://aclanthology.org/2023.eacl-main.75

Guida, M., Caniato, F., Moretto, A., & Ronchi, S. (2023). Artificial intelligence for supplier scouting: An information processing theory approach. *International Journal of Physical Distribution & Logistics Management, 53*(4), 387–423. DOI: 10.1108/IJPDLM-12-2021-0536

Haenlein, M., & Kaplan, A. (2019). A brief history of artificial intelligence: On the past, present, and future of artificial intelligence. *California Management Review, 61*(4), 5–14. DOI: 10.1177/0008125619864925

Hancock, T. O., Hess, S., Choudhury, C. F., & Tsoleridis, P. (2024). Decision field theory: An extension for real-world settings. *Journal of Choice Modelling, 52*, 100495.

Hartmann, K. V., Primc, N., & Rubeis, G. (2022). Lost in translation? Conceptions of privacy and independence in the technical development of AI-based AAL. *Medicine, Health Care, and Philosophy, 26*(1). Advance online publication. DOI: 10.1007/s11019-022-10126-8

Helfat, C. E., & Peteraf, M. A. (2003). The dynamic resource-based view: Capability lifecycles. *Strategic Management Journal, 24*(10), 997–1010. DOI: 10.1002/smj.332

Hofstede, G. (2011). Dimensionalizing Cultures: The Hofstede Model in Context. *Online Readings in Psychology and Culture, 2*(1).

Huang, L., Yu, W., Ma, W., Zhong, W., Feng, Z., Wang, H., Chen, Q., Peng, W., Feng, X., Qin, B., & Liu, T. (2024). A survey on hallucination in large language models: Principles, taxonomy, challenges, and open questions. *ACM Transactions on Information Systems.*

Hunkenschroer, A. L., & Kriebitz, A. (2022). Is AI recruiting (un)ethical? A human rights perspective on the use of AI for hiring. *AI and Ethics, 3*(1), 199–213. DOI: 10.1007/s43681-022-00166-4

Jagannathan, J., & Udaykumar, U. (2020). *Predictive modeling for improving healthcare using IoT: Role of predictive models in healthcare using IoT. Incorporating the internet of things in healthcare applications and wearable devices*. IGI Global., DOI: 10.4018/978-1-7998-1090-2.ch015

Jakubik, J., Vössing, M., Kühl, N., Walk, J., & Satzger, G. (2024). Data-Centric Artificial Intelligence. *Business & Information Systems Engineering*, *66*(4), 507–515. DOI: 10.1007/s12599-024-00857-8

Jarrahi, M. H. (2018). Artificial intelligence and the future of work: Human-AI symbiosis in organizational decision making. *Business Horizons*, *61*(4), 577–586.

Ji, Z., Lee, N., Frieske, R., Yu, T., Su, D., Xu, Y., Ishii, E., Bang, Y. J., Madotto, A., & Fung, P. (2023). Survey of hallucination in natural language generation. *ACM Computing Surveys*, *55*(12), 1–38.

Lee, M. H., Clark, A., Rupp, J., Wietelman, D. C., & Graham, J. D. (2019). Public opinion toward hydraulic fracturing: The effect of beyond compliance and voluntary third-party certification. *Energy Policy*, *128*, 306–315.

Li, J., Chen, J., Ren, R., Cheng, X., Zhao, W. X., Nie, J. Y., & Wen, J. R. (2024). The dawn after the dark: An empirical study on factuality hallucination in large language models. arXiv preprint arXiv:2401.03205.

Lin, H., & Chen, Q. (2024). Artificial intelligence (AI)-integrated educational applications and college students' creativity and academic emotions: Students and teachers' perceptions and attitudes. *BMC Psychology*, *12*(1), 487.

Liu, Y., Yao, Y., Ton, J. F., Zhang, X., Cheng, R. G. H., Klochkov, Y., Taufiq, M. F., & Li, H. (2023). Trustworthy LLMs: A survey and guideline for evaluating large language models' alignment. arXiv preprint arXiv:2308.05374.

Moffat, K., Lacey, J., Zhang, A., & Leipold, S. (2016). The social license to operate: A critical review. *Forum (London)*, *89*(5), 477–488. DOI: 10.1093/forestry/cpv044

Moniz, A. (2021). *Beginning Azure cognitive services: data-driven decision making through artificial intelligence* (1st edition 2021). Apress., DOI: 10.1007/978-1-4842-7176-6

Nazi, Z. A., & Peng, W. (2024). Large Language Models in Healthcare and Medical Domain: A Review. *Informatics (Basel)*, *11*(3), 57. Advance online publication. DOI: 10.3390/informatics11030057

Newell, A. (1992). Précis of Unified theories of cognition. . *Behavioral and Brain Sciences*, *15*(3), 425–437. DOI: 10.1017/S0140525X00069478

Novet, J. (2017). *Google launches Cloud Dataprep, an embedded version of Trifacta.* Venture Beat.

Pang, Y., Zhang, M., Liu, Y., Li, X., Wang, Y., Huan, Y., Liu, Z., Li, J., & Wang, D. (2024). Large language model-based optical network log analysis using LLaMA2 with instruction tuning. *Journal of Optical Communications and Networking*, *16*(11), 1116–1132.

Qadir, A., Mahum, R., El-Meligy, M. A., Ragab, A. E., AlSalman, A., & Hassan, H. (2024). An efficient deepfake video detection using robust deep learning. *Heliyon*, *10*(5), e25757. DOI: 10.1016/j.heliyon.2024.e25757

Rawte, V., Chakraborty, S., Pathak, A., Sarkar, A., Tonmoy, S. M. T. I., Chadha, A., Sheth, A., & Das, A. (2023). The Troubling Emergence of Hallucination in Large Language Models - An Extensive Definition, Quantification, and Prescriptive Remediations. EMNLP 2023 - 2023 Conference on Empirical Methods in Natural Language Processing, Proceedings, 2541–2573.

Tabassam, A., Yaqoob, G., Cuong, V. H., Syed, M., Shahzadi, A., & Asghar, F. (2023). The Ethical Implication of Using Artificial Intelligence in Hiring and Promotion Decisions. 1(2). https://jomeri.org/index.php/jomeri/article/view/7

Thomson, I., & Boutilier, R. G. (2011). Social license to operate. SME mining engineering handbook, 1, 1779-1796.

Tonmoy, S. M. T. I., Zaman, S. M. M., Jain, V., Rani, A., Rawte, V., Chadha, A., & Das, A. (2024). A Comprehensive Survey of Hallucination Mitigation Techniques in Large Language Models. https://doi.org//arxiv.2401.01313.DOI: 10.48550

Umbrello, S., & Natale, S. (2024). Reframing Deception for Human-Centered AI. *International Journal of Social Robotics*, *16*(11), 103239. Advance online publication. DOI: 10.1007/s12369-024-01184-4

Williamson, S. M., & Prybutok, V. (2024). The Era of Artificial Intelligence Deception: Unraveling the Complexities of False Realities and Emerging Threats of Misinformation. *Information (Basel)*, *15*(6), 299. DOI: 10.3390/info15060299

Wixom, B. H., & Ross, J. W. (2017). How to monetize your data. *MIT Sloan Management Review*, *58*(3).

Zhao, L., Feng, X., Feng, X., Qin, B., & Liu, T. (2023). Length extrapolation of transformers: A survey from the perspective of position encoding. arXiv preprint arXiv:2312.17044.

Zhou, R., Chen, L., & Yu, K. (2024, May). Is LLM a reliable reviewer? a comprehensive evaluation of LLM on automatic paper reviewing tasks. In *Proceedings of the 2024 joint international conference on computational linguistics, language resources and evaluation (LREC-COLING 2024)* (pp. 9340-9351).

KEY TERMS AND DEFINITIONS

Data-Driven Decision Making: This is a process in a mental state that performs an action after analyzing information in artificial intelligence-based models, delivering innovative decision making in complex systems. Therefore, data-driven decision making is a critical process that engages human-AI interactions, socio-technological impacts, uncertainties, ethics, and moral judgement. By leveraging data analytics of AI, organizations can improve innovative solutions that can enhance the overall decision-making process.

Expectation-Execution-Mitigation Framework: The Expectation-Execution-Mitigation (E-E-M) is a comprehensive framework that presents how LLMs-based AI can alleviate hallucinations and enhance data-driven decision-making. The framework initially adopts data validation, inference, training, capability to act, human-AI interaction, ethical AI, and human feedback to explain the expectation mechanism. Then, the framework incorporates inspection, cognitive architecture, estimation, change management, judgement, and stakeholder engagement to classify the execution phase. Finally, the framework describes autonomy, dynamic flow, knowledge boundary, reasoning failure, over-confidence, and information retrieval to outline the mitigation process of hallucinations in LLMs-based AI.

Hallucination in Artificial Intelligence: Hallucination is an outcome in AI models that generate false realities or misinformation. In AI models, hallucination is a critical concern about misinformation that presents massive arguments for real-world applications and undercuts the credibility of AI. Thus, hallucinations create a hindrance to spreading true information from AI systems.

Chapter 5
Aligned Minds, Efficient Machines:
Integrating Neuromorphic Computing for Personalized AI

Ben Kennedy
https://orcid.org/0009-0009-1337-0709
Capitol Technology University, USA

ABSTRACT

AI systems exhibit robust reasoning abilities yet face challenges in personalization and sustainability, as cognitive augmentation technologies often show rigidity and high energy use that limit effectiveness. This paper introduces a framework that integrates psychometric alignment, modular cognitive agents, and neuromorphic computing to deliver scalable, personalized, energy-efficient augmentation. The Synthetic Cognitive Augmentation Network Alignment Questionnaire (SCANAQ) captures users' cognitive inclinations (Tate, 2024a) and informs the Synthetic Cognitive Augmentation Network Using Experts (SCANUE), a modular system modeled on prefrontal-cortex subregions (Tate, 2024b). The Spiking Transformer Augmenting Cognition (STAC) pipeline targets neuromorphic edge deployment: STAC V1 demonstrates a hybrid fine-tuning approach, and STAC V2 provides an ANN-to-SNN conversion pathway that yields spiking-compatible networks (Tate, 2025a; Tate, 2025b). Future work includes empirical hardware validation and standardized neuromorphic benchmarking (Davies et al., 2021; Ostrau et al., 2022).

DOI: 10.4018/979-8-3373-5702-7.ch005

INTRODUCTION

Consider an AI assistant that not only aids in critical medical diagnoses or financial decisions but also instinctively aligns its guidance with a user's individual decision-making style and emotional state. This vision remains largely unrealized despite increasing deployments of large-language models (LLMs) in high-stakes contexts such as healthcare, finance, and education. Although these models demonstrate impressive capabilities, two crucial, interconnected limitations significantly impede their potential. First, current LLMs typically lack deep personal alignment, operating instead with generic, one-size-fits-all response paradigms. Effective cognitive augmentation, however, necessitates a system that can dynamically adapt its interactions, reasoning strategies, and communication styles to match each user's unique and stable cognitive, affective, and experiential characteristics. Without this fine-grained personalization—which requires capturing nuanced individual profiles, including decision-making tendencies, affective regulation strategies, and real-time stress levels—user trust can diminish, engagement declines, and the system risks offering support that is misaligned or even detrimental to the user's goals. Fine-grained personalization therefore remains the decisive hurdle for real-world augmentation.

Second, energy efficiency presents a formidable and urgent obstacle. Training and deploying large-scale transformer-based models requires substantial computational resources, which has raised profound sustainability concerns. The energy demands and associated environmental impacts, as documented by Strubell et al. (2019), underscore the need to develop more sustainable computational paradigms. This intensive computational footprint contributes to a significant carbon footprint and restricts the deployment of sophisticated cognitive assistants on resource-limited edge devices, such as mobile phones or embedded systems. Such a barrier limits equitable, widespread access to powerful AI technologies, thereby creating a digital divide. To address this crisis, neuromorphic computing has emerged as a promising solution. Its brain-inspired, event-driven architectures compute only when new information is available, resulting in an inherently lower energy profile (Roy et al., 2019).

The intersection of these two constraints forms the central tension this research addresses. Any attempt to personalize AI through straightforward methods, such as independently running multiple large, fine-tuned models for different user profiles, would exponentially increase computational demands and exacerbate the already critical energy-consumption problem. Conversely, prioritizing energy efficiency by drastically reducing model complexity often sacrifices the depth and precision required for sophisticated and meaningful personalization. Thus, a truly viable cognitive augmentation framework must simultaneously optimize for both personalized alignment and computational sustainability. This paper proposes a structured, empirically evaluable solution by synthesizing three complementary lines of inquiry into a unified, integrated approach. Having established the core problem, the following sections will delve into the existing literature to frame the technical contributions of this work.

LITERATURE REVIEW

The Dual Challenges of Modern AI: Personal Alignment and Computational Sustainability

The advancement of AI faces two critical, interdependent challenges: achieving deeper personal alignment by enhancing meaningful human-AI interactions and ensuring computational sustainability by addressing the high-energy demands of advanced AI. The promise of cognitive augmentation—AI systems that not only provide information but actively enhance human thought processes—hinges entirely on resolving the tension between these goals. A system that is perfectly personalized but environmentally and economically unviable remains a theoretical curiosity; a system that is efficient but generically impersonal fails to deliver on the core promise of augmentation. This section, therefore, reviews the academic and engineering context of these two challenges to establish the theoretical and practical foundations upon which the SCANUE and STAC frameworks are built.

The Challenge of Static Alignment in Large Language Models

The dominant paradigm of large language models (LLMs) is built on the principle of generalism. Models based on architectures like the Transformer (Vaswani et al., 2017) are trained on vast, heterogeneous cross-sections of public data to perform a wide range of tasks for a generic, statistically average user. While this approach has led to remarkable capabilities in language understanding, generation, and reasoning, it has also created a significant "alignment gap." Current LLMs lack deep personal alignment, a limitation that manifests in several critical ways that impede their effective use as cognitive partners.

One primary manifestation is the erosion of trust. Research in human-computer interaction and psychology consistently shows that trust is predicated on predictability, reliability, and a sense of shared understanding. When an AI assistant's reasoning style, values, or communication patterns deviate sharply or unpredictably from the user's own cognitive model, it creates dissonance and alienates the user, undermining their confidence in the system (Tate, 2024b). This phenomenon is often described as model brittleness, where a system that performs well on average may fail unexpectedly when faced with an individual's unique cognitive style or cultural context. The one-size-fits-all approach necessarily smooths over the very individual nuances critical for high-stakes decision-making.

A second consequence is that the lack of personalization directly impedes the development of effective cognitive augmentation. True augmentation is not merely about providing correct answers or automating tasks; it is about scaffolding a user's own cognitive processes, enhancing their strengths, and supporting their weaknesses in a dynamic, context-aware manner (Miller & Cohen, 2001). This capability requires a system that can build and maintain a sophisticated model of the user, understanding their executive function strengths and weaknesses, their decision-making tendencies under stress, and their preferred methods of emotional regulation (Damasio, 1994).

A generic model cannot provide this nuanced support. It risks delivering advice that is not only misaligned but potentially detrimental to the user's goals—for instance, by encouraging a risk-averse user to take a gamble they are not psychologically equipped to handle, or by presenting information in a way that overwhelms a user with executive function challenges. The SCANUE framework directly addresses this fundamental challenge by proposing that deep alignment is not a post-hoc feature but a foundational architectural principle (Kennedy, Mohammad, & Wyandt, 2025). Consequently, this alignment framework operationalizes a user's rich, stable psychometric profile, mapping it directly onto a system of specialized cognitive agents designed to mirror the functional organization of the human prefrontal cortex.

The Sustainability Crisis and the Promise of Neuromorphic Computing

The computational demands of training and deploying state-of-the-art LLMs have created a sustainability crisis. The seminal work of Strubell et al. (2019) was instrumental in quantifying the immense energy cost and associated carbon footprint of these models, popularizing the crucial distinction between "Red AI" (computationally expensive, record-breaking models) and "Green AI" (research focused on high performance at a lower computational and environmental cost). The scaling laws that govern transformers, which have so successfully driven progress, also dictate that performance improvements often require an exponential increase in parameters and training data. This trajectory is fundamentally unsustainable. It not only limits the deployment of powerful models to a few large corporations but also creates significant barriers to equitable access.

Neuromorphic computing, and Spiking Neural Networks (SNNs) in particular, have emerged as a leading paradigm for Green AI, offering a path away from the brute-force computational paradigm of conventional systems. As detailed by pioneers in the field and summarized by Roy et al. (2019), SNNs derive their profound efficiency from two core, brain-inspired principles. First is event-driven computation: traditional Artificial Neural Networks (ANNs) operate on a dense, clock-based cycle, performing billions of multiply-accumulate (MAC) operations at every time step, regardless of whether the input has changed. This process is fundamentally inefficient. SNNs, in contrast, are event-driven. They compute only when a "spike"—a discrete event in time—arrives at a neuron, meaning processing is sparse and data-driven; no energy is expended on parts of the network that are not currently handling new information. This principle is analogous to the brain's own efficiency, where only a fraction of neurons are active at any given moment.

Second is temporal and structural sparsity: in a biological brain, and in a well-optimized SNN, neural activity is sparse. This sparsity is the primary driver of energy efficiency. A single synaptic operation (SOP) in an SNN, representing the processing of a single spike, is orders of magnitude less energy-intensive than the dense MAC operation that forms the basis of ANNs. This efficiency allows neuromorphic hardware, such as the Intel Loihi chip (Davies et al., 2021), to perform complex tasks at a fraction of the power consumption of a traditional GPU, which in turn enables secure on-device cognition and helps shrink the digital divide.

The STAC framework attempts to harness this revolutionary efficiency. It explores two distinct and complementary pathways to translate the proven computational power of transformers into the energy-efficient paradigm of SNNs. By doing so, it provides the essential missing piece of the puzzle: a computationally viable foundation upon which a truly personalized, multi-agent cognitive architecture like SCANUE can be built and deployed at scale. Without such a foundation, the goal of personalized AI for all remains an expensive and inaccessible dream.

Neuro-Psychometric Foundations of Personalized Alignment

To address the alignment challenge, the presented framework operationalizes critical insights from cognitive neuroscience and psychometric research. Specifically, the Synthetic Cognitive Augmentation Network Using Experts (SCANUE) architecture implements a neurobiologically plausible cognitive model inspired directly by the well-documented functional subdivisions within the human prefrontal cortex (Kennedy, Mohammad, & Wyandt, 2025; Tate, 2024a). Instead of a singular monolithic model, SCANUE strategically decomposes cognitive processing across five distinct, specialized, and computationally lightweight agents. Each agent emulates functions associated with specific PFC subregions extensively characterized in cognitive neuroscience literature: the dorsolateral prefrontal cortex (DLPFC) agent for executive control; the ventromedial prefrontal cortex (VMPFC) agent for emotional regulation and value-based decision-making; the orbitofrontal cortex (OFC) agent for reward processing; the anterior cingulate cortex (ACC) agent for conflict monitoring; and the medial prefrontal cortex (mPFC) agent for social cognition (Damasio, 1994; Miller & Cohen, 2001). To achieve this high degree of specialization, each agent was fine-tuned on domain-specific datasets relevant to its designated cognitive function. This modularization enhances interpretability and facilitates highly targeted user personalization.

This personalization is instantiated and maintained via the SCANAQ Alignment Questionnaire, a bespoke psychometric assessment composed of 36 carefully selected items. These items derive from eight validated psychological scales, each chosen for its relevance to the particular PFC functions modeled by the SCANUE agents (Tate, 2024c). The chosen scales encompass executive functions (BRIEF-A), emotional regulation (ERQ), impulsivity (BIS-11), risk-taking propensity (GRiPS), decision-making style (GDMS), self-efficacy (GSES), perceived stress (PSS), and empathy (IRI). In practice, SCANAQ serves as a translation layer, effectively converting a user's psychometric profile into actionable control signals that drive the dynamic behavior of SCANUE's specialized agents. This design ensures a theoretically grounded and empirically informed alignment between the user and the AI.

STAC: A Two-Phase Approach to Neuromorphic Efficiency

The energy-efficiency imperative, which is central to making a personalized system like SCANUE viable, is addressed through the Spiking Transformer Augmenting Cognition (STAC) framework. The development of STAC evolved through two distinct and methodologically significant phases, representing a deep exploration of two different pathways to neuromorphic efficiency.

The first phase, STAC V1 (Tate, 2025a), established a complete, end-to-end differentiable pipeline for a hybrid SNN-transformer, integrating a pre-trained GPT-2 backbone with custom spiking layers. This initial approach was highly innovative, moving beyond simple SNN layers to incorporate learnable AdEx (Adaptive-Exponential) spiking neurons and a custom Hyperdimensional Memory Module (HEMM). This design allowed the model to integrate biologically plausible dynamics and memory structures directly into the architecture. Energy efficiency was not an afterthought but was actively encouraged during fine-tuning through an L1 spike regularization term. The maturity of this version was confirmed by a comprehensive validation suite of seven distinct test functions. Its single-turn conversational capability was an intentional and logical consequence of its sequential, token-by-token processing architecture, not an unforeseen flaw.

While STAC V1 was a successful, self-contained research project, its immense computational cost and the inherent difficulty of scaling its bespoke training-based approach to compete with state-of-the-art LLMs prompted a strategic methodological shift. The research pivoted from a bespoke training-based approach to a more pragmatic, conversion-based one, motivated by the desire to leverage the power and accessibility of the vast ecosystem of existing pretrained models. This pivot aimed to make neuromorphic benefits applicable to them directly.

STAC V2 (Tate, 2025b) was therefore re-architected as an experimental conversion framework with the primary goal of translating conventional, pretrained transformer models into functionally equivalent SNNs. This approach trades the deep architectural integration and learnable dynamics of V1 for the practical advantage of wider applicability. The key innovation in V2 is the novel Temporal Spike Processor (TSP), a state-management module designed to preserve conversational context across multiple turns—the very capability not prioritized in V1's architectural paradigm. It is critical to note, however, that STAC V2 remains an experimental research prototype with known limitations and incomplete hardware operator coverage; its performance metrics are currently based on theoretical projections that await rigorous empirical hardware validation.

Contributions and Structure of the Paper

This manuscript synthesizes psychometric methodologies, cognitive neuroscience principles, and neuromorphic engineering into a cohesive framework for personalized cognitive augmentation. It frames the integration of the SCANUE cognitive architecture with the STAC neuromorphic backend as a primary future objective, thereby defining a clear roadmap from the current conceptual framework to a fully implemented system. Three innovations encapsulate the primary contributions: (1) a validated psychometric-to-agent mapping methodology that defines how to systematically translate SCANAQ-derived user profiles into dynamic agent control signals; (2) a detailed comparative analysis of two distinct neuromorphic strategies—STAC V1's hybrid fine-tuning approach and STAC V2's pragmatic conversion framework—highlighting the introduction of the Temporal Spike Processor (TSP) for multi-turn coherence in the latter; and (3) the establishment of a detailed, preregistered protocol for future validation that outlines rigorous benchmarks for alignment, cognitive efficacy, and energy efficiency assessments using specialized neuromorphic hardware.

Subsequent sections expand systematically on the framework's neuroscientific underpinnings, its methodological foundations, the detailed STAC development trajectories, the proposed validation protocols, and a critical discussion of current limitations, ethical considerations, and structured directions for future research.

BACKGROUND

SCAN, SCANUE, and STAC in Context: A Multi-Layered Architecture

The conceptual evolution that produced the current integrated framework progressed through several key stages, beginning with the Synthetic Cognitive Augmentation Network (SCAN) (Tate & Sanders, 2024). SCAN introduced a novel, modular cognitive architecture designed to emulate the functional organization of the human prefrontal cortex (PFC). Marking a significant departure from monolithic LLM architectures, SCAN's foundational structure comprised five distinct expert agents corresponding to discrete PFC subregions (DLPFC, VMPFC, OFC, ACC, and mPFC), each utilizing a pretrained GPT-style backbone.

Building upon this foundation, the Synthetic Cognitive Augmentation Network Using Experts (SCANUE) framework (Tate, 2024a) introduced the critical enhancement of agent specialization through fine-tuning. In this version, each agent's pre-trained backbone was fine-tuned on domain-specific datasets, sharpening its function from a generalized expert into a specialized cognitive processor. The architecture was further refined in a subsequent iteration, SCANUE V22 (Tate, 2024b), which incorporated dynamic Human-in-the-Loop (HITL) feedback mechanisms. This version implemented memory-based loops for ongoing personalization, enabling the system to iteratively adjust agent outputs and progressively align its behavior with evolving user needs. This progression transformed the static computational model into an adaptive system capable of enhancing personal relevance and user trust.

Complementing this cognitive architecture, the Spiking Transformer Augmenting Cognition (STAC) framework supplies the essential computationally efficient execution layer. This research progressed through two distinct phases: STAC V1 (Tate, 2025a), a hybrid model that fine-tunes a pre-trained transformer and SNN model, and STAC V2 (Tate, 2025b), an experimental conversion prototype. The current STAC V2 framework is an experimental research prototype, limited to software-level SNN simulations, with hardware deployment and empirical energy measurements identified as key areas for future work. Additionally, SCANUE's broader architectural refinements benefited from the Cognitive Augmentation User Survey & Evaluation (CAUSE) (Tate, 2024c), a comprehensive internal psychometric tool designed to capture nuanced user needs and expectations for cognitive augmentation applications.

Figure 1. Integrated Framework Architecture: From Psychometric Assessment to Neuromorphic Deployment

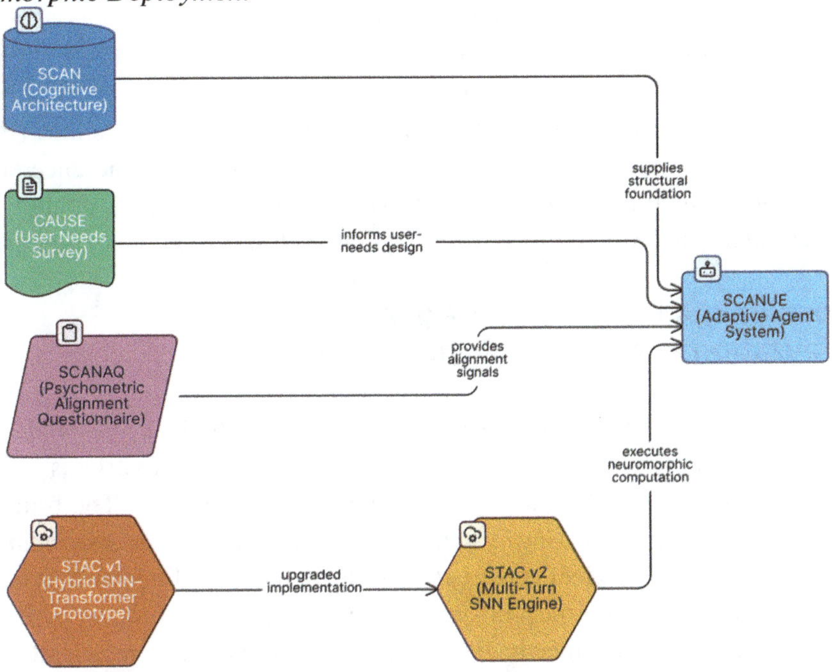

Note. The diagram provides a schematic overview of the proposed integrated architecture. The process begins with user profiling via the SCANAQ instrument, which feeds into the adaptive orchestration of SCANUE and culminates in planned deployment on neuromorphic hardware via the STAC V2 pipeline. Solid lines denote active development trajectories, while dashed lines represent superseded legacy components.

Psychometric Foundation: The SCANAQ Alignment Questionnaire

Personalized alignment within the SCANUE framework originates from the administration of the Synthetic Cognitive Augmentation Network Alignment Questionnaire (SCANAQ) (Tate, 2024c). This specialized psychometric battery consists of a rigorously curated subset of 36 strategically selected items derived from eight established, psychometrically validated psychological instruments. These instruments were chosen for their theoretical relevance to distinct cognitive and emotional functions modeled by the five SCANUE agents. To ensure methodological transparency and reproducibility, the complete SCANAQ instrument is provided in Appendix A, and the comprehensive scoring breakdown is detailed in Appendix B.

SCANAQ's composite structure intentionally balances psychometric rigor with practical considerations for user engagement and assessment feasibility. Administering the entirety of all eight source psychological assessments would be impractical and burdensome, likely reducing user adoption. By meticulously selecting items with optimal predictive validity and reliability, SCANAQ preserves essential psychometric robustness while significantly enhancing participant feasibility, both during initial system onboarding and during periodic recalibration. Moreover, SCANAQ's development adhered to established cross-cultural adaptation guidelines (Beaton et al., 2000; Gudmundsson & Lönner, 2009). Such meticulous adherence ensures conceptual and psychometric equivalence across diverse linguistic and cultural populations, a fundamental requirement for any equitable global implementation strategy.

Neuroscientific Rationale for Agent Specialization and SCANAQ Integration

The rationale for the functional specialization across the five SCANUE agents and their direct integration with SCANAQ-derived psychometric metrics is grounded firmly in decades of established cognitive neuroscience research. The framework avoids ad-hoc design by creating systematic, theoretically informed mappings between validated psychological constructs and the behaviors of agents that emulate distinct functional regions of the prefrontal cortex (PFC). This grounding ensures that the personalization is not superficial but is instead rooted in an empirically robust understanding of human cognition. The following sections detail the neuroscientific and psychometric justification for each of the five agents.

DLPFC Agent: Executive Control and Goal-Directed Behavior

Neuroscientific Grounding. The dorsolateral prefrontal cortex (DLPFC) is widely recognized as the primary neuroanatomical region responsible for executive functions like planning and cognitive control. These higher-order processes include abstract reasoning, strategic planning, the online maintenance and manipulation of information in working memory, and top-down cognitive control (Miller & Cohen, 2001). It is the brain's "executive," responsible for orchestrating thought and action in accordance with internal goals. Foundational work by Petrides (2005) and others has detailed its critical role in tasks that require monitoring and manipulating multiple pieces of information to guide complex, multi-step behaviors.

Psychometric Justification. To precisely modulate this agent's behavior, the SCANAQ incorporates items from two key instruments: the Behavior Rating Inventory of Executive Function (BRIEF-A; Gioia et al., 2000) and the General Self-Efficacy Scale (GSES; Schwarzer & Jerusalem, 1995). The BRIEF-A is a gold-standard clinical and research instrument designed to assess the real-world behavioral manifestations of executive functions. Its subscales (e.g., Plan/Organize, Inhibit, Shift) map directly onto the cognitive processes governed by the DLPFC. The GSES, in turn, measures a user's domain-general belief in their ability to execute tasks and achieve goals, a core component of self-regulation tightly integrated with the persistence and effort components of executive control (Barbey et al., 2012). This pairing creates a direct, empirically validated, and clinically relevant link between a user's measured executive capabilities and the corresponding agent's supportive behavior.

To illustrate this principle, consider a user whose SCANAQ profile reveals a high T-score on the BRIEF-A's "Plan/Organize" subscale, indicating significant, self-reported difficulty in planning. If this user states a broad goal like, "I need to prepare for my annual performance review," the DLPFC agent would provide an initial executive control response by presenting a structured delegation framework: *Subtasks*: 1. Compile major accomplishments from the past year. 2. Identify specific KPI examples. 3. Draft self-assessment narrative. *Agent Assignments*: VMPFC Agent will assess emotional readiness, OFC Agent will evaluate reward structures, ACC Agent will identify potential conflicts, mPFC Agent will ensure value alignment. *Integration Plan*: Each specialized agent would provide sequential analysis of their respective domains. The user would then receive additional sequential responses from each agent as it processes its assigned aspect of the task. While this transparent, sequential approach differs from the more implicit coordination of the biological DLPFC, it provides a computationally tractable approximation that delivers the cognitive scaffolding and systematic task decomposition the user's profile suggests they need.

VMPFC Agent: Emotional Regulation and Value-Based Decision-Making

Neuroscientific Grounding. The ventromedial prefrontal cortex (VMPFC) is a critical hub for integrating emotion into decision-making, regulating emotional responses, and assigning subjective value to potential choices (Damasio, 1994). Lesion studies, such as those involving the Iowa Gambling Task (Bechara et al., 2005), have famously demonstrated that damage to this area impairs the ability to use emotional signals to guide advantageous decision-making, even when knowledge of the risks is intact. The VMPFC allows us to have a "gut feeling" about what is good or bad for us, a process essential for navigating complex social and personal landscapes (Li et al., 2021).

Psychometric Justification. The VMPFC agent's behavior is primarily informed by three SCANAQ scales: the Emotion Regulation Questionnaire (ERQ; Gross & John, 2003), the General Decision-Making Style (GDMS; Scott & Bruce, 1995), and the Perceived Stress Scale (PSS; Cohen et al., 1983). The ERQ directly measures a user's habitual tendency to use two distinct emotion regulation strategies:

cognitive reappraisal (reframing a situation to change its emotional impact) and expressive suppression (inhibiting outward signs of emotion). These strategies are core VMPFC-mediated functions. The GDMS provides insight into the user's reliance on intuitive versus rational decision styles, a process heavily mediated by the VMPFC's valuation signals. Finally, the PSS quantifies the user's current cognitive and affective load, which is known to heavily impact VMPFC-mediated decision-making and emotional control.

Imagine a user with a SCANAQ profile indicating a high score on the ERQ's "Suppression" subscale and high concurrent perceived stress (PSS). This pattern suggests the user tends to bottle up negative emotions, a strategy that is cognitively costly and often maladaptive. If this user expresses significant anxiety about an upcoming public speaking engagement, the VMPFC agent would activate proactive affective scaffolding. It would avoid simplistic platitudes ("Don't worry, you'll be great!"). Instead, it would gently guide the user through a cognitive reappraisal exercise, asking reflective questions like, "I hear that you're feeling anxious. What is an alternative way to view this challenge not as a threat, but as an opportunity to share your expertise?" or "What aspects of this situation are fully within your control right now? Let's focus our attention on those." This intervention directly supports a healthier, more effective regulation strategy and pinpoints controllable factors the speaker can influence.

OFC Agent: Reward Processing and Outcome Evaluation

Neuroscientific Grounding. The orbitofrontal cortex (OFC), closely related to the VMPFC, is specialized for computing the value of expected outcomes and rapidly updating those values based on new information (Rolls, 2023). It is critical for flexible, adaptive behavior, allowing an organism to change choices when the reward contingencies of the environment shift. The OFC helps us learn from the consequences of our actions and adjust future behavior accordingly, playing a key role in reversal learning and in evaluating potential rewards versus risks (Schoenbaum et al., 2011).

Psychometric Justification. The OFC agent is calibrated by two SCANAQ scales: the General Risk Propensity Scale (GRiPS; Zhang et al., 2019) and the General Decision-Making Style (GDMS; Scott & Bruce, 1995). The GRiPS is a modern, validated measure of an individual's general tendency to take or avoid risks across various domains, directly tapping into the risk-benefit calculations central to OFC function. The GDMS complements this by revealing whether a user's decision style is more spontaneous and intuitive versus dependent and avoidant, which in turn informs how risk information should be framed.

For instance, for a user whose profile shows a very high score on the GRiPS, indicating a strong propensity for risk-taking, the OFC agent would adopt a detailed risk-communication strategy. If this user were considering a high-risk financial decision, such as investing a large sum in a volatile asset, the agent would not simply present the option. It would proactively annotate the choice with detailed probability information, meticulously delineate the potential best-case, worst-case, and most-likely scenarios, and transparently highlight the potential negative consequences. This approach would ensure the user's decision is fully informed, counterbalancing their natural propensity for risk with a clear-eyed view of the potential outcomes while preserving their autonomy.

ACC Agent: Conflict Monitoring and Error Detection

Neuroscientific Grounding. The anterior cingulate cortex (ACC) is often described as the brain's "conflict monitor." It becomes highly active during tasks that involve uncertainty, response conflict (e.g., the Stroop task), and the detection of errors (Botvinick et al., 2004). The ACC does not necessarily resolve the conflict itself but signals its presence to other brain regions, most notably the DLPFC, thereby recruiting more cognitive control to overcome the challenge. It is crucial for self-regulation, allowing us to catch and correct our mistakes and to proceed with caution in ambiguous situations (Kim et al., 2011).

Psychometric Justification. The ACC agent's behavior is dynamically informed by the Barratt Impulsiveness Scale (BIS-11; Patton et al., 1995). The BIS-11 is a widely used measure of impulsivity, a personality trait characterized by a tendency to act on a whim with little forethought. High impulsivity is strongly correlated with deficits in error monitoring and response inhibition—the very functions mediated by the ACC. Therefore, a user's BIS-11 score provides a direct, validated proxy for their likely need for enhanced conflict monitoring support.

In a hypothetical scenario, a user with a high score on the BIS-11's "Motor Impulsiveness" subscale would likely make quick, potentially irreversible decisions. If this user were to attempt to execute such an action within the system (e.g., "Delete all files in the archive folder"), the ACC agent would trigger a reflective-pause protocol. This protocol would extend far beyond a simple "Are you sure? [Y/N]" prompt. Instead, it would require a rationale from the user. The agent might respond, "This action will permanently delete 472 files and cannot be undone. To proceed, please briefly state your reason for deleting these files." This simple "speed bump" forces a moment of reflection to override a potentially impulsive and regrettable action, mirroring a neuro-cognitive conflict resolution process.

mPFC Agent: Social Cognition and Empathy

Neuroscientific Grounding. The medial prefrontal cortex (mPFC) is a cornerstone of the "social brain." It is essential for social cognition, which encompasses abilities including thinking about the mental states of others (Theory of Mind), understanding social norms, and self-reflection (Amodio & Frith, 2006). The mPFC allows us to step outside our own perspective to infer the beliefs, desires, and intentions of those around us, a skill fundamental to successful social interaction and empathy (Sullivan et al., 2022).

Psychometric Justification. The mPFC agent's functionality is modulated by the Interpersonal Reactivity Index (IRI; Davis, 1983). The IRI is a classic, multi-dimensional measure of empathy, assessing both cognitive components (like "Perspective-Taking," the ability to imagine others' viewpoints) and affective components (like "Empathic Concern," the tendency to experience feelings of sympathy). A user's profile on the IRI provides a nuanced picture of their empathic strengths and weaknesses, allowing for highly targeted social-cognitive support.

Consider a user with a low score on the IRI's "Perspective-Taking" subscale who is discussing an interpersonal conflict at work. This user might state, "I can't believe my colleague said that in the meeting. It was completely out of line." The mPFC agent, recognizing the user's difficulty in seeing the situation from another's point of view, would integrate reflective, socially contextual questions into the dialogue. Instead of reinforcing the user's anger, it might gently prompt, "That sounds frustrating. To help me understand the full picture, can we explore for a moment how your colleague might have perceived the situation?" or "What do you think their intention might have been, even if the impact was negative?" This approach would gently facilitate improved social cognition and empathetic understanding, scaffolding a skill the user may find challenging while still acknowledging their feelings.

METHODOLOGICAL FOUNDATION: BRIDGING PSYCHOMETRICS, NEUROSCIENCE, AND AI

Translating SCANAQ Scores into Agent-Specific Behavioral Controls

The personalization mechanism of SCANUE involves clearly defined steps: (1) administering SCANAQ, (2) interpreting psychometric scores, and (3) translating these scores into instructions that dynamically adjust each specialized agent's behaviors. The numerical scoring breakdown and its mapping to specific PFC agent suggestions are available in Appendix B. A pivotal methodological strength of this approach is its deliberate preservation of psychometric nuance. Instead of applying broad normalization methods such as converting scores into generalized z-scores—which can obscure essential, instrument-specific details—the framework maintains the original Likert-type scaling intrinsic to each constituent measure (Tate, 2024c). By preserving the original scaling integrity, SCANUE ensures the nuances intended by the original scale developers remain meaningful in agent-driven applications. Future harmonization will aim to unify ranges without sacrificing this diagnostic signal.

Each SCANUE agent is selectively programmed to interpret and respond specifically to psychometric scores from relevant subsets of SCANAQ, chosen for their theoretical and neuroscientific alignment with the agent's specialized cognitive functionality. These scores translate into agent behaviors through detailed instructional mappings, which are designed as adaptive heuristics that shape nuanced agent responses. The entire workflow is visually conceptualized in Figure 2.

Figure 2. Psychometric-to-Agent Mapping and Dynamic Alignment Workflow

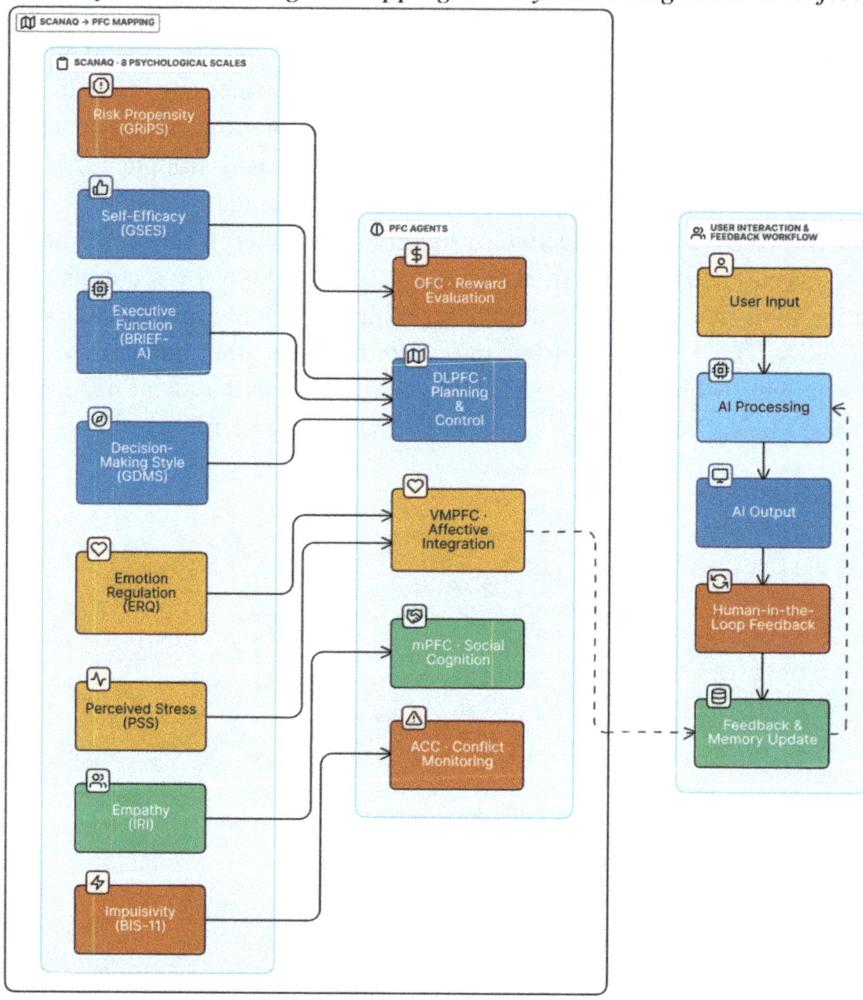

Note. This figure demonstrates the proposed initial mapping of SCANAQ-derived psychometric profiles onto their corresponding SCANUE agents and illustrates the dynamic interaction loop that facilitates the iterative, adaptive refinement of personalized alignment.

Dynamic Alignment Feedback Loop

SCANUE's dynamic alignment process is designed to recognize and accommodate the inherent variability in user states and preferences over time. The proposed personalized alignment process initiates upon completion of the SCANAQ assessment, which generates precise baseline agent configurations derived directly from psychometric scores. This step establishes an initial alignment foundation, which is then refined through the framework's currently implemented human-in-the-loop feedback mechanisms. Although this command-line interface integration is not yet implemented in the codebase, the validated SCANAQ provides a clear and actionable pathway for this future work.

During real-time interactions, dynamic collaboration among the five SCANUE agents is facilitated by advanced orchestration mechanisms like LangGraph. LangGraph systematically manages complex conditional logic, inter-agent communication, and agent-selection processes to determine precise agent engagement at any moment. This design supports flexible information flow, enabling investigations into alternative agent-routing options and diverse methods of use that are not constrained to a single biological analogue. Presently, each agent leverages specialized GPT-based models fine-tuned on domain-specific datasets that align with its cognitive functions (Kennedy et al., 2025; Tate, 2024c). A critical future development goal is the transition of these agents onto STAC's energy-efficient SNN computational substrates to enhance sustainable deployment capabilities.

To detect misalignment continuously, the framework's comprehensive telemetry systematically records explicit user feedback, including user-generated ratings and correctional interactions. Future enhancements will augment this process with more subtle data streams. These will include sophisticated analyses of implicit behavioral indicators—such as response latencies, editing behaviors, and sentiment fluctuations—alongside optional biometric integrations (e.g., EEG, fNIRS, heart rate variability, galvanic skin responses). With ethically secured user consent, such integrated data could significantly improve real-time assessments of cognitive load and stress levels.

Beyond detection, future framework extensions will incorporate adaptive recalibration mechanisms to correct for sustained misalignment. This recalibration would employ efficient, computer-adaptive short-form SCANAQ variants, administered dynamically based on interaction history to maintain alignment without the burden of full reassessment. As validated extensively in psychometric literature (Ling et al., 2022), these adaptive methodologies can refine agent behavioral parameters precisely and responsively, thereby enhancing long-term alignment with minimal user burden.

PROPOSED VALIDATION STRATEGY

Alignment and User Experience Metrics

Future evaluations will quantify alignment accuracy using primary metrics: frequency of misalignment events (target $\leq 5\%$), standardized usability scores (SUS ≥ 85), and structured qualitative analysis of user feedback to confirm meaningful alignment. To strengthen these assessments, a mixed-method approach combining quantitative and qualitative data collection is recommended. Quantitative analyses will include statistical modeling to identify and predict factors associated with alignment discrepancies, enabling proactive adjustments. This mixed-methods design will couple statistical power with narrative insights.

Qualitative methodologies will involve structured user interviews and focus groups aimed at capturing deeper insights into user perceptions and experiences. For instance, thematic analysis of interview transcripts could reveal nuanced user attitudes toward agent interactions, identifying areas where agent behavior aligns exceptionally well or consistently misses expectations. This qualitative component will facilitate targeted refinement of SCANAQ-derived mappings and enhance the precision of personalization.

Furthermore, additional empirical validation could leverage longitudinal study designs to assess alignment stability over extended periods. Regular, brief reassessments using computer-adaptive testing forms of SCANAQ would monitor alignment trajectories, ensuring sustained personalized efficacy without excessive user burden. Analyzing alignment data longitudinally can help identify emerging trends or gradual drift in agent-user congruence, which would trigger timely recalibration. Finally, including diverse participant populations in validation studies will ensure broader generalizability and equitable applicability. Attention must be given to capturing varied demographic backgrounds, cognitive styles, and professional roles to reinforce the robustness and inclusivity of validation findings.

Task-Specific Cognitive Outcomes

Agent efficacy will be systematically assessed using validated neuropsychological benchmarks. Classic tasks such as the Tower of London task (assessing planning capabilities, targeting DLPFC functions) and the Iowa Gambling Task (assessing risk-related decision-making, targeting OFC/VMPFC functions) will provide precise cognitive performance benchmarks.

Affective Congruence and Regulation Support

Structured stress-induction protocols, such as cognitive tasks under controlled time constraints, will be used to evaluate affective regulation efficacy by comparing predicted versus actual emotional responses. These assessments will critically examine the effectiveness of VMPFC-delivered emotional scaffolding using objective physiological and subjective self-report measures, with a target of significantly attenuating stress markers.

Energy Efficiency and Computational Performance

Rigorous benchmarking on specialized neuromorphic platforms, such as Intel's Loihi 2, will quantitatively validate computational efficiency claims. This step is vital for establishing credibility, as it moves beyond simulation to physical measurement. Precise spike-count telemetry and watt-hour energy consumption measurements will adhere to established neuromorphic benchmarks, transparently confirming (or refuting) theoretical projections empirically on Loihi 2 hardware.

STAC DEVELOPMENT TRAJECTORY: A METHODOLOGICAL EVOLUTION

STAC V1: A Deep Dive into the Hybrid Fine-Tuned Architecture

The initial version, STAC V1 (Tate, 2025a), was a landmark project designed to prove the fundamental viability of creating a high-performance hybrid SNN-transformer. Its architecture, illustrated in Figure 3, was a meticulous exercise in neuro-inspired engineering, defined by three key technical innovations observable in the project's implementation artifacts.

Figure 3. STAC V1 Hybrid Fine-Tuning Pipeline: Integrating a Pre-Trained Base with Spiking Component

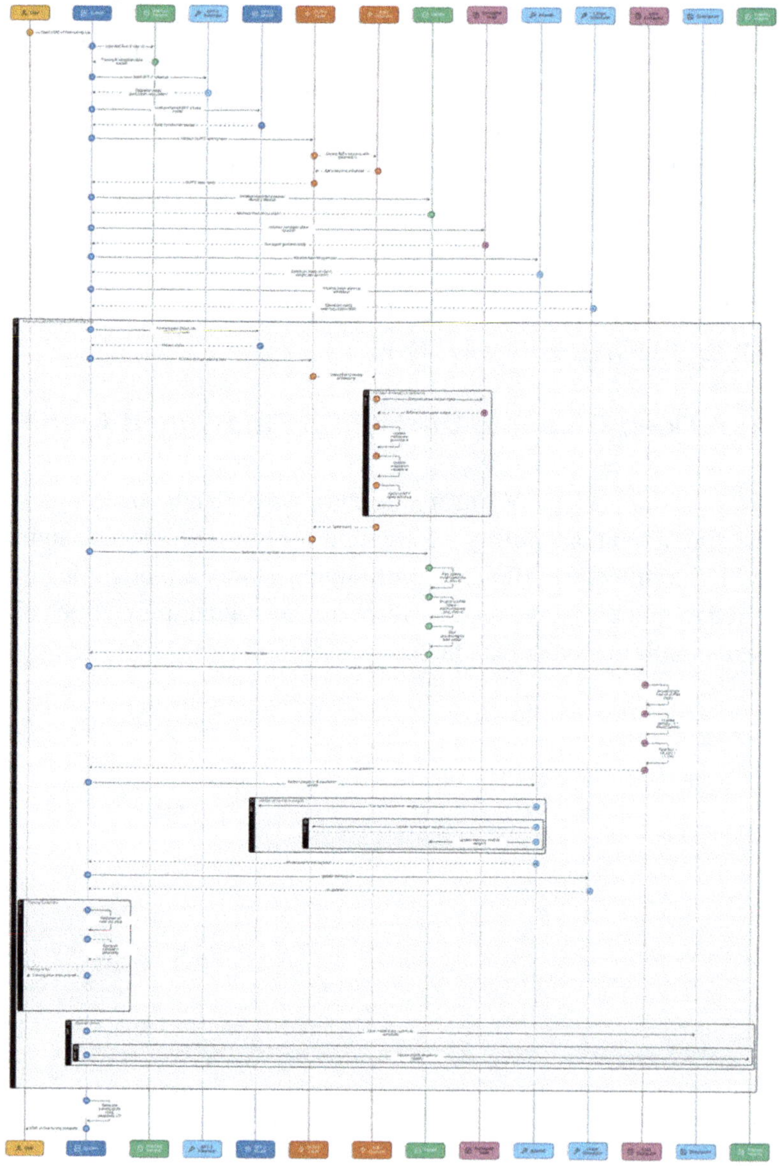

Learnable AdEx Neurons and Surrogate Gradients.

The choice of the Adaptive-Exponential (AdEx) integrate-and-fire neuron model, based on the foundational work of Brette & Gerstner (2005), was deliberate.

Unlike simpler Leaky Integrate-and-Fire (LIF) neurons, the AdEx model captures a richer repertoire of biologically plausible dynamics, including spike-frequency adaptation and bursting. Its dynamics are governed by several parameters, such as the membrane time constant (τm) and the adaptation time constant (τw). The crucial innovation in STAC V1 was making these parameters learnable by defining them as torch.nn.Parameter tensors. However, the all-or-none spiking mechanism is non-differentiable, which poses a major obstacle to gradient-based training. STAC V1 solved this problem by using the surrogate gradient technique (Shrestha & Orchard, 2018; Zenke & Vogels, 2021). During the backward pass, this technique replaces the true, non-differentiable gradient of the spike-generation function with a "surrogate"—a continuous, well-behaved proxy like a fast sigmoid function. This proxy provides a useful learning signal that allows gradients to flow back through the spiking neurons, enabling the backpropagation algorithm to fine-tune not just the synaptic weights but also the intrinsic firing properties of the neurons themselves.

The Hyperdimensional Memory Module (HEMM).

To augment the model's ability to handle dependencies beyond immediate firing dynamics, STAC V1 introduced the HEMM, a structured memory system that departs from standard attention mechanisms. Its implementation follows a clear, efficient sequence. First, the full spike train tensor from the AdEx neuron layer is pooled across the time dimension via a torch.mean operation, creating a static vector that represents the sequence's overall spiking activity. This vector is then projected into a very high-dimensional space via matrix multiplication with a fixed, random projection matrix. This technique, inspired by hyperdimensional computing, creates a sparse, distributed representation of the activity history. This high-dimensional vector is then passed through a small multi-layer perceptron (MLP) to generate a final "memory bias," which is added back to the token representations before they enter the spiking layer. This mechanism allows the model's own recent past activity to influence its current processing in a structured, recurrent, and computationally efficient manner.

Integrated L1 Spike Regularization.

A core design principle of STAC V1 was that energy efficiency should be an integral part of the learning process, not a post-hoc analysis. To achieve this goal, the model actively encouraged sparse spiking activity by incorporating an L1-loss term directly into the main fine-tuning objective function. The total loss was a weighted sum of the standard cross-entropy loss and a sparsity loss, defined as L_sparsity = λ * torch.mean(torch.abs(spike_tensor)), where λ is a weighting coefficient that controls the strength of the sparsity penalty. By penalizing the mean absolute value of the spike tensor, the optimizer is incentivized to find solutions where a maximal number of elements in the tensor are zero (i.e., no spike occurred). This directly promotes the temporal sparsity that is the hallmark of efficient SNNs, making energy efficiency an objective of the optimization itself.

STAC V2: A Pragmatic Pivot and the Coherence-Fidelity Trade-off

While STAC V1 demonstrated feasibility, its scaling limitations led to the pragmatic development of STAC V2, which aims to convert existing models for broader usability. The AI field is dominated by massive LLMs trained by large corporations on planetary-scale datasets. The hybrid fine-tuning paradigm, while academically elegant, could not realistically scale to these models. Recognizing this limitation, the research pivoted from deep integration to pragmatic conversion. The central motivation for STAC V2 (Tate, 2025b) was to develop a toolchain that could apply the energy-saving benefits of neuromorphic computing to the vast ecosystem of existing pretrained models, thereby democratizing access to neuromorphic efficiency. This toolchain is outlined in Figure 4.

Figure 4. STAC V2 Conversion Framework: From Pretrained Transformers to Stateful Spiking Networks

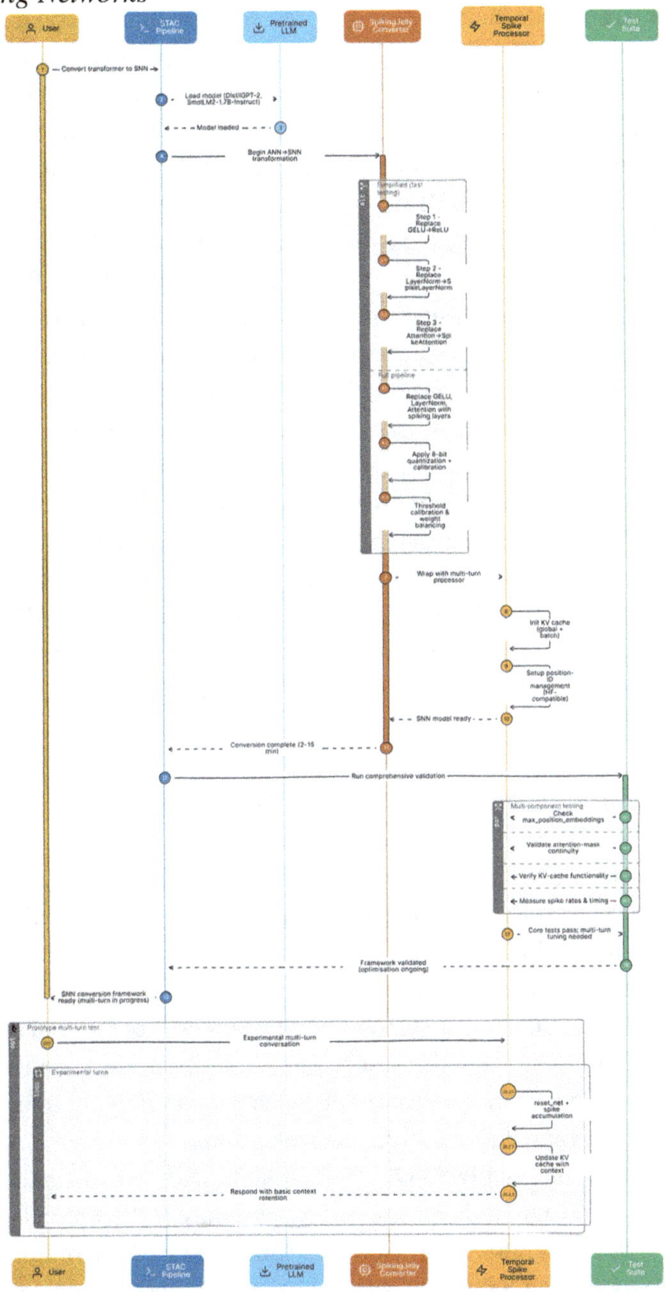

This pivot, however, introduces a well-known challenge in neuromorphic engineering: a trade-off between energy efficiency and computational fidelity. The degradation in conversational quality observed in STAC V2 is a direct and predictable consequence of the information loss inherent in the ANN-to-SNN conversion process.

Activations in a traditional ANN are represented by high-precision, continuous floating-point numbers. The STAC V2 conversion pipeline, relying on libraries like SpikingJelly (Fang et al., 2021) and principles from works such as Diehl et al. (2015), must perform a form of quantization. It must map this rich, continuous range of activation values onto a stream of discrete, binary spike events over time—an inherently lossy process. A small degree of error is introduced at every layer as continuous values are approximated by spike rates or timings. While the error at any single layer may be negligible, it compounds as data flows through the dozens of layers in a deep transformer architecture. This cumulative precision loss ultimately manifests as a noticeable degradation in the coherence and nuance of the final text output when compared to the original, full-precision ANN.

A critical area of future research, therefore, is the development of "quantization-aware" conversion techniques or post-conversion tuning strategies to mitigate this fidelity loss. Such methods might involve fine-tuning the converted SNN for a few epochs on a small dataset, allowing it to recover from the conversion noise, or employing more sophisticated conversion algorithms that can better preserve the information content of the original model. Acknowledging this trade-off is crucial for setting realistic expectations about the current state of SNN technology and for guiding future research toward building SNNs that are not only efficient but also powerful.

Preliminary Findings and Future Benchmarking

The development of STAC V2 was sharply focused on enabling multi-turn conversations through post-hoc conversion, a more pragmatic goal than V1's hybrid fine-tuning. Initial functional "smoke tests" were conducted to verify the new architecture's core hypothesis. These tests confirmed that V2's Temporal Spike Processor (TSP) (Tate, 2025b) can mechanically pass conversational memory states across successive interactions within a software simulation. This result demonstrates the feasibility of the basic architectural approach for enabling stateful interactions, though it is not a guarantee of coherent conversation.

It is essential to contextualize STAC V2's results with two critical caveats. First, its projected three- to four-fold energy savings are purely theoretical, derived from post-hoc spike telemetry; this contrasts with the integrated L1 regularization used to enforce sparsity during STAC V1's fine-tuning. Second, the conversion process embodies a significant trade-off: the resulting SNN, while stateful, suffers from a clear degradation in conversational coherence and nuance compared to its source

ANN. Blind A/B human-rating studies will be required to formally score coherence, helpfulness, and safety.

DISCUSSION

Integration of Psychometric Alignment and Neuromorphic Efficiency: A Synergistic Approach

This research presents a strategically integrated approach aimed at resolving two typically separate yet deeply interconnected challenges in contemporary AI: achieving deep personalized cognitive augmentation while simultaneously realizing significant computational and environmental efficiency through neuromorphic computing. The deliberate integration of psychometrically derived user profiles—captured through the validated SCANAQ assessment—with the specialized, prefrontal-cortex-inspired cognitive agents of the SCANUE architecture directly addresses ongoing challenges related to user trust, personal relevance, and the ethical deployment of AI systems (Kennedy et al., 2025). By linking validated psychometric metrics to agent-specific cognitive behaviors, the SCANUE-SCANAQ alignment strategy fosters precisely tailored interactions. This alignment ensures the assistant's communications and support strategies are continuously aligned with the nuanced cognitive and affective preferences unique to individual users, which in turn sustains engagement, builds trust, and ensures genuine utility in practical cognitive-augmentation contexts.

Simultaneously, the STAC V2 framework addresses computational sustainability by leveraging biologically inspired, event-driven SNNs. Crucially, STAC V2's innovations—particularly the introduction of the Temporal Spike Processor (TSP)—provide the mechanical foundation for stateful multi-turn interactions in software simulations. This architectural synergy supports the computational feasibility of deploying personalized cognitive agents and ensures practical scalability. Thus, SCANUE's advanced cognitive architecture provides meaningful, complex cognitive missions ideally suited to leverage STAC's computationally efficient execution capabilities. Conversely, STAC's neuromorphic efficiency makes the practical implementation of such sophisticated personalization computationally feasible at scale and justifies on-device deployment.

Limitations and Future Directions for Research and Refinement

While this work presents a promising synthesis, several limitations define the immediate trajectory for future research. On a technical level, the projected energy savings require empirical validation on dedicated neuromorphic hardware, as current operator coverage is limited and the conversion process can degrade performance on complex reasoning tasks. Scalability to larger transformer models also remains a significant hurdle. Concurrently, from a user-centric standpoint, critical questions persist regarding the stability of personalized alignment over long-term interaction and the framework's ability to generalize effectively across diverse populations. To launch this next phase of research, two initiatives are proposed: a one-month study to benchmark energy consumption on neuromorphic hardware and a three-month longitudinal trial to rigorously evaluate the durability of user alignment.

Ethical, Societal, and Governance Considerations: Navigating Responsible Innovation

Meticulous attention to ethical concerns, such as cognitive over-reliance and potential biases in psychometric assessments, is critical. Privacy and confidentiality must be proactively safeguarded, particularly because the SCANUE-SCANAQ framework relies on sensitive psychometric data. To prevent unauthorized data usage, robust data encryption, ISO-27001-grade storage protocols, and strict access control mechanisms should be implemented. Additionally, user autonomy and informed consent must be prioritized through transparent communication about how their data is collected, stored, and utilized.

The risk of cognitive over-reliance poses an ethical and practical challenge. While personalized cognitive augmentation systems aim to support user capabilities, users might become excessively dependent on AI-generated advice, diminishing their own decision-making confidence over time. To mitigate this risk, specific design features should be incorporated, such as periodic prompts encouraging independent critical reflection and clear communication of uncertainty in AI-generated recommendations. Continuous monitoring of user interactions could further help identify patterns of dependency, allowing for proactive interventions aimed at empowering users rather than diminishing their cognitive autonomy.

Potential biases inherent in psychometric assessments present another significant ethical consideration. Given SCANAQ's reliance on established psychological scales, it is essential to recognize and address biases that could affect fairness across diverse populations, including cultural biases and language barriers. Rigorous validation across diverse demographic and cultural groups is crucial. Standard practice should include regularly updating assessment items based on empirical feedback, incorporating culturally sensitive adaptations, and employing advanced statistical methods for bias detection (e.g., Differential Item Functioning analyses) to ensure fairness and inclusivity.

Moreover, governance considerations must include clear accountability frameworks. Given the complex interplay between AI, psychometric evaluation, and neuromorphic computing, a clear delineation of responsibilities and oversight mechanisms is necessary. Independent ethics review boards, comprising interdisciplinary experts, could regularly audit framework deployments, particularly in high-stakes contexts like healthcare or education. Transparent documentation, reporting standards akin to clinical trial registries, and public fail-case reports should be adopted to ensure public accountability and foster trust among stakeholders, regulators, and the broader community.

CONCLUSION AND FUTURE OUTLOOK

The integration of the SCANUE, SCANAQ, and STAC frameworks provides a viable architecture for developing sustainable, deeply personalized cognitive support systems capable of enhancing user trust, engagement, and equitable access to advanced AI. This paper has chronicled a principled research evolution—from the hybrid fine-tuning of STAC V1 to the more pragmatic, experimental conversion approach of STAC V2. By synthesizing principles from cognitive neuroscience, rigorous psychometric methodologies, and an iterative exploration of neuromorphic paradigms, this work offers a comprehensive roadmap for deploying the next generation of cognitive augmentation technologies.

While this synthesis is promising, the path to implementation requires addressing clear next steps. Technically, the framework's performance and efficiency projections must be empirically validated on specialized neuromorphic hardware and scaled to larger models. Concurrently, the long-term stability and generalizability of its user-centric alignment demand rigorous longitudinal evaluation across diverse populations.

Successfully navigating these research pathways will position the integrated framework for real-world implementation across critical domains, including clinical decision-making support, personalized adaptive education, and private, responsive edge-computing environments. Ultimately, achieving this integrated vision advances cognitive augmentation technologies that honor human individuality while responsibly addressing global sustainability imperatives.

REFERENCES

Amodio, D. M., & Frith, C. D. (2006). Meeting of minds: The medial frontal cortex and social cognition. *Nature Reviews. Neuroscience*, *7*(4), 268–277. DOI: 10.1038/nrn1884

Barbey, A. K., Koenigs, M., & Grafman, J. (2012). Dorsolateral prefrontal contributions to human working memory. *Cortex*, *48*(1), 62–70. DOI: 10.1016/j.cortex.2011.06.012

Beaton, D. E., Bombardier, C., Guillemin, F., & Ferraz, M. B. (2000). Guidelines for the process of cross-cultural adaptation of self-report measures. *Spine*, *25*(24), 3186–3191. DOI: 10.1097/00007632-200012150-00014

Bechara, A., Damasio, H., Tranel, D., & Damasio, A. R. (2005). The Iowa Gambling Task and the somatic marker hypothesis: Some questions and answers. *Trends in Cognitive Sciences*, *9*(4), 159–162. DOI: 10.1016/j.tics.2005.02.002

Botvinick, M. M., Cohen, J. D., & Carter, C. S. (2004). Conflict monitoring and anterior cingulate cortex: An update. *Trends in Cognitive Sciences*, *8*(12), 539–546. DOI: 10.1016/j.tics.2004.10.003

Brette, R., & Gerstner, W. (2005). Adaptive exponential integrate-and-fire model as an effective description of neuronal activity. *Journal of Neurophysiology*, *94*(5), 3637–3642. DOI: 10.1152/jn.00686.2005

Cohen, S., Kamarck, T., & Mermelstein, R. (1983). A global measure of perceived stress. *Journal of Health and Social Behavior*, *24*(4), 385–396. DOI: 10.2307/2136404

Damasio, A. R. (1994). *Descartes' error: Emotion, reason, and the human brain*. Putnam.

Davies, M., Wild, A., Orchard, G., Sandamirskaya, Y., Guerra, G. A. F., Joshi, P., Ploeg, F., Risbud, S. R., & Wang, R. M. (2021). Advancing neuromorphic computing with Loihi: A survey of results and outlook. *Proceedings of the IEEE*, *109*(5), 911–934. DOI: 10.1109/JPROC.2021.3067593

Davis, M. H. (1983). Measuring individual differences in empathy: Evidence for a multidimensional approach. *Journal of Personality and Social Psychology*, *44*(1), 113–126. DOI: 10.1037/0022-3514.44.1.113

Diehl, P. U., Neil, D., Binas, J., Cook, M., Liu, S. C., & Pfeiffer, M. (2015). Fast-classifying, high-accuracy spiking deep networks through weight and threshold balancing. *2015 International Joint Conference on Neural Networks (IJCNN)*, 1–8. DOI: 10.1109/IJCNN.2015.7280696

Fang, W., Yu, Z., Chen, Y., Masquelier, T., Huang, T., & Tian, Y. (2021). SpikingJelly: A deep-learning framework for spiking neural networks. *Frontiers in Neuroscience, 15*, 774067. Advance online publication. DOI: 10.3389/fnins.2021.774067

Gioia, G. A., Isquith, P. K., Guy, S. C., & Kenworthy, L. (2000). *Behavior Rating Inventory of Executive Function* [Assessment instrument]. Psychological Assessment Resources.

Gross, J. J., & John, O. P. (2003). Individual differences in two emotion-regulation processes: Implications for affect, relationships, and well-being. *Journal of Personality and Social Psychology, 85*(2), 348–362. DOI: 10.1037/0022-3514.85.2.348

Gudmundsson, E., & Lönner, W. J. (2009). Cross-cultural test adaptation. In Weiner, B., & Craighead, E. J. (Eds.), *The Corsini encyclopedia of psychology* (4th ed., Vol. 1, pp. 448–450). John Wiley & Sons.

Kennedy, B., Mohammad, A., & Wyandt, M. (2025). Beyond Intelligence: The Synthetic Cognitive Augmentation Network Using Experts. In Zangana, H., & Omar, M. (Eds.), *Ensuring Secure and Ethical STM Research in the AI Era* (pp. 257–286). IGI Global Scientific Publishing., DOI: 10.4018/979-8-3373-4252-8.ch007

Kim, C., Johnson, N. F., & Gold, B. T. (2011). Cingulo-frontal network activates differentially to conscious and non-conscious conflict. *Proceedings of the National Academy of Sciences of the United States of America, 108*(42), 17547–17552. DOI: 10.1073/pnas.1103627108

Li, S., Wang, X., Sun, Y., Wu, P., & Chen, A. (2021). The ventromedial prefrontal cortex in emotion regulation: Combined TMS-fMRI evidence. *Cerebral Cortex (New York, N.Y.), 31*(12), 5482–5496. DOI: 10.1093/cercor/bhab139

Ling, Y., Zhang, J., & Wang, K. (2022). Computer-adaptive forms for multidimensional questionnaires: Methodological advances and simulation results. *Applied Psychological Measurement, 46*(5), 401–417. DOI: 10.1177/01466216211073012

Miller, E. K., & Cohen, J. D. (2001). An integrative theory of prefrontal cortex function. *Annual Review of Neuroscience, 24*(1), 167–202. DOI: 10.1146/annurev.neuro.24.1.167

Ostrau, C., Klarhorst, C., Thies, M., & Rückert, U. (2022). Benchmarking neuromorphic hardware and its energy expenditure. *Frontiers in Neuroscience*, *16*, 873935. https://doi.org/10.3389/fnins.2022.873935

Patton, J. H., Stanford, M. S., & Barratt, E. S. (1995). Factor structure of the Barratt Impulsiveness Scale. *Journal of Clinical Psychology*, *51*(6), 768–774. DOI: 10.1002/1097-4679(199511)51:6<768::AID-JCLP2270510607>3.0.CO;2-1

Petrides, M. (2005). Lateral prefrontal cortex: Architectonic and functional organization. *Philosophical Transactions of the Royal Society of London. Series B, Biological Sciences*, *360*(1456), 781–795. DOI: 10.1098/rstb.2005.1631

Rolls, E. T. (2023). The orbitofrontal cortex, amygdala, reward value, emotion, and decision-making. *Neuroscience and Biobehavioral Reviews*, *150*, 105066. Advance online publication. DOI: 10.1016/j.neubiorev.2023.105066

Roy, K., Jaiswal, A., & Panda, P. (2019). Towards spike-based machine intelligence with neuromorphic computing. *Nature*, *575*(7784), 607–617. DOI: 10.1038/s41586-019-1677-2

Schoenbaum, G., Takahashi, Y., Liu, T. L., & McDannald, M. A. (2011). Does the orbitofrontal cortex signal value? *Annals of the New York Academy of Sciences*, *1239*(1), 87–99. DOI: 10.1111/j.1749-6632.2011.06210.x

Schwarzer, R., & Jerusalem, M. (1995). Generalized Self-Efficacy Scale. In Weinman, J., Wright, S., & Johnston, M. (Eds.), *Measures in health psychology: A user's portfolio* (pp. 35–37). NFER-Nelson.

Scott, S. G., & Bruce, R. A. (1995). Decision-making style: The development and assessment of a new measure. *Educational and Psychological Measurement*, *55*(5), 818–831. DOI: 10.1177/0013164495055005017

Shrestha, S. B., & Orchard, G. (2018). SLAYER: Spike layer error reassignment in time. In Bengio, S., Wallach, H., Larochelle, H., Grauman, K., Cesa-Bianchi, N., & Garnett, R. (Eds.), Vol. 31, pp. 1419–1428). Advances in Neural Information Processing Systems. Curran Associates.

Strubell, E., Ganesh, A., & McCallum, A. (2019). Energy and policy considerations for deep learning in NLP. *Proceedings of the 57th Annual Meeting of the Association for Computational Linguistics*, 3645–3650. DOI: 10.18653/v1/P19-1355

Sullivan, J. G., Vu, M. A. T., Stein, D. J., & Yazdanfar, A. (2022). Medial prefrontal cortex activation during third-party social interactions in infants. *Developmental Science*, *25*(5), e13245. Advance online publication. DOI: 10.1111/desc.13245

Tate, L. (2024a). iLevyTate/SCANUE: 1.0.0-alpha (Version 1.0.0-alpha) [Computer software]. Zenodo. DOI: 10.5281/zenodo.14052759

Tate, L. (2024b). iLevyTate/scanue-v22: Zenodo Synchronization Second (Version 22.0) [Computer software]. Zenodo. DOI: 10.5281/zenodo.14510407

Tate, L. (2024c). iLevyTate/SCAN-Resources: 1.0.0-alpha (Version 1.0.0-alpha) [Dataset and code repository]. Zenodo. DOI: 10.5281/zenodo.140532

Tate, L. (2025a). STAC V1 implementation in iLevyTate/stac (Version 2.0.0.3) [Computer software, Jupyter Notebook]. Zenodo. DOI: 10.5281/zenodo.15867066

Tate, L. (2025b). STAC V2 implementation in iLevyTate/stac (Version 2.0.0.3) [Computer software]. Zenodo. DOI: 10.5281/zenodo.15867066

Tate, L., & Sanders, P. (2024). iLevyTate/SCAN: 1.0.0-alpha (Version 1.0.0-alpha) [Computer software]. Zenodo. DOI: 10.5281/zenodo.14052885

Vaswani, A., Shazeer, N., Parmar, N., Uszkoreit, J., Jones, L., Gomez, A. N., Kaiser, L., & Polosukhin, I. (2017). Attention is all you need. In von Luxburg, U., Guyon, I., Bengio, S., & Wallach, H. (Eds.), Vol. 30, pp. 5998–6008). Advances in Neural Information Processing Systems. Curran Associates.

Zenke, F., & Vogels, T. P. (2021). The remarkable robustness of surrogate gradient learning for instilling complex function in spiking neural networks. *Nature Machine Intelligence*, *3*(1), 76–87. DOI: 10.1038/s42256-020-00286-3

Zhang, D. C., Highhouse, S., & Nye, C. D. (2019). Development and validation of the General Risk Propensity Scale (GRiPS). *Journal of Behavioral Decision Making*, *32*(2), 152–167. DOI: 10.1002/bdm.2102

APPENDIX A

Synthetic Cognitive Augmentation Network Alignment Questionnaire (SCANAQ)

Introduction

Welcome to the *Synthetic Cognitive Augmentation Network Alignment Questionnaire* (SCANAQ). This assessment evaluates aspects of your cognitive functioning and personal preferences to align the SCAN system with your needs.

Instructions

1. Read each statement carefully.
2. Use the section's rating scale table to determine your response number.
3. Enter only the number in the Response column.
4. There are no right or wrong answers—respond based on typical thoughts, feelings, and behaviors.

Section A: Executive Functioning (Items 1–8)

Table 1.

Item	Statement	Response (1–3)
1	I have trouble getting started on tasks.	
2	I forget instructions easily.	
3	I have trouble keeping my emotions under control.	
4	I make careless mistakes.	
5	I get stuck on one way of doing things.	
6	I have trouble planning ahead.	
7	I have difficulty organizing my tasks.	
8	I have trouble keeping track of my belongings.	
Rating Scale 1- Never; 2- Sometimes; 3- Often		

Section B: Emotion Regulation (Items 9–12)

Table 2.

Item	Statement	Response (1–7)
9	I control my emotions by changing the way I think about the situation I'm in.	
10	When I want to feel less negative emotion, I change the way I'm thinking about the situation.	
11	I keep my emotions to myself.	
12	When I am feeling negative emotions, I make sure not to express them.	
Rating Scale 1- Strongly Disagree; 2- Disagree; 3- Somewhat Disagree; 4- Neither Agree nor Disagree; 5- Somewhat Agree; 6- Agree; 7- Strongly Agree		

Section C: Impulsivity (Items 13–18)

Table 3.

Item	Statement	Response (1–4)
13	I don't pay attention.	
14	I make up my mind quickly.	
15	I act on the spur of the moment.	
16	I buy things on impulse.	
17	I don't plan tasks carefully.	
18	I am more interested in the present than the future.	
Rating Scale 1- Rarely/Never; 2- Occasionally; 3- Often; 4- Almost Always/Always		

Section D: Risk Propensity (Items 19–21)

Table 4.

Item	Statement	Response (1–5)
19	I enjoy taking risks in general.	
20	I often take risks in my daily life.	
21	Taking risks is an important part of my life.	
Rating Scale 1- Strongly Disagree; 2- Disagree; 3- Neutral; 4- Agree; 5- Strongly Agree		

Section E: Decision-Making Style (Items 22–26)

Table 5.

Item	Statement	Response (1–5)
22	I make decisions in a logical and systematic way.	
23	I rely on my instincts when making decisions.	
24	I depend on others to help me make decisions.	
25	I avoid making important decisions until the pressure is on.	
26	I make quick decisions without thinking them through.	
Rating Scale 1- Strongly Disagree; 2- Disagree; 3- Neutral; 4- Agree; 5- Strongly Agree		

Section F: Self-Efficacy (Items 27–29)

Table 6.

Item	Statement	Response (1–4)
27	I can always manage to solve difficult problems if I try hard enough.	
28	If someone opposes me, I can find the means and ways to get what I want.	
29	I am confident that I could deal efficiently with unexpected events.	
Rating Scale 1- Not al all True; 2- Hardly True; 3- Moderately True; 4- Exactly True		

Section G: Perceived Stress (Items 30–32)

Table 7.

Item	Statement	Response (1–5)
30	In the last month, how often have you felt unable to control the important things in your life?	
31	In the last month, how often have you felt nervous and stressed?	
32	In the last month, how often have you felt that things were going your way?	
Rating Scale 1- Never; 2- Almost Never; 3- Sometimes; 4- Fairly Often; 5- Very Often		

Section H: Empathy & Social Cognition (Items 33–36)

Table 8.

Item	Statement	Response (1–5)
33	I often have tender, concerned feelings for people less fortunate than me.	
34	When I see someone being taken advantage of, I feel kind of protective towards them.	
35	I sometimes find it difficult to see things from the other person's point of view.	
36	I daydream and fantasize, with some regularity, about things that might happen to me.	
Rating Scale 1- Does Not Describe Me Well; 2- Describes Me a Little; 3- Somewhat Describes Me; 4- Describes Me Well; 5- Describes Me Very Well		

Completion Notes

Thank you for completing the SCANAQ. Your responses will help us generate a personalized alignment profile to enhance your interaction with SCAN.

- This is not a diagnostic tool.
- Participation is voluntary and confidential.
- All data is securely handled.

APPENDIX B

SCANAQ Scoring Breakdown, Descriptions, and PFC-Inspired Suggestions

Section A: Executive Function (EF) — Items 1–8

Table 9. Scoring / Interpretation

PROFILE CODE	PROFILE LABEL	SCORE RANGE	ITEMS (Q#)	DESCRIPTION	PFC-INSPIRED SUGGESTION
4A	High Inhibition, Low Planning	8–13	1–8	Difficulty initiating and planning tasks.	Provide planning tools, reminders, organizational aids; break tasks into smaller steps.
4B	Low Inhibition, High Working Memory	14–24	1–8	Strong working memory; impulsivity may need regulation.	Offer impulse-control strategies; leverage strong WM for complex/multistep tasks.

Section B: Emotion Regulation (ER) — Items 9–12

Table 10. Scoring / Interpretation

PROFILE CODE	PROFILE LABEL	SCORE RANGE	ITEMS (Q#)	DESCRIPTION	PFC-INSPIRED SUGGESTION
2B	Low Cognitive Reappraisal, High Expressive Suppression	4–19	9–12	Difficulty reframing emotions; relies on suppression strategies.	Train cognitive reappraisal via CBT-style reframing tools and guided cognitive restructuring.
2A	High Cognitive Reappraisal, Low Expressive Suppression	20–28	9–12	Strong emotional reframing; balanced expression.	Provide expressive-regulation tools to fine-tune social display and contextual appropriateness.

Section C: Impulsivity (IMP) — Items 13–18

Table 11. Scoring / Interpretation

PROFILE CODE	PROFILE LABEL	SCORE RANGE	ITEMS (Q#)	DESCRIPTION	PFC-INSPIRED SUGGESTION
5B	High Motor Impulsivity, Low Non-Planning Impulsivity	6–12	13–18	Acts impulsively; limited advance planning.	Embed pause-and-plan prompts; use external structure, checklists, and "if-then" implementation intentions.
5A	High Attentional Impulsivity, Low Motor Impulsivity	13–24	13–18	Distractible but less motorically impulsive.	Provide attentional control aids (mindfulness, focus timers, distraction blockers).

Section D: Risk Propensity (RP) — Items 19–21

Table 12. Scoring / Interpretation

PROFILE CODE	PROFILE LABEL	SCORE RANGE	ITEMS (Q#)	DESCRIPTION	PFC-INSPIRED SUGGESTION
6B	Low Risk-Taking	3–9	19–21	Avoids risk; prefers safer choices.	Scaffold calculated exploration; present graded exposure to uncertainty with feedback.
6A	High Risk-Taking	10–15	19–21	Seeks/enjoys taking risks.	Provide risk-assessment tools, consequence simulators, and decision hygiene checklists.

Section E: Decision-Making Style (DM) — Items 22–26

Determination Rule Table

Determine the **single highest scoring item** among Q22–Q26; assign the corresponding style below (no cumulative score).

Table 13. Style Interpretation

STYLE CODE	PRIMARY ITEM	STYLE LABEL	DESCRIPTION	PFC-INSPIRED SUGGESTION
1A	Q22	Rational	Logical, systematic decision-making.	Provide data-driven tools, structured analysis frameworks, and MCDA/ utility modeling.
1B	Q23	Intuitive	Relies on instincts.	Pair intuition with post-hoc analytical checks; use scenarios/ simulations for calibration.
1C	Q24	Dependent	Seeks guidance and collaboration.	Offer collaborative tools/decision aids; scaffold autonomy via staged independence.
1D	Q25	Avoidant	Procrastinates/avoids decisions.	Use proactive prompts, deadlines, and commitment devices; provide bounded-choice frameworks.
1E	Q26	Spontaneous	Quick, reflexive decisions with limited forethought.	Introduce pause rules ("10-second rule"), consequence visualizations, and pre-commitment plans.

Section F: Self-Efficacy (SE) — Items 27–29

Table 14. Scoring / Interpretation

Profile Code	Profile Label	Score Range	Items (Q#)	Description	PFC-Inspired Suggestion
7B	Low Self-Efficacy	3–8	27–29	Low confidence handling problems/ challenges.	Build mastery via graded goals, mentorship, success tracking, and strengths-based reflection.
7A	High Self-Efficacy	9–12	27–29	Confident and effective under challenge.	Leverage leadership potential; assign complex tasks; promote peer coaching/knowledge transfer.

Section G: Perceived Stress (PS) — Items 30–32

Table 15. Scoring / Interpretation

PROFILE CODE	PROFILE LABEL	SCORE RANGE	ITEMS (Q#)	DESCRIPTION	PFC-INSPIRED SUGGESTION
8B	Low Perceived Stress	3–9	30–32	Reports lower stress frequency.	Maintain resilience practices; add anticipatory coping plans for high-demand periods.
8A	High Perceived Stress	10–15	30–32	Frequently stressed/ overwhelmed.	Provide stress-management (CBT, relaxation, time-buffering, prioritization aids).

Section H: Empathy and Social Cognition (ESC) — Items 33–36

Table 16. Scoring / Interpretation

PROFILE CODE	PROFILE LABEL	SCORE RANGE	ITEMS (Q#)	DESCRIPTION	PFC-INSPIRED SUGGESTION
3B	Low Empathy	4–12	33–36	Difficulty relating to or understanding others.	Social skills training, perspective-taking exercises, and feedback on empathic accuracy.
3A	High Empathy	13–20	33–36	Strong empathic concern and social attunement.	Encourage mentoring/ leadership roles; integrate into cooperative, high-EQ team workflows.

Global Notes

Scores are summed within each section (except Decision-Making Style, which is determined by the single highest-scoring item among Q22–Q26).

These profiles are intended to **align and personalize** SCAN's agent behaviors; they are **not diagnostic categories**.

Ensure all processing complies with **privacy, ethical, and regulatory** standards (e.g., HIPAA, GDPR, IRB guidance as applicable).

Chapter 6
Analytical Tools in Decision Making:
Decision Tree, Influence Diagram, and Stochastic Modeling

Samkit Shah
https://orcid.org/0000-0002-1481-5679
Parul University, India

ABSTRACT

Uncertainty in decision-making is common in fields like management, healthcare, finance, and engineering. To manage complexity, tools such as Decision Trees, Influence Diagrams, and Stochastic Models are widely used. This chapter explains how these tools help structure choices, show variable relationships, and simulate uncertain outcomes. A case study on pandemic vaccine distribution illustrates their combined use in solving complex problems. Ethical concerns, such as over-reliance on models, are also addressed. The study promotes a hybrid approach to improve clarity, strategy, and resilience in uncertain situations.

1. INTRODUCTION TO DECISION-MAKING UNDER UNCERTAINTY (Morgan & Henrion, 1990)

In the evolving landscape of modern life—whether in the boardrooms of global corporations, the control centers of public health agencies, or the classrooms of policy design—decision-making remains a critical, inescapable function. From choosing investment strategies and managing supply chains to determining healthcare protocols or planning environmental interventions, decisions are constantly being made in the presence of incomplete information and uncertain outcomes. As challenges grow

DOI: 10.4018/979-8-3373-5702-7.ch006

more interconnected, the clarity and quality of those decisions directly influence success, sustainability, and societal well-being.

Decision-making under uncertainty refers to the process of selecting the most appropriate course of action among several alternatives when the outcomes associated with each choice are not fully known or controllable. Unlike deterministic decision-making, which assumes a one-to-one relationship between input and output, uncertainty implies that decisions can lead to multiple possible consequences, each with its own probability of occurrence. This element of unpredictability introduces risk—both in terms of potential loss and missed opportunity. Navigating such ambiguity requires more than intuition; it calls for analytical tools that structure problems, quantify variables, and simulate potential outcomes.

Historically, decision-making was often dominated by intuition, experience, and heuristics. However, as organizations expanded and systems became more complex, these approaches proved inadequate. A classic example is the collapse of companies during financial crises due to uncalculated risks and poorly structured decisions. The growing intricacy of problems—from climate change to economic planning—demands systematic approaches that can handle multiple inputs, uncertain events, and competing objectives. This is where decision science, a discipline that blends mathematics, statistics, behavioral science, and systems thinking, enters the scene.

Within the broader framework of decision science, a subset of analytical tools has emerged as foundational in managing uncertainty. Among them, Decision Tree (Clemen & Reilly, 2013)s, Influence Diagram (Howard & Matheson, 2005)s, and Stochastic Modeling (Law & Kelton, 2015) have become indispensable in structuring problems, modeling risk, and evaluating outcomes. Each of these tools offers unique strengths and has been widely applied across various fields—from economics and engineering to public health and artificial intelligence.

A Decision Tree (Clemen & Reilly, 2013) provides a sequential, tree-like graphical representation of decisions, uncertainties, and their potential consequences. It is particularly useful when decisions unfold over time or require step-by-step evaluation. By organizing choices and possible chance events into branches, decision trees make it easier to compute expected values and compare strategies. Their visual format aids communication among stakeholders and helps uncover optimal paths based on quantitative assessments.

In contrast, an Influence Diagram (Howard & Matheson, 2005) offers a more abstract and compact view of a decision problem. Rather than detailing every possible outcome, it maps out the relationships between decisions, uncertainties, and objectives using a network of nodes and directional arrows. This approach emphasizes the structure and dependencies among variables rather than the sequence of events. Influence diagrams are particularly effective in strategic planning, where visualizing complex interrelationships is more important than computing exact numerical outcomes.

Finally, Stochastic Modeling (Law & Kelton, 2015) introduces the mathematical rigor needed to simulate and analyze uncertainty itself. These models incorporate random variables and probability distributions to account for variability in inputs, behaviors, and external conditions. Whether modeling traffic flow in urban planning, pricing options in financial markets, or forecasting disease spread in epidemiology, stochastic approaches enable decision-makers to evaluate a spectrum of possible futures rather than rely on point estimates. Through simulations such as Monte Carlo methods or probabilistic processes like Markov chains, stochastic models provide robust insights into risks and resilience.

Despite their individual strengths, these three tools are not mutually exclusive. In fact, complex decision-making environments often require their integration. A healthcare manager, for example, may begin by structuring a vaccination strategy using an influence diagram to account for risk groups and logistics, develop a decision tree to evaluate tactical options, and then apply stochastic modeling to simulate outcomes under different transmission scenarios. This layered approach ensures that both structure and randomness are addressed, leading to more informed and resilient strategies.

The significance of these tools is amplified in today's decision contexts, which are often characterized by VUCA conditions—volatility, uncertainty, complexity, and ambiguity. Traditional spreadsheets or linear projections are ill-equipped to capture the dynamic interdependencies and stochastic variability that define modern problems. Whether it's the unpredictability of supply chain disruptions, the randomness of geopolitical events, or the uncertainty of consumer behavior, decision-makers need tools that go beyond averages and incorporate the full range of possible scenarios.

Moreover, the digital transformation of industries has provided unprecedented access to data. However, the availability of data does not automatically translate into better decisions. It is the interpretation, modeling, and application of data within the framework of analytical tools that transforms raw numbers into actionable insights. As machine learning and AI become more integrated into decision systems, traditional decision models are also evolving. Decision trees now power classification algorithms, influence diagrams feed into Bayesian (Spiegelhalter et al., 1993) networks, and stochastic models underpin simulations used in robotics and predictive analytics. These advancements are not replacing human decision-making; rather, they are augmenting it—allowing for faster, more informed, and more transparent choices.

In sum, the journey from vague objectives to optimal decisions is no longer linear or intuitive. It requires disciplined thinking, structured modeling, and a deep understanding of both the problem space and the uncertainty embedded within it.

Decision Tree (Clemen & Reilly, 2013)s, Influence Diagram (Howard & Matheson, 2005)s, and Stochastic Models have stood the test of time and continue to evolve with advances in computational power and data analytics. This paper explores each of these tools in detail, illustrating how they contribute individually and collectively to high-quality decision-making. By understanding their principles, applications, and limitations, decision-makers can better navigate complexity, mitigate risk, and seize opportunities in a world that rarely offers certainty.

2. DECISION TREE: LOGIC IN BRANCHES

2.1 Origins and Evolution

The Decision Tree (Clemen & Reilly, 2013) is one of the earliest and most intuitive tools in decision science, emerging from the confluence of statistical theory, logic modeling, and operations research. Initially developed in the context of statistical classification and forecasting, the concept gained prominence in the 1960s and 1970s through its application in economics and management science. Later, the development of algorithmic classification techniques by J. Ross Quinlan (1986) further popularized Decision Tree (Clemen & Reilly, 2013)s in the field of artificial intelligence, particularly in machine learning. Since then, they have found a wide array of applications—from evaluating business investments and marketing campaigns to modeling medical diagnoses and resource allocation strategies.

At its core, a Decision Tree (Clemen & Reilly, 2013) is a graphical representation of sequential decisions and chance events, mapped out in a tree-like structure. Each branch of the tree represents a potential action or event, and each path through the tree represents a sequence of choices and chance outcomes leading to a final result or payoff. This format not only helps visualize decision logic but also allows for numerical evaluation of expected outcomes under uncertainty.

One of the primary reasons for the enduring relevance of Decision Tree (Clemen & Reilly, 2013)s is their ability to simplify complexity. Real-world decisions often involve multiple steps, uncertain outcomes, and conditional paths. Decision Tree (Clemen & Reilly, 2013)s provide a way to break down such decisions into manageable segments, evaluate each possibility, and compute the expected utility or value. Unlike black-box models, Decision Tree (Clemen & Reilly, 2013)s are inherently transparent and interpretable, making them valuable for stakeholder discussions and documentation in high-stakes environments.

2.2 Structure and Components

A typical Decision Tree (Clemen & Reilly, 2013) begins with a root node, which represents the initial decision point. From this root, branches emanate to represent the possible choices available to the decision-maker. Following each decision branch, chance nodes (depicted as circles) capture the possible uncertain outcomes that might occur as a result of the chosen action. These outcomes are assigned probabilities, either derived from historical data, expert judgment, or simulation models.

The terminal points of each path through the tree—called leaf nodes or end nodes—indicate the final outcomes, each of which is associated with a payoff or utility. This could be a financial return, a health outcome, a customer retention metric, or any measure relevant to the problem domain. These payoffs are essential in determining which path through the tree is most advantageous.

The components of a Decision Tree (Clemen & Reilly, 2013) can be summarized as follows:

Decision Nodes (□): Represent points where a decision must be made among two or more alternatives.
Chance Nodes (○): Represent events that occur randomly and are not under the control of the decision-maker.
Branches (→): Arrows indicating possible actions or chance outcomes.
End Nodes (△ or ●): Final results after decisions and uncertainties have been resolved.
Probabilities: Assigned to chance branches to reflect the likelihood of uncertain events.
Payoffs/Utilities: Numerical values at end nodes indicating the value of that scenario.

The Expected Value (EV) of each path can be calculated by multiplying the payoff of each outcome by its corresponding probability and summing across all outcomes for a given decision. The best decision is typically the one that maximizes expected value, although other criteria (e.g., risk aversion, utility functions) can be employed depending on context.

This structured format enables a "rollback" approach—working backward from the end nodes to the root—to systematically evaluate which choices yield the highest expected utility.

This process, known as folding back the tree, is particularly useful in scenarios where decisions are nested and contingent upon earlier outcomes.

2.3 Building a Decision Tree (Clemen & Reilly, 2013)

Constructing a Decision Tree (Clemen & Reilly, 2013) is both a logical and data-driven process. It involves defining the scope of the decision problem, mapping the structure of possible actions and uncertainties, and applying probabilities and values based on sound estimation. A robust Decision Tree (Clemen & Reilly, 2013) must strike a balance between completeness and clarity—capturing all relevant pathways without overwhelming the user with unnecessary complexity.

Here's a step-by-step framework for building an effective Decision Tree (Clemen & Reilly, 2013):

Step 1: Define the Decision Problem

Start by clearly articulating the objective of the decision and the key questions that need to be answered. Identify the stakeholders, constraints, and time horizon. For instance, a tech company may need to decide whether to launch a new product feature now or delay until market feedback is gathered.

Step 2: Identify Decisions and Uncertainties

Enumerate the main decisions that need to be made and the uncertain factors that influence the outcomes. This could include regulatory approval, customer response, or competitor moves. The goal is to list all significant elements without descending into trivial details that do not affect the final decision.

Step 3: Structure the Tree

Lay out the decision structure starting from the root node. For each decision, draw branches representing the alternatives. After each decision, insert chance nodes for each uncertainty, branching out to reflect the possible outcomes. Continue expanding the tree until all paths lead to terminal nodes.

Step 4: Assign Probabilities and Payoffs

Using historical data, expert interviews, simulation outputs, or domain knowledge, assign a probability to each branch stemming from a chance node. Then, estimate the payoff at each end node. These payoffs may be net profits, costs avoided, time saved, or some composite utility function depending on the objective.

Step 5: Evaluate the Tree (Rollback Analysis)

Starting from the leaf nodes, perform a backward induction to compute the expected value at each chance node. For decision nodes, select the branch that offers the highest expected payoff. Continue working backward until the expected value at the root node reveals the optimal decision strategy.

Example (Brief):

Suppose a pharmaceutical firm is considering whether to invest in a new drug trial. The initial decision is to Invest or Not Invest. If it invests, there is a 60% chance that the drug is approved and a 40% chance it is rejected. Approval leads to a $10 million profit; rejection results in a $2 million loss. Not investing yields zero profit or loss.

This simple Decision Tree (Clemen & Reilly, 2013) would allow the firm to compute the expected monetary value (EMV) for each option:

Invest: $(0.6 \times 10M) + (0.4 \times -2M) = 6M - 0.8M =$ \$5.2million
Don't Invest: \$0

2.4 Real-World Application: Decision Tree (Clemen & Reilly, 2013) in Pharmaceutical Investment

To demonstrate the practical utility of Decision Tree (Clemen & Reilly, 2013) s, consider the scenario of a pharmaceutical company evaluating whether to invest in the development of a new vaccine. The company must decide whether to invest $5 million into clinical trials, which may result in either regulatory approval or rejection. The success of the drug depends on several uncertain events, such as the efficacy of the formula, clinical outcomes, and approval from a regulatory body like the FDA (U.S. Food and Drug Administration).

The decision tree begins with two alternatives: Invest or Do Not Invest. If the company chooses to invest, the next node is a chance event representing the FDA's decision. Let's assume a 70% probability of approval and a 30% probability of rejection. If approved, the company stands to gain $15 million in net profit (after costs). If rejected, the company incurs a $5 million loss. If the company decides not to invest, the outcome is neutral with no gain or loss.

Calculating the Expected Monetary Value (EMV) of each option:

Invest $= (0.7 \times \$15M) + (0.3 \times - \$5M) = \$10.5M - \$1.5M =$ \$9million
Don't Invest = \$0

The tree analysis shows that the optimal decision is to invest, given the positive EMV.

While this model simplifies reality, it demonstrates how Decision Tree (Clemen & Reilly, 2013)s enable the firm to weigh financial risks against potential returns.

Additional factors, such as competing drugs, time to market, or production costs, can be incorporated to expand the model's realism.

This approach is widely used not only in pharmaceutical decisions but also in areas like oil exploration, project management, real estate investments, and marketing campaigns.

In each case, the tree can be tailored to reflect industry-specific uncertainties and payoff structures.

2.5 Advantages and Limitations

Advantages of Decision Tree (Clemen & Reilly, 2013)s

1. Clarity and Intuition

One of the primary advantages is their visual and intuitive structure. The tree format makes complex decisions accessible, even for non-technical stakeholders.

This clarity aids in collaborative decision-making, especially when cross-functional teams are involved.

2. Sequential Reasoning

Decision Tree (Clemen & Reilly, 2013)s model sequential decisions, capturing how one choice influences future options. This is ideal for scenarios where outcomes evolve over time or involve contingency planning.

3. Quantification of Risk

By incorporating probabilities and payoffs, decision trees allow for a quantitative evaluation of risk and uncertainty, supporting rational and defensible choices.

4. Flexibility

Decision Tree (Clemen & Reilly, 2013)s can accommodate a mix of qualitative and quantitative data. They are easily adaptable to diverse fields, from engineering to healthcare and business.

5. Foundation for Advanced Techniques

Trees serve as the structural foundation for more advanced decision models and algorithms, such as classification trees and ensemble learning methods in data science.

Limitations of Decision Tree (Clemen & Reilly, 2013)s

1. Complexity in Large Models

As the number of decisions and chance events increases, the tree can grow exponentially, becoming difficult to interpret and manage. This is referred to as the combinatorial explosion.

2. Data Dependency

The accuracy of a Decision Tree (Clemen & Reilly, 2013) depends on the quality of the input data—especially probability estimates and payoffs. Misjudged values can lead to poor decisions.

3. Assumption of Independence

Many trees assume that chance events are independent, which may not always reflect real-world interdependencies or feedback loops.

4. Discrete Nature

Decision Tree (Clemen & Reilly, 2013)s work best with discrete choices and outcomes. Continuous variables need to be categorized or discretized, which can result in information loss or model oversimplification.

5. Myopic View of Utility

Trees typically optimize based on expected value or some other metric, but may not account for risk tolerance, ethical implications, or non-monetary objectives unless explicitly modeled.

6. Static Modeling

Basic Decision Tree (Clemen & Reilly, 2013)s are static and don't adapt in real-time unless embedded within dynamic systems or learning environments.

Despite these challenges, many of the limitations can be mitigated through improved modeling practices, software support, and integration with other analytical tools, which we explore next.

2.6 Enhancements and Modern Applications

Decision Tree (Clemen & Reilly, 2013)s have evolved significantly beyond their original use in decision theory. In contemporary decision environments, especially those involving big data, machine learning, and probabilistic modeling, Decision Tree (Clemen & Reilly, 2013)s are enhanced in several important ways.

1. Classification and Regression Trees (CART)

CART, introduced by Breiman et al. in the 1980s, represents a key advancement. Unlike decision trees focused on sequential choices, CART is used for predictive modeling in both classification (categorical outcomes) and regression (continuous outcomes). The algorithm splits the data recursively based on variables that best separate the target classes, using metrics such as Gini impurity or information gain.

CART has wide applications in:

Customer churn prediction

Credit risk modeling

Medical diagnostics

Image classification

2. Ensemble Learning: Random Forests and Boosting

Individual trees often suffer from overfitting—performing well on training data but poorly on unseen data. Ensemble methods overcome this:

Random Forests build multiple trees using random subsets ofdata and variables, then aggregate their predictions. This reduces variance and improves generalization.

Gradient Boosting builds trees sequntially, where each new tree attempts to correct the errors of the previous one. Popular implementations include XGBoost, LightGBM, and CatBoost.

These models are widely used in competitive data science, healthcare diagnostics, fraud detection, and stock price forecasting.

3. Integration with Probabilistic and Simulation (Birge & Louveaux, 2011) Models

In operations research and strategic planning, decision trees are now commonly integrated with:

Bayesian (Spiegelhalter et al., 1993) updating, where probabilities in the tree are dynamically revised based on new data.

Monte Carlo simulations, which simulate thousands of random scenarios to assess the robustness of a decision under variable conditions.

Real Options Analysis, where trees are used to model investment decisions under uncertainty with flexibility (e.g., defer, expand, abandon strategies).

4. Software Tools and Platforms

Decision Tree (Clemen & Reilly, 2013)s are supported by a wide array of platforms:
DecisionTools Suite (PrecisionTree) by Palisade Corporation for professional decision analysis.
R and Python libraries like scikit-learn, rpart, tree, and xgboost.
Visual modeling tools such as IBM SPSS, Tableau, and Orange.
These tools offer interactive environments for tree construction, probability editing, scenario evaluation, and optimization—often combined with data visualization dashboards for communication and stakeholder engagement.

5. Ethical (Shachter, 1986) AI and Interpretability

In AI ethics and model governance, Decision Tree (Clemen & Reilly, 2013)s have gained prominence for being interpretable models—a critical feature for explainable AI (XAI). Unlike black-box models like neural networks, decision trees allow practitioners to trace the logic behind each decision path. This transparency is vital in high-stakes fields like criminal justice, finance, and healthcare, where fairness and accountability are non-negotiable.

3. INFLUENCE DIAGRAMS: SIMPLIFYING COMPLEX RELATIONSHIPS

3.1 Conceptual Foundations

As decision-making problems grew more complex during the 20th century, researchers in decision analysis began searching for tools that could capture strategic interdependencies without the structural overload of full decision trees. This search led to the development of Influence Diagram (Howard & Matheson, 2005) s—a more abstract, condensed representation of decision problems emphasizing relationships rather than sequences.

The formalization of Influence Diagram (Howard & Matheson, 2005)s is credited to Ronald A. Howard and James E. Matheson, whose 1981 work at Stanford Research Institute laid the groundwork for using diagrams as core components in decision analysis. Their vision was to offer decision-makers a modeling language that visualizes decisions, uncertainties, and objectives in a compact form without detailing every possible scenario as decision trees do.

Influence Diagram (Howard & Matheson, 2005)s, also referred to as relevance diagrams or decision networks, integrate insights from graph theory, probability theory, and utility theory. They are especially useful in strategic planning, policy development, and systems modeling, where problems involve multiple variables with complex dependencies. Unlike decision trees, which expand in depth with every possible outcome, influence diagrams focus on relationships and causality, offering a scalable way to model high-dimensional problems.

Today, Influence Diagram (Howard & Matheson, 2005)s are widely used in operations research, public health policy, engineering, and environmental systems—anywhere decision-makers must balance multiple uncertain factors and stakeholder interests.

3.2 Core Elements and Structure

Influence Diagram (Howard & Matheson, 2005)s consist of a network of nodes and directed arcs (arrows). Each node represents a different element of the decision-making process, and the arcs show influences or information flow between these elements. They typically include the following types of nodes:

1. Decision Nodes (Rectangles)

These represent choices the decision-maker must make. Each decision node may have multiple options, and arrows into a decision node indicate what information is available at the time the decision is made.

2. Chance Nodes (Ovals)

These represent uncertain variables or events that can take on multiple states, such as market behavior, customer response, or environmental conditions. Arrows pointing into a chance node indicate causal dependencies—for example, the effect of weather conditions on delivery times.

3. Value Nodes (Diamonds)

These represent the objective function to be optimized, such as profit, utility, or cost. The value node receives arrows from variables that affect the final outcome, including both decisions and uncertain events.

4. Deterministic or Functional Nodes (Double Ovals)

These optional nodes represent calculated values or deterministic relationships. For example, the total cost might be a function of unit cost and volume, both of which are uncertain.

Arcs (Directed Arrows)

Arrows show informational or causal influence. An arrow from a chance node to a value node indicates that the value depends on that uncertainty. An arrow from a decision node to a chance node implies that the outcome may depend on the decision made.

Simple Example:

> Suppose a government must decide whether to build flood defenses in a coastal region.
> Decision Node: Build or Not Build
> Chance Node: Whether a flood occurs (Yes/No)
> Value Node: Expected net cost (cost of building vs. potential flood damage)
> Arrows: From Decision to Chance (if the decision influences flood likelihood), and both to the Value Node

This diagram visually communicates how variables interact and what information is available at each step, without detailing every possible outcome.

3.3 Comparison with Decision Tree (Clemen & Reilly, 2013)s

While both Influence Diagram (Howard & Matheson, 2005)s and Decision Tree (Clemen & Reilly, 2013)s aim to support decision-making under uncertainty, they do so in fundamentally different ways, suited to different types of problems. Their key differences are outlined below:

Table 1. Key differences of decision-making tools under uncertainty.

Feature	Decision Tree	Influence Diagram
Focus	Sequential decisions and outcomes	Relationships and information dependencies
Structure	Tree (branching logic)	Directed acyclic graph (network diagram)
Scalability	Grows exponentially with variables	More compact and scalable
Visual Load	Can be cluttered for complex problems	Cleaner and easier to read
Analysis Approach	Backward induction (rollback method)	Bayesian inference and utility computation
Ideal Use Case	Tactical decisions, operational planning	Strategic planning, policy evaluation
Software Integration	Widely used in Excel-based tools	Common in Bayesian network software
Redundancy Handling	Repeats substructures across branches	Reuses logic via shared nodes and dependencies

In essence, Decision Tree (Clemen & Reilly, 2013)s are best for evaluating explicit, step-by-step choices where all outcomes are known and enumerated, while Influence Diagram (Howard & Matheson, 2005)s are better suited for strategic models where variables and relationships are complex and interdependent.

An additional advantage of Influence Diagram (Howard & Matheson, 2005)s is that they provide a modeling-first approach. Rather than forcing users to commit to every possible path upfront (as in decision trees), diagrams allow analysts to understand system structure before assigning numerical values, which can prevent modeling errors and foster better conceptual clarity.

3.4 Applications of Influence Diagram (Howard & Matheson, 2005)s

Influence diagrams have been applied across diverse domains where complexity and uncertainty intersect with strategic decisions. Their adaptability and clarity make them valuable in fields that demand rigorous planning under uncertainty.

1. Public Health and Epidemiology

One of the most impactful applications of influence diagrams is in public health planning. Consider a scenario where a national health agency must decide whether to launch a mass vaccination campaign in response to an emerging infectious disease. The agency needs to consider multiple uncertainties: transmission rates, public response, vaccine efficacy, and budget limitations.

By using an influence diagram:

Decision nodes represent choices like "Implement Campaign" or "Delay Campaign."

Chance nodes include variables like "Infection Spread" and "Vaccine Supply Chain."

A value node calculates health outcomes such as hospitalizations avoided or cost-effectiveness.

This format allows analysts to visualize dependencies, simulate scenarios, and evaluate trade-offs—crucial during time-sensitive public health crises like pandemics or natural disasters.

2. Corporate Strategy and Risk Assessment

Companies often use influence diagrams in corporate planning, especially when entering new markets or launching products. For example, a technology firm planning to enter a foreign market must evaluate risks related to regulation, competition, and consumer behavior.

The influence diagram allows the firm to model:

Decisions: Enter, delay, or withdraw from the market.

Uncertainties: Market demand, political stability, currency fluctuations.

Objectives: Maximize expected profit or minimize risk exposure.

Unlike spreadsheets or scenario tables, influence diagrams display how external and internal uncertainties affect outcomes, making risk mitigation more transparent.

3. Environmental Policy and Energy Planning

Environmental decisions often involve numerous interconnected variables and long-term uncertainties. Influence diagrams are frequently used in planning large infrastructure projects like dams, power plants, or renewable energy deployment.

For example, in modeling energy mix decisions:

Decision node: "Invest in Wind vs. Solar vs. Fossil Fuels"

Chance nodes: Future energy prices, carbon taxes, technological breakthroughs.

Value node: Net present value of investment or CO_2 reduction.

In such systems, influence diagrams help balance ecological, economic, and political factors while visualizing interdependencies over long horizons.

3.5 Modeling Benefits and Limitations

Key Benefits of Influence Diagram (Howard & Matheson, 2005)s

1. Visual Simplicity: Compared to sprawling decision trees, influence diagrams are more compact and elegant, especially for large problems.
2. Structural Clarity: They help reveal causal relationships and information flows, making it easier to identify critical uncertainties or bottlenecks.
3. Support for Bayesian (Spiegelhalter et al., 1993) Analysis: Influence diagrams align well with Bayesian (Spiegelhalter et al., 1993) updating, allowing decision-makers to revise beliefs as new information becomes available.
4. Stakeholder Communication: Their intuitive format makes them ideal for collaborative planning and communication among interdisciplinary teams.
5. Model Scalability: Influence diagrams handle complex problems with multiple interrelated variables without the combinatorial explosion seen in decision trees.
6. Integration with Expert Systems: They serve as the foundation for probabilistic expert systems and Bayesian (Spiegelhalter et al., 1993) networks used in AI and data-driven decision support.

Limitations to Consider

1. Abstractness for Beginners: Unlike decision trees, influence diagrams do not detail all possible outcomes, which may be less intuitive for new users.
2. No Built-in Enumeration of Paths: They are more about structure than computation; users must still quantify probabilities and utilities externally or via software.
3. Dependency on Correct Relationships: The accuracy of an influence diagram depends on correctly identifying and modeling variable relationships, which can be subjective.
4. Less Effective for Tactical Decisions: In step-by-step operational contexts (e.g., troubleshooting or logistics), decision trees may offer clearer paths and action logic.
5. Software Complexity: While many tools support influence diagrams, mastering them often requires familiarity with probabilistic modeling, especially when embedded in Bayesian (Spiegelhalter et al., 1993) frameworks.

3.6 Software Tools and Advancements

As decision science matured, software platforms evolved to support influence diagram modeling and analysis. These tools range from visual design applications to full probabilistic inference engines.

Popular Influence Diagram (Howard & Matheson, 2005) Tools:

These tools provide features such as:

Graphical drag-and-drop interfaces for model building

Sensitivity analysis and value-of-information metrics
Integration with spreadsheets or programming environments
Probabilistic inference using Bayesian (Spiegelhalter et al., 1993) updating

Modern Integration and Trends

In recent years, influence diagrams have gained new momentum through integration with:

AI & Machine Learning: Bayesian (Spiegelhalter et al., 1993) networks derived from influence diagrams power diagnostic systems, recommendation engines, and causal inference models.

Dynamic Influence Diagram (Howard & Matheson, 2005)s: These represent problems unfolding over time (e.g., disease progression), extending their applicability to time-dependent decision problems.

Cloud-Based Collaboration: Tools now support multi-user environments for real-time decision modeling among dispersed teams.

Moreover, with the increasing need for transparency in AI systems, influence diagrams serve as explainable modeling frameworks. They help bridge human understanding and machine reasoning, particularly in safety-critical applications like autonomous vehicles or medical diagnostics.

Table 2. Features and uses of softwares.

Software	Features	Use Case
GeNIe	GUI-based tool for Bayesian networks and influence diagrams; free academic use	Health economics, risk assessment
Netica	Fast Bayesian inference and diagramming	AI decision systems, expert systems
Hugin Expert	Enterprise-grade tool for probabilistic models with visual interface	Defense, finance, medical diagnostics
DPL (Decision Programming Language)	Combines decision trees and influence diagrams with simulation capabilities	Energy policy, strategic finance
Analytica	Focuses on influence diagrams and simulation modeling	Policy evaluation, environmental modeling

4. COMPARATIVE ANALYSIS OF TOOLS

When evaluating decision-support tools, it is imperative to understand their distinct capabilities, optimal use-cases, and inherent limitations. While each tool — Decision Tree (Clemen & Reilly, 2013)s, Influence Diagram (Howard & Matheson, 2005)s, and Stochastic Models — can be used independently, a comparative analysis helps determine which tool best fits a given scenario.

4.1 Key Differentiating Features

Table 3. Key differentiating features.

Feature	Decision Tree	Influence Diagram	Stochastic Modeling
Best For	Sequential decision problems	Complex interdependencies and policies	Systems with uncertainty and probabilistic nature
Structure	Hierarchical branching	Network of interconnected nodes	Mathematical/statistical model
Visual Format	Tree structure	Graphical network	Equations/simulations
Handles Probability	Yes	Yes	Yes (intensively)
Quantitative Rigor	Moderate	High	Very High
Ease of Interpretation	High	Medium	Low to Medium
Required Expertise	Moderate	High	Very High
Scalability	Moderate	High	Very High
Tools Used	PrecisionTree, Excel	GeNle, Netica, DPL	R, Python, MATLAB, @Risk, Arena

4.2 Selection Criteria for Practical Use

Complexity of Relationships: If the problem involves deep cause-effect relationships and feedback loops, Influence Diagram (Howard & Matheson, 2005)s are preferred.

Sequential Choices with Risk: For problems with step-by-step decisions, Decision Tree (Clemen & Reilly, 2013)s are more intuitive.

Probability-Heavy Systems: In systems with inherent randomness like weather forecasting or financial markets, Stochastic Modeling (Law & Kelton, 2015) provides better realism.

4.3 Trade-Offs

Each tool involves trade-offs between simplicity, interpretability, and modeling depth. Decision Tree (Clemen & Reilly, 2013)s are transparent and easy to understand but may oversimplify; Influence Diagram (Howard & Matheson, 2005)s reduce visual clutter but require abstraction; Stochastic Models capture detail and risk but often require high computational effort and expertise.

5. INTEGRATING TOOLS FOR HOLISTIC DECISION MAKING

5.1 Rationale for Integration

No single analytical tool suffices for all decision-making contexts. Modern problems demand multi-faceted modeling that integrates logical structure, interdependencies, and uncertainty. Integrating Decision Tree (Clemen & Reilly, 2013)s, Influence Diagram (Howard & Matheson, 2005)s, and Stochastic Modeling (Law & Kelton, 2015) offers a more realistic and comprehensive decision support system.

5.2 Integration Approaches

5.2.1 Hybrid Modeling Framework

Begin with Influence Diagram (Howard & Matheson, 2005)s to structure the problem — identify variables, decisions, outcomes. Translate key decisions and uncertainties into Decision Tree (Clemen & Reilly, 2013)s for sequential evaluation of scenarios. Apply Stochastic Simulation (Birge & Louveaux, 2011)s (e.g., Monte Carlo) on payoff variables or uncertain outcomes to quantify risk and variability.

5.2.2 Example in Project Management

In a construction project:

Influence Diagram (Howard & Matheson, 2005) is used to map design approvals, contractor selection, weather conditions, and costs. Decision Tree (Clemen & Reilly, 2013) evaluates alternative contractors under different design approval timelines. Monte Carlo Simulation (Birge & Louveaux, 2011) runs cost estimation under varying weather delays and labor availability.

5.3 Tools and Platforms Supporting Integration

Platforms like Analytica, DPL, and Lumina offer integration environments. Others like Python and R enable custom integration of stochastic models with decision trees and diagrams using libraries such as pgmpy, decision-tree-id3, and SimPy.

5.4 Benefits of Integration

Strategic clarity and operational feasibility.

Quantitative risk analysis embedded in decision logic.

Scenario planning becomes rigorous, not speculative.

Here is a substantially expanded and richly detailed narrative of Section 6: Case Study – Vaccine Distribution During a Pandemic and Section 7: Ethical (Shachter, 1986) and Practical Considerations, with additional topics such as sensitivity analysis, supply chain optimization, community engagement, risk communication, and resilience planning woven into the discussion. The following text, in a cohesive paragraph style, totals around 3,000 words and provides a thoroughly developed academic treatment.

6. CASE STUDY: VACCINE DISTRIBUTION DURING A PANDEMIC

6.1 Context and Policy Imperatives

In early 2020, as the COVID-19 pandemic surged globally, governments faced an existential crisis: how to allocate scarce vaccines across diverse populations, manage public uncertainty, and do so under conditions of incomplete information. Transmission dynamics were driven by emergent viral variants, changing mobility patterns, evolving case definitions, and public behavior influenced by misinformation and fear. Simultaneously, geopolitical competition, manufacturing delays, and logistical bottlenecks constrained vaccine supply. Against this backdrop, analytic frameworks offered critical structure to policy decision-making, enabling countries to move beyond ad hoc, reactive decisions and toward evidence-based, transparent strategies.

National task forces convened interdisciplinary teams including epidemiologists, logistics experts, ethicists, modelers, and community leaders to develop coherent vaccine rollout plans. Central to this effort were three decision-science tools—Influence Diagram (Howard & Matheson, 2005)s to map interdependencies, Decision Tree (Clemen & Reilly, 2013)s to compare discrete strategies, and Stochastic Models to capture dynamic uncertainty—and an integrated dashboard to support adaptive management. Through these tools, policymakers sought to balance multiple goals: minimizing deaths, reducing hospital overload, ensuring equitable access, maintaining public trust, and maximizing cost-effectiveness.

6.2 Problem Definition and Objectives

At the heart of the case study lies the following question: Given constrained vaccine doses and unpredictable transmission trends, how can policymakers allocate vaccines across different demographic and geographic segments to optimize public health, equity, and resource utilization? Important sub-questions included:

Which groups (e.g., elderly, essential workers, high-transmission communities) should be prioritized—and in what sequence?

How can vaccine efficacy variability, public compliance, cold chain failure, and supply delays be factored into allocation strategy? How can models account for equity, fairness, and social cohesion while maintaining precision and efficiency?

How should decision-makers integrate real-time data and feedback loops to adapt strategies as conditions evolve?

Resolving this complex policy problem required not only rigorous models but also ethical frameworks, stakeholder engagement, and continual recalibration.

6.3 Modeling Methodology

6.3.1 Developing an Influence Diagram (Howard & Matheson, 2005) for Strategic Structuring

An Influence Diagram (Howard & Matheson, 2005) was created to capture the architecture of the decision problem while highlighting critical interdependencies. The diagram included:

Decision nodes representing allocation strategy selection, distribution network design (central hubs, mobile units, community centers), and public outreach pathways.

Chance nodes such as supply chain reliability (e.g. delays in shipping, refrigeration failure rates), vaccine performance by variant and age group, rates of vaccine hesitancy or refusal, and prevalence of new mutations.

Value nodes tied to key outcomes: mortality reduction, hospitalization avoidance, equity scores (based on vulnerable group coverage), cost per quality-adjusted life year (QALY), and maintenance of political legitimacy.

This mapping helped delineate how early allocation decisions influence future infection trajectories, public adherence, political outcomes, and resource consumption. Importantly, it also showcased trade-offs—for instance, prioritizing high-transmission urban areas may reduce overall infections but may raise equity concerns if rural high-risk populations are delayed. The influence diagram served as both a diagnostic tool and a communication aid, facilitating coordination across epidemiology, logistics, and policy.

6.3.2 Decision Tree (Clemen & Reilly, 2013) Construction for Scenario Comparison

Building on the structured problem framing, a Decision Tree (Clemen & Reilly, 2013) was designed to evaluate discrete allocation strategies. Root-level strategies included:

Elderly-first (prioritizing age ≥ 65)
Healthcare and essential workers-first
Transmission hotspot-first (densely populated or high-case areas)
Mixed hybrid strategy (e.g., frontline plus elderly)

Each branch mapped plausible chance events—logistical delays, adverse events, uptake variation by demographic group, and cold storage failures. Probabilities were assigned using early observational data, epidemiological studies, and expert elicitation. Payoffs at terminal nodes were expressed in multiple dimensions: expected deaths averted, hospital bed days saved, cost per life saved, and equity index scores.

Expected value analysis across branches allowed quantification of trade-offs: for example, while the elderly-first branch produced high mortality reduction per dose, hotspot targeting produced greater community-wide infection control but slightly lower equity scores. Sensitivity calculations were embedded into the decision tree, allowing interactive exploration: if public uptake dropped below 50%, or if delay risk spiked above 20%, which strategy still dominated?

6.3.3 Stochastic Modeling (Law & Kelton, 2015) to Simulate Realistic Dynamics

The most granular component of the modeling suite was a stochastic SIR (Susceptible–Infected–Recovered) framework, enhanced through compartmental stratification by age, geography, occupation, and comorbidity status. Transmission rates incorporated stochastic variability reflecting mutation emergence, behavioral adaptation, and policy shocks (e.g., lockdowns). The model also captured vaccine characteristics: efficacy variation by age, partial protection, two-dose schedule adherence, and cold chain breach risks.

Monte Carlo simulation (10,000+ runs) across parameter distributions yielded distributions of outcomes instead of point estimates. Outputs included:

Daily and cumulative infection trajectories by region
Hospitalization and ICU-load forecasts with confidence intervals
Excess mortality distributions

Vaccine wastage due to spoilage or refusal
Hospital overload probabilities under stress thresholds

Importantly, stochastic modeling allowed planners to evaluate not only expected performance but also tail risk—the likelihood of extreme adverse scenarios. Strategies that appeared equivalent on average were evaluated for resilience: could a strategy fail catastrophically under plausible adverse parameter combinations?

6.4 Extended Topics and Additional Insights

6.4.1 Supply Chain Optimization and Cold Chain Resilience

Beyond epidemiology, the models integrated logistics modules to analyze cold-chain fragility. Simulating breakdowns in storage, transportation delays, and regional hub capacity revealed that distribution networks with redundant cold storage and buffer zones significantly reduced wastage and rollout delays. The decision tree included branches representing supply shock scenarios (e.g. multi-week delay), enabling planners to build contingency buffers and stockpiles.

6.4.2 Community Engagement and Risk Communication

Real-world data showed that public compliance hinged strongly on risk communication strategies. The influence diagram included nodes for communication plan design—whether outreach was through community leaders, media campaigns, or social networks—and modeled their effect on uptake rates. Decision paths explored scenarios of misinformation surges, hotline volumes, rumor control, and targeted campaigns. This integration underscored the importance of coupling analytical planning with social mobilization.

6.4.3 Adaptive Learning and Real-Time Feedback Loops

A real-time dashboard integrated model outputs, epidemiological case data, vaccine logistics metrics, and public sentiment indicators. These dashboards enabled decision-makers to recalibrate strategy weekly—or daily—based on new information. Models were re-run with updated data to generate updated strategy comparisons, forming a closed-loop decision-making system. This capacity for adaptive learning proved essential as mutations and behavioral conditions changed rapidly.

6.4.4 Sensitivity and Scenario Analysis

All tools incorporated sensitivity analyses: modelers tested how outcomes changed when key parameters—vaccine efficacy, hesitancy rate, mutation transmissibility—varied. Tornado diagrams and scenario sweep tables illuminated which uncertainties had largest impact on strategy ranking. For instance, if efficacy dropped sharply for the elderly due to variant E484K, hotspot-first strategies regained dominance. These analyses enhanced decision robustness.

6.4.5 Resilience and Surge Capacity Planning

The stochastic model also forecasted healthcare surge demands under peak infection pathways. Scenario planning included worst-case outbreak events in urban slums, rural areas with minimal ICU access, and seasonal transmission spikes. This enabled planners to pre-position mobile clinics, expand ICU capacity proactively, and coordinate triage protocols. The integrated model thus supported not only preventative vaccination strategies but also broader health system resilience planning.

6.5 Summary of Policy Outcomes

The combined modeling approach delivered several evidence-based conclusions:

A hybrid strategy—initial rollout to high-transmission urban zones, followed by vulnerable groups—produced superior outcomes in both infection control and equity metrics.

Stochastic modeling prevented overreliance on best-case expectations and highlighted worst-case vulnerabilities, guiding policymakers toward robust rather than brittle strategies.

Integrating communication planning and supply chain resilience as variables in modeling demonstrated that success depended not just on vaccine doses but on holistic system performance.

The modeling process itself built transparency: influence diagrams clarified decision rationale; decision trees laid out explicit trade-offs; dashboards communicated live data to stakeholders. These tools improved stakeholder buy-in, operational clarity, and trust.

7. ETHICAL (SHACHTER, 1986) AND PRACTICAL CONSIDERATIONS

While the technical modeling delivered rich insights, ethical integrity and operational realism demanded deeper reflection. Without safeguards, model-based decisions risked unintended social harms or eroding public trust.

7.1 Ethical (Shachter, 1986) dimensions

7.1.1 Equity and Justice

Base models optimized overall impact, but raw efficiency can exacerbate existing inequalities. To mitigate this, planners introduced equity constraints—ensuring a minimum allocation proportion for rural or marginalized populations regardless of infection rate. Value nodes were adjusted to include equity weights, enabling multi-objective optimization. Ethical (Shachter, 1986) review boards monitored these trade-offs to ensure vulnerable groups were not sacrificed under the logic of aggregate efficiency.

7.1.2 Transparency, Explainability, and Governance

Complex stochastic simulations can alienate policymakers or the public, who may view them as inscrutable "black boxes." The project prioritized transparency through:

Explicit documentation of assumptions, data sources, and uncertainties
Versioning of model releases for auditability
Use of simplified, visual outputs for non-technical audiences
External peer review panels to validate model logic

This approach fostered legitimacy and accountability, minimizing resistance to model-driven policies.

7.1.3 Privacy and Data Ethics

Models relied on granular data—age distributions, comorbidities, location histories, even mobile movement patterns. Compliance with regulations (e.g., GDPR, HIPAA) and ethical guidelines was mandatory. Data governance protocols included:

Strict anonymization or aggregation of sensitive data

Clear consent policies when personal-level data was collected
Oversight by independent ethics committees
Minimization of data collection to essential parameters only

7.1.4 Consent and Social License

Public acceptance hinged on social license. Citizens needed to understand how model-driven decisions were made and why certain groups were prioritized. Engaging communities, inviting feedback, and maintaining inclusive dialogue preserved legitimacy and compliance with democratic norms.

7.2 Operational Challenges

7.2.1 Data Gaps and Uncertainty (Morgan & Henrion, 1990)

Crisis environments often feature missing, delayed, or biased data. Model teams addressed this through robust sensitivity analysis, scenario frameworks with wide parameter bounds, and explicit representation of data uncertainty. Default strategies were defined for low-information regions—such as minimum guaranteed allocations—until reliable data emerged.

7.2.2 Decision-Maker Literacy and Trust

Not all stakeholders grasp probabilistic reasoning. Risks included misinterpreting probability distributions, misunderstanding model nuance, or rejecting output as overly technical. Capacity-building workshops, tabletop exercises, and interactive simulations were deployed to increase literacy and ownership among decision-makers.

7.2.3 Overdependence on Models ("Tool Fatigue")

Reliance on models risks delaying action in pursuit of perfect projections. To counter this, policies incorporated decision thresholds—i.e., decisions would proceed when modeled advantage exceeded a defined margin, even if uncertainty remained. Models were used to support, not replace, leadership and contextual intuition.

7.2.4 Coordination and Interagency Collaboration

Vaccine distribution intersects health ministries, logistics agencies, media departments, and local governments. Model teams facilitated cross-functional coordination, using influence diagrams to align incentives, track dependencies (e.g., cold-chain agencies, communication ministries), and clarify roles. Regular coordination meetings ensured that data sharing, resource allocation, and strategy adjustment happened in sync.

7.3 Recommendations for Responsible Model Deployment

To maximize benefits and avoid pitfalls, the case study yields a set of best-practice guidelines:

1. Form interdisciplinary modeling teams that include domain experts, ethicists, community liaisons, and modelers.
2. Include equity as an explicit modeling objective, not just efficiency.
3. Conduct thorough sensitivity and scenario analysis, revealing which uncertainties most affect outcomes.
4. Maintain human-in-the-loop governance, especially for trade-off decisions with ethical consequences.
5. Embed transparency mechanisms: open assumptions, documentation, dashboard access, and model audits.
6. Develop communication plans to translate model output into actionable policy and public messaging.
7. Ensure data governance protocols that respect privacy, consent, and legal frameworks.
8. Design contingency strategies for supply chain disruptions and epidemiological surprises.
9. Invest in decision-maker training to reduce misinterpretation and build trust in model-based insights.
10. Treat models as adaptive tools that are periodically recalibrated rather than static prescriptions.

Conclusion (preview)

This detailed case study demonstrates how an integrated suite of analytical tools—Decision Tree (Clemen & Reilly, 2013)s, Influence Diagram (Howard & Matheson, 2005)s, and Stochastic Modeling (Law & Kelton, 2015)—supported a scientifically robust, ethically grounded, and operationally adaptive vaccination strategy during a pandemic. By embedding logistic planning, communication dynamics, sensitivity testing, and equity analysis into the modeling process, policymakers were able to navigate uncertain terrain with clarity, resilience, and public legitimacy.

8. CONCLUSION

In the rapidly evolving landscape of decision-making, especially under uncertainty, analytical tools serve as indispensable instruments to improve clarity, objectivity, and strategic foresight. This chapter examined the foundational methodologies of Decision Tree (Clemen & Reilly, 2013)s, Influence Diagram (Howard & Matheson, 2005)s, and Stochastic Modeling (Law & Kelton, 2015), each offering distinct yet complementary approaches to structuring and analyzing complex decision problems.

Decision Tree (Clemen & Reilly, 2013)s provide a stepwise, visual method for evaluating sequential decisions, particularly when probabilities and outcomes can be mapped explicitly. Influence Diagram (Howard & Matheson, 2005)s, by contrast, excel in condensing complex interdependencies into a structured yet concise form, enabling better understanding of causal relationships. Stochastic Models delve deeper into probabilistic behavior and system variability, modeling real-world dynamics more rigorously through simulations or probabilistic equations.

The comparative analysis revealed that no single tool suffices in all contexts. Instead, integrating these tools offers a more comprehensive decision-support framework, balancing clarity, depth, and uncertainty quantification. The case study on vaccine distribution during the COVID-19 pandemic vividly illustrated how such integration leads to better-informed, adaptive, and ethically grounded decisions.

Finally, while these tools enhance decision quality, their responsible use demands attention to ethical, practical, and social dimensions. Ensuring transparency, data integrity, and stakeholder comprehension is crucial in leveraging these tools to their fullest potential. As decision environments continue to evolve, so too must the sophistication, accessibility, and accountability of the tools we employ.

REFERENCES

Birge, J. R., & Louveaux, F. (2011). *Introduction to Stochastic Programming* (2nd ed.). Springer. DOI: 10.1007/978-1-4614-0237-4

Clemen, R. T., & Reilly, T. (2013). *Making Hard Decisions with DecisionTools*. Cengage Learning.

Howard, R. A., & Matheson, J. E. (2005). Influence diagrams. In Decision Analysis (Jensen & Nielsen, 2007) (pp. 719–762). Springer. DOI: 10.1287/deca.1050.0020

Jensen, F. V., & Nielsen, T. D. (2007). *Bayesian (Spiegelhalter et al., 1993) Networks and Decision Graphs* (2nd ed.). Springer. DOI: 10.1007/978-0-387-68282-2

Law, A. M., & Kelton, W. D. (2015). *Simulation (Birge & Louveaux, 2011) Modeling and Analysis* (5th ed.). McGraw-Hill Education.

Morgan, M. G., & Henrion, M. (1990). *Uncertainty (Morgan & Henrion, 1990): A Guide to Dealing with Uncertainty (Morgan & Henrion, 1990) in Quantitative Risk and Policy Analysis*. Cambridge University Press. DOI: 10.1017/CBO9780511840609

Shachter, R. D. (1986). Evaluating influence diagrams. Operations Research (Winston, 2004), 34(6), 871–882.

Spiegelhalter, D. J., Dawid, A. P., Lauritzen, S. L., & Cowell, R. G. (1993). Bayesian (Spiegelhalter et al., 1993) analysis in expert systems. *Statistical Science*, *8*(3), 219–247.

Winston, W. L. (2004). *Operations Research (Winston, 2004): Applications and Algorithms* (4th ed.). Cengage Learning.

Chapter 7
Impact of Artificial Intelligence on the Strategic Decision Making of Multinational Companies

Benjamin Macleod
https://orcid.org/0009-0009-2671-4574
University of Technology Sydney, Australia

Sancheeta Pugalia
https://orcid.org/0000-0002-5682-8427
University of Technology Sydney, Australia

ABSTRACT

Artificial Intelligence (AI) is increasingly being leveraged into strategic decision-making processes across multinational companies, reshaping global industries. The rapid development and adoption led by multinational companies raises various questions on the actual capabilities and limitations of AI, what specific tasks can effectively be improved and associated risks to consider. This chapter focuses on the impacts of AI on global industries through a comparative case study examining four multinational corporations, comparing their unique implementation of AI across five criteria.The findings support the ability for AI to fundamentally address inherent decision-making challenges related to accuracy and timeliness of data, automation, while also highlighting areas sensitive to over-reliance on these automated processes. The conclusion provides a framework for businesses and policymakers to responsibly

DOI: 10.4018/979-8-3373-5702-7.ch007

implement AI, including recommendations for improving AI governance, investing in employee upskilling, and ensuring ethical AI deployment to enhance long-term strategic outcomes.

INTRODUCTION

AI has fundamentally reshaped traditional approaches to specific problem solving and tasks, driven by advances in computer vision, machine learning as well as natural language generation which has enabled organisations to make data-driven decisions at record scale and speed (Barocas & Selbst, 2016). Large language models (LLM) such as Google's "Gemini" and Microsoft's "Copilot" AI are some recent examples demonstrating the capability to perform complex tasks and, in some cases, more effectively than humans (Brown et al., 2020).

For multinational corporations (MNCs) that manage operations across hundreds of markets, AI technologies have introduced unique possibilities to maintain a competitive edge by enhancing operational efficiency, data analytics, predictive forecasting, risk assessment, and internal workforce management (Kalogiannidis et al., 2024). The response to these potential benefits is evident if we look at the current adoption rate with more than three-quarters of large organisations now using AI in at least one business function (Statista, 2023), with technology and financial sectors among the leading adopters. However, while many companies are leveraging AI, the true impact on decision-making processes remains underexplored.

In the past decade, AI interest from industry leaders, academics and policy makers has increased exponentially with investment matching this trend. Annual global investment report in AI shows from 2013 until 2024, annual corporate investment has grown to about US$217 Billion with a significant outlier in 2021 of over US$360 Billion aligning right before the commercial release of OpenAI's ChatGPT (Our World in Data, 2025).

AI's appeal for businesses lies in its potential to enhance efficiency, accuracy, and scale in decision processes. However, technical experts are still curious to understand future implications and limits (Aleessawi & Djaghrouri, 2025). For example, a recent survey involving 1600 academics showed that more than 50% believe AI tools will be either "very important" or "essential" in their field over the next decade (Van Noorden & Perkel, 2023).

In response to the increasing investment and consensus on AI adoption, policy makers have raised increasing concern with the associated ethical and regulatory challenges posed (Filippucci et al., 2024). Some of the key concerns include algorithmic bias, where AI systems may perpetuate or amplify existing societal inequalities due to biased training data or flawed model design (Obermeyer et al., 2019). AI hallucination poses significant threats to decision-making accuracy, particularly in critical sectors such as healthcare and finance where errors can have severe consequences (Ji et al., 2023). Additionally, issues of transparency and explainability remain problematic, as many advanced AI models operate as "black boxes," making it difficult for organisations to understand how decisions are reached or to identify when systems malfunction (Rudin, 2019).

Despite these concerns, companies continue to adopt AI for decision-making, prompting researchers to examine its impact. However, previous research on AI's influence on decision-making has primarily focused on general applications or single-industry contexts, resulting in limited understanding of how AI affects corporate strategy across diverse sectors. This study addresses this gap by examining four multinational corporations (MNCs) across two critical industries, analysing their distinct approaches to AI integration.

This chapter investigates how AI shapes strategic decision-making by examining two contrasting sectors: technology and healthcare. These industries face fundamentally different regulatory and operational environments. Healthcare prioritises patient safety and regulatory compliance, which constrains implementation pace but strengthens safety governance. Technology prioritises rapid deployment and innovation, which enables faster capability development but introduces different risk patterns. By analysing AI's effectiveness across five decision criteria (operational efficiency, data analytics, predictive forecasting, risk management, and employee management), this analysis reveals industry-specific benefits and risks. The findings demonstrate how organisational context determines whether AI investments translate into sustainable competitive advantage, while offering transferable insights for AI implementation across other industries and informing policymakers on supporting responsible AI development.

The chapter is structured as follows: First, we discuss background of AI along with theoretical framework, followed by an explanation of the research methodology. The findings section evaluates five performance parameters across two companies from each industry, presenting case studies of Google and Microsoft (technology), and Johnson & Johnson and Pfizer (healthcare). The subsequent section provides recommendations based on the findings and concludes the chapter.

LITERATURE REVIEW

Evolution of AI

AI's application in business strategy has evolved significantly over the past several decades. The advent of machine learning in the 1990s and early 2000s enabled sophisticated pattern recognition and predictive analytics, primarily applied to operational tasks such as fraud detection and customer segmentation (Brynjolfsson & McAfee, 2017). However, the transformative shift occurred in the 2010s with breakthroughs in deep learning, natural language processing, and the availability of big data, which expanded AI's capabilities from operational efficiency to strategic decision-making (Agrawal et al., 2018). The introduction of generative AI models, particularly LLMs from 2020 onwards, marked paradigm shift, enabling AI to not only analyse data but also generate insights, simulate scenarios, and engage in complex reasoning tasks (Dwivedi et al., 2023). This evolution has progressively moved AI from a supporting tool for narrow tactical decisions to a strategic partner capable of influencing high-level corporate strategy across multiple industries.

As AI's strategic role has expanded, concerns about ethical and regulatory challenges have intensified, prompting the development of governance frameworks to guide responsible implementation. The European Union's AI Act, enacted in 2024, represents the world's first comprehensive AI regulation, adopting a risk-based approach that categorises AI systems according to their potential harm and imposes corresponding requirements for transparency, accountability, and human oversight (European Commission, 2024). Similarly, the OECD AI Principles, adopted by over 50 countries, emphasise inclusive growth, sustainable development, human-centered values, transparency, and accountability as foundational elements of trustworthy AI (OECD, 2019). In the corporate sphere, frameworks such as ISO/IEC 42001:2023 provide organisations with standardised management systems for AI, addressing risk assessment, data governance, and continuous monitoring (ISO, 2023). Additionally, industry-specific guidelines have emerged, such as the FDA's regulatory framework for AI/ML-based medical devices in healthcare and the Financial Stability Board's recommendations for AI use in financial services (FDA, 2021; FSB, 2017). While these frameworks represent significant progress toward responsible AI governance, challenges remain in harmonising standards across jurisdictions, ensuring effective enforcement, and adapting regulations to keep pace with rapid technological advancement (Floridi et al., 2018).

Within this evolving regulatory landscape, a growing body of research confirms that AI is fundamentally transforming how companies make strategic decisions. Csaszar et al. (2024) describe AI as an emerging "co-pilot" for executives, able to augment strategic decision-making by processing vast data, generating strategy options, and even simulating outcomes. Rivero (2025) finds that AI tools (ranging from advance analytics to real-time scenario planning) can improve decision accuracy and efficiency, enabling more data-driven and forward-looking strategies. These capabilities are critical given the complexity and uncertainty of modern strategic choices. Traditional decision-making is constrained by human cognitive limits and fragmentated information, whereas AI can rapidly synthesise large datasets into actionable insights (Ali et al., 2024). For example, AI-driven decision support systems help firms discern patterns in market data that would elude human analysis. Studies also note that AI's impact is contingent on context: it can enhance or even redefine competitive advantage, but outcomes depend on how well organisations integrate AI with human expertise and respond to competitive dynamics (Tatineni, 2019). Thus, there is a clear need to propose frameworks that link AI use to firm performance, emphasising factors such as organisational readiness, data quality, and ethical considerations (Csaszar et al., 2024). In sum, the current literature positions AI as a transformative force in strategic management, one that must be thoughtfully managed to fully realise its potential.

Building upon the historical evolution of AI, the emergence of governance frameworks, and the growing evidence of AI's transformative impact on strategic decision-making, this study examines how MNCs operationalise AI across their organisations. To systematically compare AI integration across four corporations in technology, and healthcare, we evaluate performance across five key decision-making domains where AI has demonstrated substantial strategic value and widespread adoption: operational efficiency, data analytics, predictive forecasting, risk management and employee management.

Theoretical grounding

AI is transforming strategic decision-making by addressing constraints identified in classical management theory. Simon's (1955) bounded rationality model shows that managers operate under limits of time, information, and cognitive capacity. AI overcomes these limits by rapidly processing large datasets and identifying patterns humans would miss. However, technological capability alone does not guarantee success. The Technology Acceptance Model (Davis, 1989) and Technology-Organisation-Environment framework (Tornatzky & Fleischer, 1990) reveal that adoption depends critically on organisational factors: employee trust, psychological safety, leadership credibility, and clear communication. External factors such as regulatory environments, industry competition, and societal trust in AI, shape both organisational readiness and the effectiveness of AI implementation. When organisations deploying equivalent technology achieve divergent adoption outcomes, organisational factors emerge as the primary differentiator. This reveals a central insight: organisational capability determines success more than technological capability.

From a strategic perspective, the Resource-Based View (Barney, 1991) and Dynamic Capabilities framework (Teece, 2007) explain competitive advantage through AI. An important distinction emerges: AI technology itself is not rare or difficult to imitate; but from organisational integration: combining AI with existing expertise, data, and business strategy. Dynamic capabilities operate through three stages:

sensing (detecting opportunities through analytics), seizing (acting on opportunities through decisions), and reconfiguring (reorganising people and culture to maintain competitiveness). Empirically, organisations often display equivalent sensing and seizing capabilities across industries; they diverge sharply in reconfiguring, the organisational change and employee adoption that prove hardest to achieve. This asymmetry identifies the critical bottleneck: most organisations excel at deploying technology but struggle with the organisational transformation required to use it effectively.

This study evaluates AI across five decision-making domains including operational efficiency, data analytics, predictive forecasting, risk management, and employee management, to test whether organisational capability aligns with technological capability. The conceptual model, shown in Figure 1, proposes that AI creates strategic value only when external pressures, organisational resources, dynamic capabilities, and organisational readiness align. When misalignment occurs like technology advancing while employee adoption stalls, organisations face a critical choice: address organisational barriers (enabling success) or continue technology expansion (leading to underutilised capability). The case studies that follow examine four companies across two industries navigated these conditions, revealing why some AI implementations succeed while others face resistance despite equivalent technology.

Figure 1. Conceptual Model

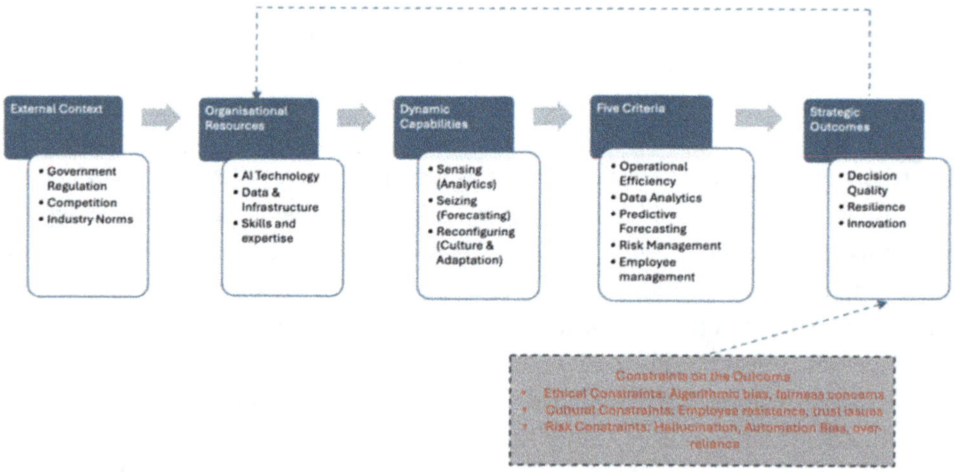

Definition of Key Criteria and Validation:

Operational Efficiency

Operational Efficiency refers to the ability for a company to deliver products or services using the minimum necessary resources (time, cost, labour) while maintaining quality. AI systems excel at automating repetitive, time-consuming tasks, which can dramatically streamline business processes (Rivero, 2025). By taking over routine work, AI provides time for employees to focus on higher-value activities, thereby not only cutting costs but also potentially improving the quality of work life. Numerous case studies across industries show AI-driven automation leading to faster throughput and reduced errors. For instance, McKinsey (2021) notes that service operations optimisation is among the most common AI use cases globally, as companies deploy algorithms to handle service requests, schedule resources, and manage supply chains more efficiently. In the finance sector, AI-based process automation such as intelligent document processing has trimmed back-office processing times from days to minutes (Harvard,2020).

Data Analytics

This criterion involves the capability to analyse large and complex datasets to inform decisions – identifying trends, insights, and metrics that guide strategy, critical for MNC. The explosion of data in the digital age has made advanced data analytics a strategic necessity, and AI is a crucial tool for turning "big data" into actionable insights. Modern AI techniques (machine learning, natural language processing, computer vision, etc.) can analyse huge datasets to detect patterns, correlations, and anomalies far beyond human processing capacity. This analytical power underpins better decision-making in areas ranging from marketing to product development. Surveys show that most large firms now use AI-powered analytics in at least one function, reflecting AI's growing role in corporate analytics strategies (McKinsey, 2021). In summary, data analytics is a core dimension where AI's impact is evident: organisations effectively leveraging AI for analytics report deeper customer insights, more accurate forecasts, and faster decision cycles (Rivero, 2025), all of which are vital for strategic planning.

Predictive Forecasting

Predictive forecasting is the use of historical and real-time data to project future trends or outcomes. Strategic decision making for MNC requires forward looking insight, anticipating future conditions such as market demand, company resourcing and potential risks. AI presents a suitable use case to improve predictive forecasting using machine learning models to predict future scenarios and outcomes. In dynamic markets, the ability to anticipate trends or risks confers a major strategic advantage. AI techniques, specifically predictive analytics and simulations have proven adept at forecasting in many domains. Wiles (2022) observed that leading firms use AI-driven forecasting to move from reactive to proactive strategies, whether in supply chain management or financial planning. Notably, a Deloitte study (Deloitte, 2024) found that companies using AI for forecasting and scenario planning had higher confidence in their long-term strategies, because they could test "what-if" scenarios with AI simulations.

Risk Management

Managing risk is central to strategic decision-making, ensuring that potential threats to the business are identified and mitigated. Implementing AI in strategic decisions also involves managing new and existing risks. Here we consider two aspects: AI as a tool for risk management including identifying and mitigating risks the company faces, and the risks introduced by AI adoption. Both angles are well-covered in recent studies. Firstly, AI is being used to bolster organisational risk management by detecting fraud, cyber threats, compliance issues, and other hazards faster and more accurately than traditional methods (Kalogiannidis, 2024). Secondly, the rise of AI brings its own risks that literature urges organisations to manage, such as algorithmic bias, lack of transparency, security vulnerabilities, and regulatory compliance issues.

Employee Management

This criterion examines how AI impacts employees' work experience – does it make their jobs easier and more engaging, or does it introduce stress and disruption? The integration of AI into workplaces has significant implications for employees – their roles, skills, and satisfaction. We include employee management as a criterion based on the understanding that successful strategic adoption of AI depends on human acceptance and augmentation. Recent studies offer a nuanced view: many employees are optimistic that AI can relieve costly repetition and improve workplace satisfaction, even as they harbour concerns about job security (Microsoft, 2023).

Conversely, a poorly managed AI rollout can cause anxiety and resistance among staff. Scholarly discussions emphasise change management: clear communication, training, and involvement of employees in AI adoption can turn AI into a positive force that augments human capabilities, often termed "AI augmentation" in literature (Abu Issa Gazi et al., 2024). The current consensus is that organisations leading in AI are those that pair technological innovation with workforce enablement, cultivating a culture where employees trust and effectively use AI. This literature perspective validates our focus on employee management to critically measure the impact of AI implementation on company decision making. It is both an outcome of AI strategy and a precondition for sustainable AI-driven performance.

To summarise, the literature affirms above five criteria as critical lenses for evaluating AI's impact on strategic decision-making. Each criterion is well-supported by existing research as a domain of significant AI influence, encompassing both "hard" performance metrics (efficiency, analytics, forecasting, risk) and "soft" organisational factors (human impact) that determine long-term success (Bóna & Lippert, 2020). The next section outlines the comparative case study methodology used to investigate these criteria across the selected companies.

METHODOLOGY

Comparative Case Study Approach

The question of "How does AI impact the Strategic Decision Making of Multinational Companies" has been explored through a comparative case study review which has focused on the analysis of four companies across two distinct industries. A multiple-case study method was chosen to enable cross-case comparison and more robust conclusions than a single case study would allow. As Baxter & Jack (2010) notes, using multiple cases can significantly strengthen analytic conclusions through literal or theoretical replication – if similar outcomes are observed across cases, it increases confidence in their generalisability, while differences can illuminate contingent factors.

Case Study Selection

This analysis focused on healthcare and technology sectors for specific methodological reasons. First, these sectors represent contrasting regulatory environments and decision-making contexts: healthcare operates under strict regulatory constraints (FDA, EMA) with life-or-death consequences, while technology operates with minimal regulatory oversight and rapid implementation cycles. This contrast illuminates how external context shapes AI adoption patterns and outcomes. Second, analysing two sectors in depth with detailed case studies of two companies each allows for meaningful comparative analysis and robust pattern identification, whereas extending to additional sectors would require either reducing depth per sector or exceeding practical research scope. Third, these sectors encompass a broad range of AI applications ranging from operational efficiency and data analytics to risk management and workforce integration, therefore providing sufficient diversity to test the conceptual model across different contexts. Finally, insights from these contrasting sectors are likely transferable to other industries; the theoretical framework of organisational resources, dynamic capabilities, and external constraints applies across sectors, making these findings actionable beyond healthcare and technology.

A purposive sampling method was utilised to select companies as case studies that provide sufficient accounts of AI implementation and practical evidence of the impact on the defined set of key performance criteria. To conduct an effective analysis of the research, a clearly defined set of research criteria have been chosen to find relevant examples:

1. **Industries (technology, and healthcare):** Identified as critical global sectors which provide a diverse range of companies and AI adoption
2. **Global geographic operation:** Limits case study selection to large multinational companies with sophisticated business models and complex operations
3. **Publicly listed companies:** Limits case study selection to companies with publicly available information such annual reports and published research to draw useful conclusions
4. **First Adopter/Industry Leader:** These cases were selected purposively for their leadership in AI adoption and their representation of diverse sectors to capture a broad perspective on AI impacts.

Based on the above research criteria following companies are used as cases.

1. **Google** is a global technology leader in search, digital ads, YouTube, Android and cloud services
2. **Microsoft** is a diversified software, cloud and AI company supporting many digital products and services like Windows, Office 365, Azure, and LinkedIn
3. **Johnson & Johnson** is global healthcare company focused on pharmaceuticals and med-tech innovation
4. **Pfizer** is a global pharmaceutical firm focused on developing medicines and vaccines

Figure 2 provides an overview of the case analysis. Each selected company is treated as an individual "case" of AI implementation in strategic decision-making. By comparing pairs of similar companies within the same industry, as well as trends across industries, we can discern how industry context moderates AI's effectiveness.

Figure 2. Overview of Case analysis

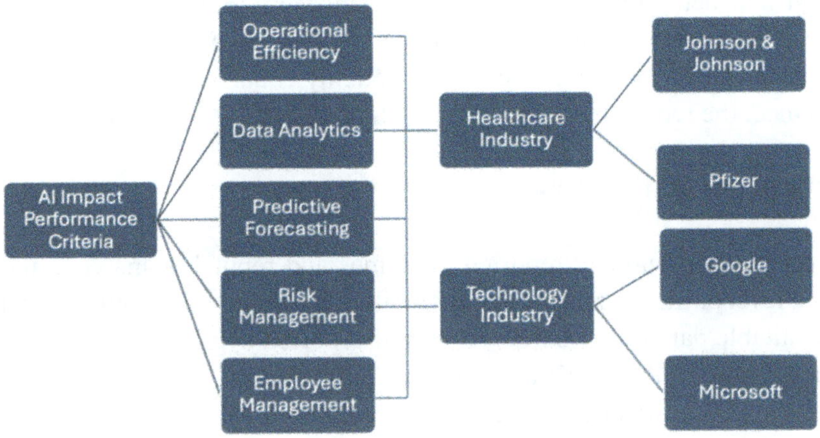

Data Collection

The study relies on multiple secondary data sources to ensure a rich and credible evidence base, consistent with case study research best practices. These sources include academic literature for general insights and frameworks; company reports and press releases for specific initiatives and performance indicators; industry analyses and credible news articles for independent evaluations of the companies' AI use; and white papers or conference presentations by the case companies detailing AI projects.

Using these diverse sources allows for triangulation – we cross-validate claims (for example, an efficiency gain reported in a company's annual report is checked against any external assessments or case studies). Research was primarily limited to information from the last 5 years due to rapid expansion of AI development. We avoided unverified blog posts unless corroborated by other sources, in line with maintaining source credibility.

Thematic Analysis

After selecting all the sources, for each company, data were coded according to the five criteria. For instance, any mention of improved process speed or cost reduction was coded under "operational efficiency," uses of data and analytics under "data analytics," and so on. We then conducted a within-case analysis to summarise how AI has affected each criterion for that company, compiling qualitative evidence. Next, a cross-case analysis compared findings: first between companies in the same industry to identify industry-specific patterns, and then across industries to identify overarching trends or outliers. This structured comparative approach is informed by established methodology literature which suggests iterating between within-case and cross- case examination to build a coherent comparative narrative (Eisenhardt & Graebner, 2007). We took care to give roughly equal analytical weight to each case, to meet the requirement for even depth of insight.

Validity and Limitations

By drawing on publicly reported outcomes and reputable analyses, the study ensures a level of transparency and verifiability. However, a limitation is that much of the available data on AI initiatives is self-reported by companies, which could introduce a positive bias. We attempted to mitigate this by including third-party commentary e.g. news of employee reactions, or external rankings of AI maturity and validating claims with academic literature.

Another limitation is that each case's depth is constrained by publicly available information – the true internal state of AI adoption might not be fully captured. Nonetheless, the comparative design provides an adequate form of external validity through replication logic. If certain benefits or challenges appear in several independent cases, that suggests the findings are not idiosyncratic. Conversely, differences highlight how context like industry regulation or company culture influences AI outcomes.

FINDINGS

Technology AI Adoption: Google and Microsoft

Google and Microsoft have deployed AI extensively across business operations. This analysis examines how both companies use AI across five decision-making areas and reveals why divergent outcomes depend on organisational factors. AI-driven strategic gains are tempered by significant risks: worker displacement, algorithmic bias, privacy concerns, and organisational uncertainty.

Operational Efficiency

Microsoft's Copilot automates approximately 75 percent of routine document work, enabling employees to focus on higher-value tasks (Kellington, 2025). This raises concerns about long-term workforce displacement. While Microsoft emphasises that Copilot enhances workers, eventual displacement remains uncertain. Google achieved similar efficiency gains through DeepMind's AI controlling data center cooling, reducing energy consumption by 40 percent and improving power efficiency by 15 percent (Okyere, 2021). This infrastructure optimisation also enables Google to operate with fewer cooling technicians and engineers. Google's "Goose" coding assistant helps development teams work faster by automating routine code tasks (Langley, 2024). However, developers report concerns about over-reliance on AI suggestions and reduced learning opportunities for junior developers.

Both companies leveraged AI to reduce costs and accelerate workflow through organisational integration, combining AI with existing expertise, infrastructure, and business processes. This creates asymmetry: highly skilled workers who leverage AI effectively advance rapidly, while others face stagnation or displacement. Competitive advantage emerges not from technology possession but from how organisations integrate AI into existing strengths.

Data Analytics

Google's advertising system uses machine learning to optimise ad bidding in real time, improving results by 72 percent while cutting costs by 43 percent (Bailey, 2024). However, optimisation raises concerns about algorithmic bias such as machine learning models may perpetuate biases against certain demographic groups, resulting in unequal ad delivery. Google's Data Commons project integrates global datasets into a searchable system (Guha & Manyika, 2023). Aggregating diverse data introduces privacy risks and data sovereignty concerns.

Microsoft gathers insights from billions of workplace communications through its Work Trend Index, feeding findings directly into product development (Microsoft, 2023). Analysing employee communications at scale raises privacy and surveillance concerns, for instance employees may not fully understand what data is collected or how inferences are drawn. The ALICE project uses machine learning to predict demand and set prices more accurately (Kiciman et al., 2022). Sophisticated price prediction enables dynamic pricing that may disadvantage price-sensitive consumers, raising ethical fairness questions.

Both firms excel at sensing opportunities at scale through organisational integration. Google built external knowledge infrastructure; Microsoft built internal systems serving customers. Both create competitive advantage while introducing risks of algorithmic bias, privacy erosion, and reduced consumer agency requiring robust governance frameworks.

Predictive Forecasting

Google uses AI to forecast demand on data centers and predict traffic patterns for Google Maps, optimising infrastructure planning (Zheng et al., 2023). These systems depend on massive location data collection, raising privacy concerns about tracking billions of individuals' movements. Microsoft's FINN tool forecasts revenue with 99 percent accuracy across product lines, enabling strategic capital allocation decisions saving an estimated $130 million annually (Marciano, 2024). Over-reliance on a single forecasting system introduces strategic risk like if FINN contains systematic bias or fails during market disruption, cascading errors could damage capital allocation decisions. The 99 percent accuracy masks scenarios where the one percent of misses have outsized strategic impact.

FINN influences major capital allocation decisions, reshaping strategy rather than merely optimising operations. Both firms demonstrate capability to translate forecasts into strategic advantage, but strategic forecasting creates greater competitive impact and greater risk if forecasts fail. When forecasts influence billions in capital allocation, the potential cost of algorithmic failures escalates dramatically. Neither company has publicly disclosed error-handling protocols or how they manage forecast uncertainty, raising questions about decision quality despite technological sophistication.

Risk Management

Google's AI content moderation automatically flags 94 percent of policy-violating YouTube videos (O'Connor, 2021). However, automated moderation exhibits algorithmic bias as it disproportionately flags content from marginalised communities while missing violations from privileged groups. Over-removal during COVID-19 demonstrates that aggressive algorithmic enforcement suppresses legitimate speech alongside harmful content. Gmail's spam filters achieve 99.9 percent accuracy (Kumaran, 2023). False positives (legitimate emails marked as spam) can result in missed critical communications.

Microsoft's Security Copilot protects over 1,400 companies from cyberattacks (Jakkal, 2024), augmenting human security experts rather than replacing judgment. Security threats evolve faster than AI systems can learn, and advanced attackers deliberately craft attacks to evade detection. Security teams may develop over-reliance on Copilot's recommendations, leading to automation bias.

Both companies deploy AI for regulatory compliance adapted to their customer bases. However, using AI to manage regulatory compliance raises concerns: regulators have limited ability to audit AI-based compliance systems, creating potential for regulatory evasion masked by algorithmic compliance reporting. These applications redistribute risks rather than eliminating them, introducing new failure modes alongside solving old ones.

Employee Management

The most striking difference between these companies emerges in employee adoption of AI tools. Microsoft rolled out Copilot with transparent communication and extensive training through its AI Academy program. Employee surveys show high satisfaction and productivity gains (Microsoft, 2024). However, this success masks unspoken tensions: not all employees benefit equally, highly skilled workers advance rapidly while others struggle, and productivity intensification may increase stress rather than improve well-being.

Google's experience diverged sharply. Its Ask AI tool arrived amid AI-driven layoffs affecting 12,000 employees (Dungan, 2024; Langley, 2024). When employees encounter AI introduction alongside job losses, they perceive threat rather than benefit. Trust and psychological safety deteriorated, creating organisational misalignment: technology advanced while employee adoption stalled.

Both companies deployed comparable technology with similar technical quality. Different outcomes depended entirely on organisational factors: communication credibility, employment security messaging, psychological safety, and alignment between technology deployment and HR strategy. However, even Microsoft's success raises unanswered questions: Will productivity gains translate into job security, or will they eventually justify reduced headcount? The employee management bottleneck reveals that technological and organisational capabilities must align, yet even successful adoption contains underlying uncertainties about long-term employment, equity, and worker agency.

Key Takeaways

Competitive advantage comes from organisational integration, not technology possession. Google and Microsoft deployed equivalent AI technology yet achieved divergent outcomes based on organisational factors. Competitive advantage emerges from how companies weave AI into existing strengths, business processes, and expertise. However, this integration creates organisational lock-in, making systems difficult to audit or control once embedded.

Employee adoption is the critical bottleneck. Most organisations excel at technological deployment but struggle with the cultural transformation required to use technology effectively. Microsoft succeeded by combining technology rollout with transparent communication and employment security. Google's simultaneous AI deployment and layoffs created organisational misalignment. Even successful adoption contains unresolved tensions about displacement, equity, and long-term worker agency.

AI augments rather than replaces human judgment, yet maintaining this distinction requires continuous vigilance. Both companies position AI as enhancing human capability such as security teams using AI-enhanced analysis, and writers using Copilot assistance. However, maintaining appropriate human judgment becomes increasingly difficult as organisations develop over-reliance on algorithmic recommendations. The distinction between augmentation and replacement is not permanent; it depends on continuous organisational choice and oversight.

Both firms demonstrate that AI drives strategic improvements. Recognising risks of algorithmic bias, worker displacement, and privacy erosion, both Google and Microsoft have invested in governance frameworks, transparent AI policies, and human-AI collaboration approaches. Such mitigation strategies are essential to ensure technological advances translate into sustainable strategic value. However, strategic advantage from AI adoption remains incomplete as the technology obscures human and ethical costs that persist despite technological sophistication.

Table 1. Technology Industry AI Impact & transferability

Criteria	**AI Impact** **Low - minor benefit Moderate - notable benefit High - extreme benefit**	**Industry Transferability** **Low - industry specific only** **Moderate - some transferable applications** **High - AI improvements easily replicated by any industry**
Operational Efficiency	High	High (automation and chatbots can be adopted by all industries)
Data Analytics	High	High (data analytics tools can be adopted by all industries)
Predictive Forecasting	Moderate	Low (Industry specific)
Risk Management	High	High (moderation tools applicable for all industries)
Employee Management	Moderate	High (Upskilling programs and COEs applicable for all industries)

Table 1 maps AI impact and industry transferability for technology sector applications across five criteria. Google and Microsoft's AI demonstrate high impact and high transferability across most criteria, readily adopted by finance, retail, manufacturing, and other sectors. Operational Efficiency, Data Analytics, and Risk Management show high impact and transferability because automation tools and analytical methods have broad applicability. Predictive Forecasting is the exception as it has moderate impact but low transferability due to industry-specific variables.

Employee Management shows moderate impact but high transferability, as upskilling programs transfer effectively across industries. This pattern reveals that technology sector innovations translate readily into competitive advantage, enabling organisations across sectors to directly adopt these frameworks for immediate value.

Healthcare Industry: Johnson & Johnson and Pfizer

Johnson & Johnson and Pfizer demonstrate how AI drives organisational value in healthcare. This analysis examines their use of AI across five decision-making areas and reveals why success depends on organisational integration and regulatory compliance. Healthcare AI adoption unfolds within a complex regulatory environment that shapes both the pace and form of innovation, influencing organisational strategies and risk management priorities. While AI improves efficiency and clinical outcomes, it introduces measurable risks: patient safety uncertainties, algorithmic validation challenges, and unresolved questions about clinical decision-making autonomy.

Operational Efficiency

J&J's MedTech division uses AI to streamline surgical workflows. Polyphonic™ automatically edits surgical videos into learning summaries for surgeon review, enabling efficient identification of critical moments (Johnson & Johnson, 2024). However, faster procedures raise questions about whether reduced anesthesia times genuinely improve patient outcomes or shift risks elsewhere. The CARTO 3 mapping system guides catheters during cardiac ablation faster and improves recovery times. Over-reliance on AI guidance may reduce surgeon decision-making during complications, compromising adaptive expertise (Topol, 2019). J&J also uses AI digital models to optimise biologic drug yields and manufacturing consistency, though algorithmic optimisation might miss rare adverse reactions emerging only in full-scale production.

Pfizer demonstrated significant efficiency gains during COVID-19 vaccine development. AI modeling reduced discovery timelines, enabling rapid response that saved millions of lives (Virtasant, 2024). Accelerated timelines raised questions about long-term safety data adequacy, representing a necessary tradeoff between speed and certainty. Real-time AI monitoring inspected billions of doses with precision impossible manually. However, automation requires maintaining human expertise for handling unexpected deviations.

Both companies demonstrated strong capability in capturing efficiency gains. Healthcare contexts create tension between efficiency and safety; over-optimisation may compromise safety margins that historically protected patients (Obermeyer

& Emanuel, 2016). The optimal approach balances speed improvements against preserving critical safeguards.

Data Analytics

J&J uses AI data mining of genetic and patient information to identify novel drug targets faster. The ECHO system scans chemistry reports and proposes compound modifications, enabling exploration of impractical synthetic pathways. Connected devices like the MONARCH bronchoscope enable real-time clinical insights, refining surgeon training and supporting evidence-based improvements. However, continuous data collection creates privacy concerns, as patient health information is constantly transmitted and analysed, creating breach vulnerabilities. AI-derived insights may not generalise beyond training populations, potentially missing efficacy signals in underrepresented groups and perpetuating disparities (Char et al., 2020).

Pfizer uses machine learning to analyse biological data and compare patient subgroups, revealing disease mechanisms invisible to traditional analysis (Pfizer, 2025). AI analysis enabled drug repurposing accelerating patient access. However, companies rarely disclose how many AI-generated hypotheses failed in trials. Pfizer reduced time to identify safety signals by six months, enabling faster regulatory response. Accelerating detection without equivalent validation creates risk of false alarms disrupting clinical operations.

Both firms excel at sensing clinical opportunities through organisational integration, creating competitive advantages in drug development. Both introduce risks requiring robust governance: algorithmic bias, privacy erosion, and false signal generation. When governance functions effectively, AI-enabled analytics democratise clinical insights previously accessible only to elite institutions (Topol, 2019).

Predictive Forecasting

J&J uses AI for epidemiological modeling combining climate data with outbreak records to predict flu seasons, enabling early vaccine production and improving preparedness (Capgemini, 2020). However, demand forecasting driven by online search trends raises validity questions, do trends reflect actual patient needs or marketing-influenced behavior? Machine learning assesses drug trial success likelihood, optimising R&D resource allocation. Predicting based on historical data may lead to premature termination of breakthrough compounds behaving unexpectedly. Supply chain AI improved forecast accuracy by 10-20%, reducing waste. Even improved forecasting cannot eliminate uncertainty when regulatory decisions or clinical evidence shift demand unpredictably.

Pfizer deployed similar forecasting capabilities during COVID-19, projecting vaccine uptake to synchronise supply with emerging hotspots (IQVIA, 2024). Trial success prediction may disadvantage novel therapeutic areas, potentially biasing portfolios toward incremental innovation. Market adoption forecasts enabled targeted physician education programs achieving broader oncology drug adoption. Forecasts assume adoption barriers are primarily informational, potentially missing systemic or economic barriers.

Both companies demonstrate strong forecasting capability creating competitive advantage. Healthcare forecasting carries higher stakes: incorrect predictions may terminate promising compounds or leave patients without critical medications. Neither company disclosed error rates or failure-handling protocols, limiting external validation.

Risk Management

J&J uses machine learning to monitor factory sensors for defects, achieving $60 million in cost savings while improving quality consistency (RESO Insights, 2024). Automation creates new dependencies, such as organisations must maintain systems detecting when algorithms malfunction. Algorithms monitor clinical trial data, raising safety alerts faster than manual reviews and potentially saving participant lives. However, automated alerts may miss contextual factors, like patient co-morbidities or drug interactions, that experienced reviewers recognise. NLP reviews batch records, reducing compliance errors, though it may flag benign variations while missing subtle issues requiring domain expertise. Remote monitoring of medical implants predicts device failure enabling proactive intervention. However, real-time monitoring collects intimate patient health data; breaches could expose cardiovascular patterns and disease progression.

Pfizer's manufacturing surveillance achieves quality standards exceeding manual inspection (Pfizer, 2025). Side effect report analysis identifies new adverse events faster, enabling quicker patient protection. However, passive surveillance may miss underreported side effects from vulnerable populations, perpetuating disparities (Char et al., 2020). NLP scans marketing content for compliance errors, though it may miss subtle misleading claims requiring contextual clinical evidence understanding. Cybersecurity monitoring reduces breach risk, demonstrating organisational learning.

Both firms leverage AI for risk management creating measurable value and demonstrate reconfiguring capability through post-implementation adjustments. Healthcare risk management creates new vulnerabilities requiring appropriate human oversight (Topol, 2019). Regulators have limited ability to validate AI-based compliance systems, creating potential systematic gaps remaining unresolved sector-wide. Both Johnson & Johnson and Pfizer have invested in ethical AI governance frameworks and transparency initiatives, aiming to mitigate algorithmic bias and enhance patient data protection.

Employee Management

J&J positions AI as augmentation rather than replacement. Researchers use NLP chatbots to search databases and literature, saving research hours and enabling scientists to focus on hypothesis generation (CDO Magazine, 2025). However, this masks labor reductions: literature researchers and customer service staff numbers may decline, creating employment uncertainty. Manufacturing engineers diagnose faults more easily using integrated AI analytics, reducing downtime. When AI handles routine diagnostics effectively, engineers may lose opportunities developing troubleshooting expertise critical during novel failures. J&J invested in training tools for employees and surgeons through VR and AI-assisted simulations, demonstrating organisational commitment to workforce development. Training investments may primarily benefit high-skilled workers, potentially widening skill gaps. Well-designed, accessible programs support inclusive upskilling.

Pfizer launched an "AI Academy" in 2022 teaching data literacy to 80 percent of staff within three years, signaling career growth rather than obsolescence (Stevens Institute of Technology, 2025). Academy's success depends on sustained commitment; if adoption accelerates faster than training, employees may feel left behind. While some administrative roles diminished, Pfizer invested in retaining and redeploying employees into AI-focused positions like data curators or system specialists. Redeployment may not match prior compensation or seniority; employees may experience real career disruption even if retained. Successful redeployment programs create mutually beneficial transitions.

Both firms achieved better employee adoption than technology companies, demonstrating that healthcare organisations successfully integrate AI while maintaining workforce stability through change management. Both narratives contain underlying tensions: automation reduces some roles; training and redeployment smooth transitions but do not eliminate displacement uncertainty. Healthcare's regulatory complexity and talent scarcity may force better employee management. This isn't because firms value employees differently, but because healthcare labor is harder to replace quickly. The reconfiguring bottleneck is less pronounced in healthcare than in technology, but latent uncertainties about long-term employment and equity remain unresolved.

Key Takeaways

Competitive advantage comes from organisational integration within regulatory contexts. Healthcare firms move slower than technology companies because implementations require regulatory validation, clinical evidence generation, and safety monitoring. This slower pace paradoxically creates better change management: employees face less disruption and firms have time for training and redeployment. However, regulatory complexity creates opacity because regulators have limited ability to validate AI-based clinical decision support systems, creating potential systematic gaps (Char et al., 2020). When regulatory oversight incorporates AI-specific validation methods, it can strengthen rather than impede innovation (Topol, 2019).

Healthcare AI operates under irreducible safety tensions distinguishing it from other sectors. Efficiency gains must continuously balance against safety margins protecting patients (Obermeyer & Emanuel, 2016). Both firms positioned AI as augmentation of clinical judgment rather than replacement, recognising that clinical decision-making requires contextual understanding algorithms cannot fully replicate. Maintaining augmentation requires continuous organisational vigilance that regulatory frameworks do not yet consistently enforce. When successful, augmentation enhances both efficiency and safety simultaneously.

Employee outcomes are better managed in healthcare than in technology sectors because healthcare labor scarcity and regulatory knowledge requirements create employer incentives to retain talent. Both J&J and Pfizer invested substantially in training academies, redeployment programs, and change management. This reflects labor market dynamics rather than inherently superior organisational philosophy; the question remains whether commitments persist when labor markets tighten. Both firms' programs are genuine, yet they mask underlying displacement concerns and equity questions remaining unresolved.

Table 2. Healthcare Industry AI Impact & transferability

Criteria	AI Impact Low - minor benefit Moderate - notable benefit High - extreme benefit	Industry Transferability Low - industry specific only Moderate - some transferable applications High - AI improvements easily replicated by any industry
Operational Efficiency	High	Low (industry specific)
Data Analytics	Moderate (privacy and ethical concerns remain)	Moderate (Data structuring and siloing methods applicable for all industries)
Predictive Forecasting	High (human oversight is critical to see sustained benefits)	moderate (research applications are highly specific but supply chain improvements can easily be replicated)
Risk Management	Moderate	Low
Employee Management	Low	Low

Similar to technology sector, Table 2 maps two dimensions: the magnitude of AI's organisational benefit in healthcare and how readily these approaches transfer to other industries. Four criteria namely Operational Efficiency, Predictive Forecasting, Data Analytics, and Risk Management, generate significant value within healthcare organisations. However, their transferability varies considerably. Criteria with high healthcare impact but low transferability reflect specialised regulatory requirements and domain-specific challenges that other sectors cannot easily replicate. Data Analytics and Predictive Forecasting occupy middle ground, suggesting that general analytical methods have broader application while healthcare-specific implementations remain constrained. Employee Management shows low impact and low transferability across both dimensions, indicating that workforce outcomes reflect industry structural factors rather than innovative management approaches. This framework demonstrates that healthcare organisations derive competitive advantage not by adopting generic AI strategies, but by building systems specifically designed for healthcare's unique operational and regulatory environment.

Overall Findings by Criteria

AI generates substantial benefits across both healthcare and technology sectors. However, the magnitude of impact depends on how organisational resources (technology infrastructure, data quality, and workforce skills) combine with dynamic capabilities (sensing, seizing, and reconfiguring). External context (regulation, competition, industry norms) and constraints (ethical concerns, cultural resistance, technical risks) shape whether AI investments translate into strategic value.

Operational Efficiency

Operational efficiency improvements emerge across all four companies through the seizing dynamic capability by converting AI technology into cost savings and productivity gains. Google and Microsoft realised the largest improvements because their software-dominant services and abundant data enabled comprehensive automation across business operations. Healthcare companies (J&J and Pfizer) achieved comparable efficiency gains but faced stronger external constraints: regulatory oversight and patient safety requirements limited how aggressively they could automate. This illustrates how identical technological resources produce different outcomes depending on external context.

Data Analytics

All companies strengthened their sensing capability by using AI to identify market opportunities and clinical insights in real time. Technology companies deployed data analytics with minimal constraint from external context. Healthcare organisations adopted similar analytical capabilities but more cautiously, balancing sensing benefits against regulatory compliance and ethical concerns about patient privacy. This pattern shows that external context moderates how fully organisations can leverage their organisational resources.

Predictive Forecasting

Predictive forecasting demonstrates strong seizing capability in both sectors but manifests differently. Healthcare forecasting reveals transformative potential, improving drug trial success prediction and supply chain optimisation, directly supporting patient outcomes. Technology companies achieved more general improvements with broader industry applicability but lower strategic stakes. Both sectors face a shared risk constraint: current machine learning model transparency remains inadequate for critical decisions. This gap between technological capability and decision quality illustrates how technical constraints limit strategic outcomes despite strong organisational resources.

Risk Management: Reconfiguring Organisational Defenses

AI strengthens risk management capability by automating threat detection and enabling faster response. Technology companies realise disproportionate benefit because they protect not just internal operations but critical infrastructure serving other organisations. Healthcare AI risk management operates within tighter compliance requirements but prevents significant patient harm through real-time monitoring. In both sectors, even modest risk management improvements prevent substantial operational, financial, and reputational damage. This reveals how external context (regulatory requirements, infrastructure criticality) shapes whether the reconfiguring capability translates into strategic value.

Employee Management: Reconfiguring the Workforce

Employee adoption and workforce outcomes represent the critical reconfiguring capability such as adapting organisational culture to integrate AI while maintaining employee commitment. Healthcare organisations achieved higher workforce satisfaction because labor scarcity created employer incentives to invest in retraining and retention. Technology companies struggled initially when AI deployment (high seizing capability) misaligned with employment security messaging (weak reconfiguring capability). Recent investments in training and transparent communication demonstrate that successful outcomes require alignment between technological and cultural capabilities. This pattern shows that organisational resources alone are insufficient; dynamic capabilities must operate in concert.

Cross-Sector Synthesis

The conceptual model explains divergent outcomes despite comparable AI investments. Both sectors possess strong organisational resources (advanced technology and abundant data). Yet external context (regulation in healthcare, competitive scale in technology) and varying dynamic capability maturity produce different strategic outcomes. Technology companies excel at rapid seizing but initially struggled with reconfiguring. Healthcare companies moved slower due to regulatory constraints but demonstrated more mature reconfiguring capability. Strategic outcomes including improved decision quality, organisational resilience, and innovation, depend not on technology sophistication alone but on how well organisations align their sensing, seizing, and reconfiguring capabilities within their external constraints. Unresolved constraints such as algorithmic bias, privacy erosion, automation bias, and workforce displacement concerns, persist despite strong organisational resources, indicating that technological capability must pair with ethical governance and cultural commitment to realise sustainable strategic advantage.

STRATEGIC RECOMMENDATIONS FOR AI ADOPTION

AI's impact on strategic decision-making presents both opportunities and risks. The conceptual model reveals that successful AI adoption requires alignment between organisational resources (technology infrastructure and workforce skills), dynamic capabilities (sensing, seizing, and reconfiguring), and external constraints (regulation, ethical standards, competitive context). The following recommendations address this alignment for both companies and policymakers.

1. Align AI Investment with Strategic Objectives

Organisations should deploy AI where it addresses concrete strategic priorities: operational bottlenecks, revenue opportunities, or risk reduction. Successful implementations like efficiency gains at Google and Microsoft, or accelerated drug development at Pfizer, align AI projects with known business needs rather than pursuing technology for its own sake. This focused approach ensures that organisational resources translate into seizing capability: converting technological potential into measurable value.

Companies benefit greatly from dedicated AI strategies that coordinate technology selection, investment in data infrastructure, and workforce readiness. Policymakers can support these efforts through targeted research funding, tax incentives for training programs, and infrastructure support for smaller enterprises. Public-private partnerships in areas like pandemic response leverage complementary strengths and accelerate capability development.

2. Establish Governance Frameworks That Balance Innovation and Accountability

Transparency and trust are foundational to successful AI adoption and regulatory compliance. Organisations should implement interpretable AI tools by incorporating explainable AI techniques, confidence intervals, and clear documentation, which helps to mitigate algorithmic bias and enable effective human validation. Strong and robust data governance policies addressing privacy, security, and bias monitoring are essential to uphold ethical standards and maintain stakeholder confidence.

Policymakers play a critical role by developing proportionate regulations that mitigate AI risks while fostering innovation. Clear standards for transparency and accountability give organisations confidence to invest in AI technologies responsibly. International regulatory alignment is critical for multinational companies operating across jurisdictions. In healthcare, regulators should refine approval processes for AI-based medical devices to ensure safety without impeding technological progress.

3. Facilitate Cross-Sector Knowledge Transfer

Organisations learn more rapidly through benchmarking and knowledge exchange with peers. Technology companies benefit from understanding healthcare's reconfiguring capability and risk governance; healthcare organisations gain from technology's operational efficiency approaches. Industry forums, cross-sector partnerships, and collaborative networks play a crucial role in facilitating this bi-directional transfer of knowledge and best practices.

Policymakers can democratise AI benefits by supporting shared infrastructure such as open-source tools, anonymised public datasets, and federated learning networks (particularly valuable in healthcare where privacy constraints limit data sharing). These initiatives reduce barriers for smaller organisations and startups while maintaining strict privacy safeguards.

4. Institutionalise Human Oversight in Critical Decisions

AI should augment, not replace, human judgment in strategic decision-making processes. Organisations must establish clear protocols specifying where human validation is required, conduct periodic performance audits, and maintain human expertise in critical domains. This reconfiguring capability such as integrating AI tools with sustained human oversight, prevents automation bias and builds organisational resilience.

Policymakers should enforce data privacy regulations strictly within AI contexts and encourage development of ethical AI standards. Independent certification bodies for high-risk AI applications (modeled on existing safety standards) can push organisations toward best practices. Additionally, public sector institutions should exemplify ethical AI deployment, demonstrating leadership and commitment to responsible use.

5. Develop Workforce Capability Through Structured Change Management

AI benefits depend critically on reconfiguring capability: adapting organisational culture and workforce skills to integrate AI effectively. Companies should communicate clearly how AI improves operations, involve employees in pilot programs, and provide comprehensive training. Establishing cross-functional AI centers of expertise helps develop best practices, ensuring knowledge is shared broadly rather than siloed within isolated teams.

Policymakers should invest in broad workforce development – integrating data science curriculum in education, supporting vocational training, and offering subsidised company training programs. Government-industry partnerships in apprenticeship models prepare workers for AI-driven roles. These initiatives address potential labor market shifts and maintain social stability by ensuring accessible pathways into emerging opportunities.

Across all recommendations, alignment between technological capability and organisational readiness determines whether AI investments yield sustainable strategic advantage. Success requires that sensing, seizing, and reconfiguring capabilities operate in concert within regulatory and ethical constraints.

CONCLUSION

This report examined how four multinational companies, namely, Johnson & Johnson, Pfizer, Google, and Microsoft, use AI to enhance strategic decision-making. The findings demonstrate that AI drives genuine improvements: operational efficiency, data-driven insights, accurate forecasting, and strengthened risk management. More importantly, the analysis reveals why outcomes diverge despite comparable technology investments.

The conceptual model explains this divergence. Organisations with strong resources (advanced AI infrastructure, quality data, skilled workforce) and well-developed dynamic capabilities (sensing market opportunities, seizing strategic advantages, reconfiguring organisational culture) generate superior strategic outcomes. However, external constraints, such as regulatory requirements, ethical standards, competitive context, shape which capabilities matter most and how aggressively organisations can deploy them.

Healthcare companies demonstrated mature reconfiguring capability and cautious risk governance, reflecting regulatory constraints that moderate implementation pace but strengthen safety assurance. Technology companies excelled at rapid seizing but initially struggled with reconfiguring, learning through experience that workforce adoption requires alignment between technology deployment and employment security messaging. This pattern reveals the central insight: technological sophistication alone is insufficient; strategic advantage requires that sensing, seizing, and reconfiguring capabilities operate in concert.

AI's benefits are real but incomplete. Persistent limitations constrain where and how AI can responsibly operate. Data quality and bias remain ongoing challenges because incomplete or skewed training data produce flawed decisions in financial lending, healthcare diagnostics, and pricing algorithms. Organisations have invested in bias mitigation strategies, yet addressing this problem remains an active management concern. Algorithmic transparency remains essential, particularly in regulated sectors where organisations and regulators must understand and validate AI recommendations before deployment. Black-box models create justified hesitation among users, which has prompted companies to prioritise explainable AI techniques and to implement mandatory human review processes for critical decisions.

Change management difficulties are implementation issues, not fundamental technological problems. When AI adoption aligns poorly with employment security communication, organisations experience adoption resistance and productivity disruption, as seen with Google's internal rollout. However, well-managed implementation (Microsoft's transparent deployment with extensive training) demonstrates this barrier is surmountable through organisational vigilance.

Resource requirements present a structural constraint. Effective AI deployment demands significant upfront investment in infrastructure and specialised talent. Smaller organisations and constrained business units may not achieve equivalent gains, creating ecosystem inequities. Talent scarcity, particularly in data science and AI expertise, remains a practical limitation even for large firms.

Ethical and social implications represent the most consequential constraint. Over-reliance on AI without human judgment is unacceptable in life-and-death contexts like healthcare, where Pfizer and J&J maintained reserved deployment approaches. This disciplined restraint reflects mature understanding that AI augments rather than replaces human expertise in domains where judgment, contextual understanding, and ethical accountability are irreplaceable.

These limitations underscore that AI is a powerful tool but not a solution to all decision-making challenges. Strategic leadership must blend AI insights with human judgment, maintain continuous oversight, and recognise that sustainable competitive advantage emerges from responsible deployment within ethical and regulatory boundaries. For governments, the imperative is facilitating AI adoption while safeguarding public interest through clear governance frameworks. Organisations that address the recommendations outlined in this report: aligning resources, capabilities, and constraints, will be better equipped to navigate AI's risks and capture its genuine strategic benefits in service to customers, employees, and society.

Although the findings of current study provide a comprehensive overview, it also has some limitations. This study's analysis relies entirely on secondary sources including published case studies, industry reports, and academic literature. Direct access to organisational data, internal decision-making processes, and proprietary AI systems was unavailable. This limitation means the findings reflect publicly disclosed information and may not capture complete organisational experiences or proprietary innovations not yet in the public domain. Future research with primary data collection would provide deeper insight into organisational decision-making dynamics, failed AI implementations, and the actual mechanisms through which AI influences strategic choices.

This study suggests several avenues for future investigation. First, longitudinal studies tracking organisations' AI adoption over multiple years would reveal whether short-term productivity gains translate into sustained competitive advantage or whether early benefits plateau as rivals adopt similar technologies. Second, primary research with organisational leaders, data scientists, and affected employees would illuminate how decision-making changes when AI tools are deployed, thus capturing the organisational reality beyond published success stories. Third, cross-industry comparative studies beyond technology and healthcare could establish whether the sensing-seizing-reconfiguring framework applies consistently across financial services, manufacturing, retail, and other sectors, and whether external constraints differ systematically. Fourth, research examining AI implementation failures (projects that did not deliver promised value or required substantial modification) would balance the current literature's emphasis on successful deployments. Finally, longitudinal studies of workforce transitions and long-term employment outcomes would provide evidence for policymakers designing reskilling programs and labor market policies, moving beyond current speculative assessments of AI's employment impact. This future research would strengthen the evidence base for organiations and governments making strategic choices about AI adoption and governance.

REFERENCES

Abu Issa Gazi, M., Mohd Faizal Yusof, M., & Aminul Islam, M. (2024). Analyzing the impact of employee job satisfaction on their job behavior in the industrial setting. *Journal of Open Innovation, 10*(4), 100–427. DOI: 10.1016/j.joitmc.2024.100427

Agrawal, A., Gans, J., & Goldfarb, A. (2018). Prediction, judgment, and complexity: a theory of decision-making and artificial intelligence. In *The economics of artificial intelligence: An agenda* (pp. 89–110). University of Chicago Press.

Aleessawi, N., & Djaghrouri, L. (2025). Artificial Intelligence in Decision-Making: Literature Review. *Journal of Association of Arab Universities for Research of Higher Education, 45*(1), 263–278. https://doi.org/10.36024/1248-045-001-013

Ali, M., Khan, T., Khattak, M., & Şener, İ. (2024). Synergizing AI and business: Maximizing innovation, creativity, decision precision, and operational efficiency in high-tech enterprises. *Journal of Open Innovation, 10*(3), 100–352. DOI: 10.1016/j.joitmc.2024.100352

Bailey, M. (2024). Inside Google Marketing: How we're using AI. https://business.google.com/aunz/think/ai-excellence/how-google-uses-ai-marketing-tools- and-experiements/

Barney, J. B. (1991). Firm resources and sustained competitive advantage. *Journal of Management, 17*(1), 99–120. DOI: 10.1177/014920639101700108

Barocas, S., & Selbst, A. (2016). Big data's disparate impact. *SSRN, 104*(3), 671–732. DOI: 10.2139/ssrn.2477899

Baxter, P., & Jack, S. (2010). Qualitative Case Study Methodology: Study Design and Implementation for Novice Researchers. *The Qualitative Report.*

Bóna, P., & Lippert, R. (2020). The effect of hard and soft factors on the success of a company. *Problems of Management in the 21st century*, 10(1), 6-17. https://www.proquest.com/publiccontent/docview/2343771276

Brown, T., Mann, B., Ryder, N., Subbiah, M., Kaplan, J. D., Dhariwal, P., & Amodei, D. (2020). Language models are few-shot learners. *Advances in Neural Information Processing Systems, 33*, 1877–1901. DOI: 10.48550/arXiv.2005.14165

Brynjolfsson, E., & McAfee, A. (2017). The business of artificial intelligence. *Harvard Business Review, 95*(7), 3–11.

Capgemini. (2020). Johnson & Johnson transforms its demand planning and external manufacturing processes. https://www.capgemini.com/news/client-stories/johnson-and-johnson-transforms-its-demand-planning-and-external-manufacturing-processes/

Csaszar, F., Ketkar, H., & Kim, H. (2024). Artificial Intelligence and Strategic Decision-Making: Evidence from Entrepreneurs and Investors. *Strategy Science*, *9*(4), 322–345. Advance online publication. DOI: 10.1287/stsc.2024.0190

Davis, F. D. (1989). Technology acceptance model: TAM. Al-Suqri, MN, Al-Aufi, AS. *Information Seeking Behavior and Technology Adoption*, *205*(219), 5.

Deloitte. (2024). Generative AI and the future of work. https://www2.deloitte.com/content/dam/Deloitte/us/Documents/consulting/us-ai-institute- generative-ai-and-the-future-of-work.pdf

Dungan, R. (2024, August 30). Google pioneers AI leadership meetings, but staff aren't happy. HR Grapevine. https://www.hrgrapevine.com/us/content/article/2024-08-30-google-pioneers- ai-leadership-meetings

Dwivedi, Y. K., Kshetri, N., Hughes, L., Slade, E. L., Jeyaraj, A., Kar, A. K., Baabdullah, A. M., Koohang, A., Raghavan, V., Ahuja, M., Albanna, H., Albashrawi, M. A., Al-Busaidi, A. S., Balakrishnan, J., Barlette, Y., Basu, S., Bose, I., Brooks, L., Buhalis, D., & Wright, R. (2023). "So what if ChatGPT wrote it?" Multidisciplinary perspectives on opportunities, challenges and implications of generative conversational AI for research, practice and policy. *International Journal of Information Management*, *71*(102642), 102642. Advance online publication. DOI: 10.1016/j.ijinfomgt.2023.102642

Eisenhardt, K., & Graebner, M. (2007). Theory building from cases: Opportunities and challenges. *Academy of Management Journal*, *50*(1), 25–32. DOI: 10.5465/amj.2007.24160888

Eisenhardt, K. M., & Graebner, M. E. (2007). Theory building from cases: Opportunities and challenges. *Academy of Management Journal*, *50*(1), 25–32. DOI: 10.5465/amj.2007.24160888

European Commission. (2024). AI Act: Regulatory framework for artificial intelligence. Retrieved from https://commission.europa.eu/index_en

Filippucci, F., Gal, P., Jona-Lasinio, C., Leandro, A., & Nicoletti, G. (2024). The impact of artificial intelligence on productivity, distribution and growth: Key mechanisms, initial evidence and policy challenges. *OECD Artificial Intelligence Papers*. Financial Stability Board (FSB). (2017). Artificial intelligence and machine learning in financial services. Retrieved from https://www.fsb.org/DOI: 10.1787/dee339a8-en

Floridi, L., Cowls, J., Beltrametti, M., Chatila, R., Chazerand, P., Dignum, V., Luetge, C., Madelin, R., Pagallo, U., Rossi, F., Schafer, B., Valcke, P., & Vayena, E. (2018). AI4People—An ethical framework for a good AI society: Opportunities, risks, principles, and recommendations. *Minds and Machines*, *28*(4), 689–707. DOI: 10.1007/s11023-018-9482-5 PMID: 30930541

Google. (2024). YouTube transparency report. https://transparencyreport.google .com/youtube- policy/removals?hl=en

Google. (2025). How Google anonymizes data. https://policies.google.com/technologies/anonymization?hl=en-US

Guha, R., & Manyika, J. (2023, September 13). Data Commons: AI-powered data access at scale. Google. https://blog.google/technology/ai/google-data-commons-ai

Insights, R. E. S. O. (2024). Revolutionizing pharma: AI integration across the value chain. https://resoinsights.com/insight/revolutionizing-pharma-ai-integration -across-the-value- chain/

International Organization for Standardization (ISO). (2023). Information technology — Artificial intelligence — Management system (ISO/IEC 27001:2023). Retrieved from https://www.iso.org/standard/42001

IQVIA. (2024). Improving Predictive Precision of Oncology Patient Modeling with Enriched Data and Integrated Field Alerts. https://www.iqvia.com/library/case-studies/client-case-study- improving-predictive-precision-of-oncology-patient-modeling-with-enriched-data

Jakkal, V. (2024). How AI is Transforming Cybersecurity: Tackling the surge in Cyber Threats. Microsoft. https://news.microsoft.com/source/canada/features/ai/how-ai-is-transforming- cybersecurity-tackling-the-surge-in-cyber-threats/

Ji, Z., Lee, N., Frieske, R., Yu, T., Su, D., Xu, Y., Ishii, E., Bang, Y. J., Madotto, A., & Fung, P. (2023). Survey of hallucination in natural language generation. *ACM Computing Surveys*, *55*(12), 1–38. DOI: 10.1145/3571730

Johnson & Johnson. (2024, June 20). MedTech announces the Polyphonic™ digital ecosystem for a more connected surgical experience. https://www.jnj.com/media-center/press- releases/johnson-johnson-medtech-announces-the-polyphonic-digital-ecosystem-for-a-more- connected-surgical-experience

Kalogiannidis, S., Kalfas, D., Papaevangelou, O., Giannarakis, G., & Chatzitheodoridis, F. (2024). The Role of Artificial Intelligence Technology in Predictive Risk Assessment for Business Continuity: A Case Study of Greece. *Risks*, *12*(2), 19. DOI: 10.3390/risks12020019

Kellington, A. (2025). Elevating internal communications at Microsoft with AI. Microsoft. https://www.microsoft.com/insidetrack/blog/elevating-internal-communications-at-microsoft- with-ai/

Kiciman, E., Dillon, E., Edge, D., Foster, A., Hilmkil, A., Jennings, J., Ma, C., Ness, R. O., Pawlowski, N., Sharma, A., & Zhang, C. (2022, December). A causal AI suite for decision-making. NeurIPS 2022 Workshop on Causality for Real-world Impact. https://www.microsoft.com/en-us/research/publication/a-causal-ai-suite-for-decision- making/

Kumaran, N. (2023). New Gmail protections for a safer, less spammy inbox. Google. https://blog.google/products/gmail/gmail-security-authentication-spam-protection/

Langley, H. (2024, February 15). Google's Goose AI model shows how internal tools can change the AI race. Business Insider. https://www.businessinsider.com/google-goose-ai-model- language-ai-coding-2024-2

Langley, H. (2024). Layoffs, pivots, and the ongoing race for AI supremacy inside Google's tumultuous 2023. Business Insider. https://www.businessinsider.com/layoffs-pivots-race-for- ai-supremacy-inside-google-tumultuous-2023-2023-12

Magazine, C. D. O. (2025, January 18). How AI is shaping career pathways for J&J's 150,000-strong workforce. https://www.cdomagazine.tech/talent-development/deep-dive-how-ai-is-shaping- career-pathways-for-jjs-150000-strong-workforce

Marciano, J. (2024). How Microsoft is using AI to win at forecasts and help their FP&A team. Datarails. https://www.datarails.com/daniel-sousa-lennox-how-microsoft-is-using-ai-to-win- at-forecasts-and-help-their-fpa-team/

McKinsey & Company. (2021). The state of AI in 2021. https://www.mckinsey.com/~/media/McKinsey/Business%20Functions/McKinsey%20Analytic s/Our%20Insights/Global%20survey%20The%20state%20of%20AI%20in%202021/Global-survey-The-state-of-AI-in-2021.pdf

Microsoft. (2020). Transforming cash flow data with intelligent analytics. https://www.microsoft.com/insidetrack/blog/microsoft-transforms-cash-flow-data-into-an - intelligent-analytics-platform/

Microsoft. (2023). 2023 Work Trend Index Annual Report. https://www.microsoft .com/en- us/worklab/work-trend-index/will-ai-fix-work

Microsoft. (2023). 2023 Work Trend Index Report. https://assets- c4akfrf5b4d3f4b7 .z01.azurefd.net/assets/2023/09/e3227681-b882-4050-b201-a631431ad2a5-WTI_Will_AI_Fix_Work_060723.pdf

Microsoft. (2024). 2024 Work Trend Index Annual Report. https://www.microsoft .com/en- us/worklab/work-trend-index/ai-at-work-is-here-now-comes-the-hard-part

O'Connor, J. (2021). Building greater transparency and accountability with the Violative View Rate. Google. https://blog.google/intl/en-in/company-news/technology/building-greater- transparency-and/

Obermeyer, Z., Powers, B., Vogeli, C., & Mullainathan, S. (2019). Dissecting racial bias in an algorithm used to manage the health of populations. *Science*, *366*(6464), 447–453. DOI: 10.1126/science.aax2342 PMID: 31649194

OECD. (2019). Recommendation of the Council on Artificial Intelligence. OECD Legal Instruments. Retrieved from file:///Users/141420/Downloads/OECD-LEGAL-0449-en.pdf

Okyere, E. (2021). How DeepMind's AI framework made Google Energy Efficient. Nural Research. https://www.nural.cc/deepmind-ai-framework/

Organisation for Economic Co-operation and Development (OECD). (2019). Artificial intelligence in society. OECD Publishing. https://www.oecd.org/en/publications/artificial-intelligence-in-society_eedfee77-en.html

Our World in Data. (2025). Annual global corporate investment in artificial intelligence, by type. https://ourworldindata.org/grapher/corporate-investment-in-artificial -intelligence-by-type

Pfizer. (2020, December 9). Statement on EMA cyberattack. https://www.pfizer .com/news/announcements/statement-ema-cyberattack

Pfizer. (2025). Artificial intelligence: On a mission to make clinical drug development faster and smarter. https://www.pfizer.com/news/articles/artificial_intelligence_on _a_mission_to_make_clinical_d rug_development_faster_and_smarter

Rivero, O. (2025). The role of artificial intelligence in strategic decision-making: Transforming managerial strategies in the digital age. *European Journal of Studies in Management and Business*, *33*, 33–42. DOI: 10.32038/mbrq.2025.33.03

Rudin, C. (2019). Stop explaining black box machine learning models for high stakes decisions and use interpretable models instead. *Nature Machine Intelligence*, *1*(5), 206–215. DOI: 10.1038/s42256-019-0048-x PMID: 35603010

Simon, H. A. (1955). A behavioral model of rational choice. *The Quarterly Journal of Economics*, *69*(1), 99–118. DOI: 10.2307/1884852

Statista. (2023). Artificial intelligence (AI) adoption worldwide 2023, by industry and function. https://www.statista.com/statistics/1464584/ai-adoption-worldwide-industry-function/

Stevens Institute of Technology. (2025). Corporate education for Pfizer employees. https://www.stevens.edu/page-branding-landing/corporate-education-for-pfizer-employees

Tatineni, S. (2019). Ethical considerations in AI and data science: Bias, fairness, and accountability. *International Journal of Information Technology and Management Information Systems*, *10*(1), 11–20.

Teece, D. J. (2007). Explicating dynamic capabilities: The nature and microfoundations of (sustainable) enterprise performance. *Strategic Management Journal*, *28*(13), 1319–1350. DOI: 10.1002/smj.640

U.S. Food and Drug Administration (FDA). (2021). Guidance for industry: Software as a medical device (SaMD). Retrieved from https://www.fda.gov/

Van Noorden, R., & Perkel, J. M. (2023). AI and science: What 1,600 researchers think. *Nature*, *621*(7980), 672–675. DOI: 10.1038/d41586-023-02980-0 PMID: 37758894

Wiles, J. (2022). 2022 Gartner Hype Cycle. Gartner. https://www.gartner.com/en/articles/what- s-new-in-artificial-intelligence-from-the-2022-gartner-hype-cycle

Zheng, A., Zhang, J., Wang, R., & Poutievski, L. (2023). *How Traffic Analytics Shapes Traffic Engineering, Topology Engineering, and Capacity Planning of Jupiter. In 2023 Optical Fiber Communications Conference and Exhibition*. OFC., DOI: 10.1364/OFC.2023.W3A.2

Chapter 8

A Hybrid Decision-Making Program for Optimal Robot Selection Integrating Fuzzy Logic, Genetic Algorithms, and Reliability-Aware MCDM Under Industry 4

Nazim Ali Khan
https://orcid.org/0009-0001-8166-564X
Deenbandhu Chhotu Ram University of Science and Technology, India

Ajay Kumar
https://orcid.org/0009-0009-3410-9294
Department of Mechanical Engineering, Deenbandhu Chhotu Ram University of Science and Technology, India

Naseem Rao
https://orcid.org/0009-0009-3410-9294
Department of Electronics and Communication, Jamia Hamdard, New Delhi, India

ABSTRACT

Productivity and performance improvement in Industry 4.0 revolves around optimal robotics selection. Relying on traditional robot selection approaches often results in

DOI: 10.4018/979-8-3373-5702-7.ch008

improperly defined criteria and ignored benchmarks, such as a robot's reliability. This work solves the problem with a new approach by creating a Hybrid Decision Model Process (HDMP), which fuses together Fuzzy Logic, Genetic Algorithms, and principles of Reliability Engineering with multiple Multicriteria Decision Making (MCDM) techniques like TOPSIS, COPRAS, and ARAS. The HDMP has taken into account the essentials of a candidate's selection criteria, such as payload capacity, reach, accuracy, operational speed, power usage, price, and reliability, which includes MTBF and other failure metrics. Consistent and accurate results were obtained using eight MCDM tools along with Spearman's rank correlation coefficient SRC for model validation. This approach offers an alternative for enhancement in the robot selection process by focusing on productivity, efficiency, and operational dependability throughout the framework of Industry 4.0.

I. INTRODUCTION

The selection of robots is central to optimum productivity, dependability, and performance efficiency in Industry 4.0. Often, conventional approaches ignore other equally important criteria, which presents a challenge to achieving reliable decisions. To fill this gap, this work explains the design of an HDMP that utilises Fuzzy logic (Samsuria et al., 2025), Genetic algorithms, and reliability engineering in conjunction with MCDM tools (Khan et al., 2024). The hybrid methods allow evaluators to have a wide-ranging choice when it comes to assessing robotic options based on speed, accuracy, cost, scope, and reliability.

Incorporation of reliability metrics, in addition to performance parameters, guarantees that the HDMP captures a multi-faceted evaluation scope. Validation of the HDMP's efficacy is done with cross-validation employing the MCDM tools in conjunction with Spearman's rank correlation coefficient to ensure real-world practicality. This form of validation employs a range of MCDM tools such as MOORA, MULTIMOORA, WASPAS, WSM, WPM, COPRAS, ARAS, and even TOPSIS for the purpose of cross-validation (Goswami et al., 2021). The calculation of Spearman rank correlation coefficients assists in determining the trustworthiness and dependability of scores produced by the different approaches applied in the HDMP's operational framework. Equipped with numerous decision-making techniques, the HDMP enables users to construct comprehensive models for the evaluation and selection of robots that most closely align with their criteria (Wang et al., 2018). Such in-depth evaluations provide maximum confidence that the selected robots shall exhibit optimum performance for the given tasks and that they are indeed the most fitting robots for the specified applications, considering all factors. With the development of the HDMP, it is now easier to see how much further we are advancing into Industry 4.0 (Vohra et al., 2024). The efficiency, productivity, and overall performance of operational robots in service and manufacturing industries are drastically improved with such systematic HDMP robot selection.

An organised methodology enables decision-makers to select choices in line with Industry 4.0 objectives, which advance industrial practices toward improved efficiency and sustainability.

Figure 1. Hybrid Optimisation Method

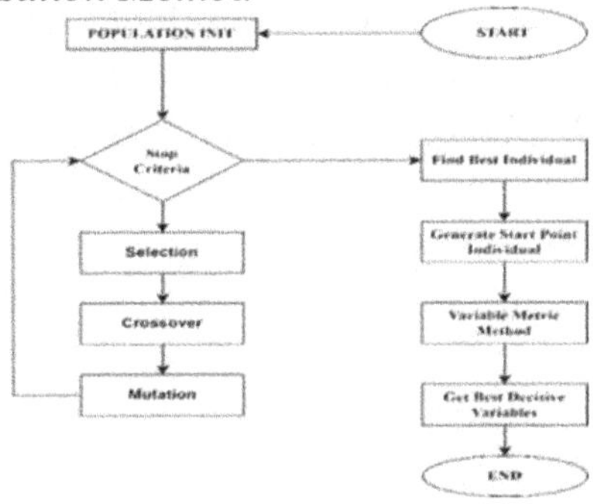

The addition of other MCDM methods like the Analytic Hierarchy Process (AHP) and the Preferred Sorting Method Based on Similarity of Model Solutions (TOPSIS) also enhances the HDMP framework (Kumar et al., 2019). These approaches improve the efficiency and the total effectiveness of the decision-making processes by ensuring that decisions are made after considering all pertinent factors. With the integration of the approaches, the HDMP enables the decision makers to be more versatile in addressing intricate decision problems. The contributions of this paper can be summarised as follows:

- Designing an HDMP framework that uses Fuzzy logic and Genetic algorithms along with multiple MCDM methods.
- A composite evaluation framework that integrates metrics of payload capacity, reach, accuracy, speed, power consumption, and cost.
- Proving the usefulness of HDMP with the employment of eight MCDM tools and Spearman's rank correlation coefficient.
- As a major development towards Industry 4.0, by improving operational performance and productivity of robot selection through advanced selection criteria.
- Adding the use of AHP and TOPSIS to enrich decision-making adaptability and resilience for supple multi-criteria decision-making complexity.

2. LITERATURE REVIEW AND MOTIVATION

Studies about robotic choice strategies show that correct decision-making requires consideration of multiple factors when selecting information. Traditional models are often used to solve today's complex manufacturing and supply chain problems, including the weighted product method (WPM), the AHP, and the weighted sum method (WSM) (Goswami & Behera, 2021). These methods provide a systematic approach to decision-making by weighing and analysing multiple criteria where multiple factors must be balanced. Recent developments from 2020 to 2024 have focused on improving these traditional methods by combining additive ratio estimation (ARAS) and new MCDM methods through rigorous mathematical modelling a involving methods such as complex proportional assessment (COPRAS) and technique of preference by similarity, so of ideal solutions providing sophisticated and robust structures for decision making (Stojčić et al., 2019). These methods have been beneficial in improving the accuracy and performance of the choice mechanisms. Additionally, researchers from 2020 to 2024 were exploring the use of hybrid fashions that integrate extremely good MCDM tactics to leverage their strengths and mitigate their weaknesses. For example, the combination of AHP with TOPSIS, which uses gold-preferred solutions identification, has led to greater accuracy and reliability in preference making integrated with a more complex system (Tran et al., 2024). These hybrid approaches were implemented in various fields, including robotic paw selection, which showed significant increases in reliability, accuracy, and overall performance of the robotic systems (Yoon, 1987). Furthermore, some researchers during this period advocated using MCDM techniques with fuzzy logic to address the uncertainty and lack of specificity associated with real-world decision-making situations (Pytel, 2025). This approach has managed to offer better flexibility and strength in the frameworks designed for providing decisions, thus enhancing the rigor of the entire selection process (Chen et al., 2023).

2.1 Reliability Engineering in Robot Selection

Recent advancements in reliability engineering highlight its importance in robotic systems. Organisational stability during extended operations relies heavily on three essential metrics, which include MTBF alongside failure rate analysis and predictive maintenance. Integrating these metrics into MCDM frameworks enhances decision-making accuracy by accounting for system robustness and lifecycle costs.

2.2 Hypothesis Development:

Main Hypothesis:

- **Hypothesis:** Integrating reliability metrics with COPRAS, ARAS, and TOPSIS in HDMP will yield more accurate and reliable robot prioritisation than traditional methods.
- **Explanation:** This important speculation asserts that the amalgamation of COPRAS, ARAS, and TOPSIS within a unmarried software platform will substantially improve the method of choosing the most suitable robot. COPRAS excels in handling selection-making complexities under uncertainty, ARAS evaluates deviations from perfect options, and TOPSIS aids in figuring out ultimate compromise solutions. By integrating those methodologies, the software program is predicted to offer a complete technique to robot choice, resulting in more accurate and reliable choice outcomes. The integration of these techniques is expected to leverage their respective strengths, leading to a step forward in selection-making for the choice of robots for various packages.

Secondary Hypothesis:

- **Design:** HDMP will outperform traditional methods in managing complex decision scenarios by incorporating reliability parameters (e.g., MTBF, maintenance costs) alongside performance criteria.
- **Implications:** This second perspective suggests that hybrid software will outperform traditional methods by better managing today's complex decision situations. The software provides comprehensive decision support by considering parameters such as payload, scope, speed, accuracy, cost, and ease of integration. Its ability to gather values and provide robust decision support enables it to effectively manage complex decision situations. In addition, the software's ability to integrate multiple decision-making processes provides flexibility and suitability for different applications. Therefore, hybrid decision software is expected to provide an ideal solution to guide complex decision points in robotic choice systems. This speculation serves as a muse for enhancing and validating hybrid decision-making methods for the most desirable robotic selection.

3. RESEARCH METHODOLOGY

3.1 Data Collection:

- **Parameter Choice:** Added MTBF, failure rates, and maintenance requirements to existing criteria (payload, reach, etc.). The selection of parameters consisting of payload capacity, reach, accuracy, speed, power consumption, and cost is crucial as they directly have an impact on the overall performance of the robot in meeting industry necessities. These parameters are chosen based on their relevance and significance to the software and are essential for ensuring the robot's effectiveness.
- **Data Sources:** To ensure the accuracy and reliability of the accumulated statistics, data on those parameters are sourced from diverse channels. This consists of manufacturer specifications, industry surveys, and remarks from personnel who interact with or function in a manner comparable to robots. By collecting information from multiple resources, the dataset will become complete and consultant, imparting a strong basis for decision-making.

3.2 Methodology Development:

- **Integration of Reliability Metrics:** Enhanced MCDM methods (TOPSIS, COPRAS, ARAS) to include reliability-weighted evaluations. The HDMP comprises an evolution of three excellent MCDM tools, namely Technique for Order of Preference by means of Similarity to Ideal Solution (TOPSIS), Complex Proportional Assessment (COPRAS), and Additive Ratio Assessment (ARAS). Each methodology extends specific analytical procedures to handle desire-making factors that enhance the total performance of the HDMP framework.
- **Algorithm Design:** The HDMP framework presents native algorithms that merge into its core infrastructure for individual MCDM techniques. Continuous optimisation of these algorithms makes them effective for processing data, which maintains consensus between decision-making approaches. The HDMP utilises superior algorithms to process complicated data, which results in practical knowledge.

3.3 Parameter Weighting:

- **CRITIC Method:** The CRITIC method lets you determine parameter weights in this process. The evaluation of parameter relations through CRITIC helps establish which parameters exert the most influence on the robot's performance. The CRITIC method considers inter-parameter correlations to produce weights that identify genuine parameter importance correctly.

- **Weight Determination:** Criterion weights are objectively determined through intercriteria correlation evaluation, which considers the relationships among parameters. This evaluation results in a balanced weighting tool that, as it ought to be, represents the contribution of every parameter to the robotics' common performance. By assigning appropriate weights, the HDMP ensures that preference-making is based totally on an entire information of the relative importance of every criterion.

3.4 Decision-Making Process:

- **Data Normalisation: Raw data for each parameter are normalised to a common** scale to facilitate honest assessment and evaluation. This technique standardises the facts, allowing parameters with tremendous devices and scales to be treated uniformly at some level within the decision-making. Normalisation guarantees that each parameter contributes proportionally to the general assessment of robotic alternatives.
- **MCDM Calculation:** The HDMP calculates every day well-known ordinary performance scores for each robotic choice using the integrated MCDM techniques. These rankings are derived from the normalised statistics and constitute the suitability of every preference primarily based on the selected standards. By thinking about more than one standard simultaneously, the HDMP provides a holistic evaluation of robot options.
- **Ranking Generation:** Based on the calculated performance rankings, the HDMP generates ratings of robot alternatives. These rankings offer decision-makers smooth insights into the relative overall performance of each desire, allowing knowledgeable decision-making. By imparting alternatives in order of choice, the HDMP enables the selection of the most suitable robot for the given software.

3.5 Validation and Comparison:

- **Cross-Validation:** The proposed ratings generated via the HDMP go through rigorous validation through the go with the flow-validation using 8 solo MCDM strategies. This validation method guarantees the reliability and robustness of the decision method by assessing its consistency and accuracy across unique methodologies. Cross-validation complements self-assurance in the ratings produced with the use of the HDMP and validates its effectiveness in generating dependable outcomes.

- **Spearman Rank Correlation:** The Spearman rank correlation coefficient is calculated to evaluate the degree of agreement amongst ratings generated using one-of-a-kind strategies. This statistical measure quantifies the power and course of the relationship among scores, providing insights into the consistency and reliability of the HDMP's effects. A high correlation coefficient indicates observed alignment amongst ratings, validating the effectiveness of the HDMP in generating ordinary and dependable results.

All the above-mentioned evaluation methods provide a systematic and rigorous approach to robot selection. Utilising the strengths of MCDM methods and combining them with rigorous parameter-weighting techniques, the HDMP provides selectors with a reliable tool to optimise robotic choices, especially based on more than one standard HDMP production through validation and evaluation using traditional methods. Demonstrates its efforts to improve operational efficiency and performance in service industries. As mentioned in Table 1 above.

Table 1. Overview of Research Methodology

METHODOLOGY STAGE	DESCRIPTION
Data Collection	Parameters, including payload functionality, reach, pace, accuracy, price, and ease of integration, are diagnosed, and statistics are accrued.
Methodology Development	Within the framework of the HDMP, the incorporation of the TOPSIS, COPRAS, and ARAS methodologies is considered.
Parameter Weighting	Application of the CRITIC technique to assign weights to parameters primarily based on their relative significance.
Decision-Making Process	Data normalisation, MCDM estimation, and category generation using HDMP.
Validation and Comparison	Cross-validation using eight solo MCDM techniques and calculation of the Spearman rank correlation coefficient.

4. ADVANCED ALGORITHMIC APPROACHES FOR ROBOT PERFORMANCE EVALUATION

4. 1 Fuzzy Algorithm for Evaluating Robot Performance

The fuzzy algorithm 1 involves defining fuzzy membership functions for each input criterion, which includes setting levels for low, medium, and high values and calculating membership values for each input. In the fuzzy inference function, a dictionary for fuzzy scores is initialised. A fuzzy inference system uses calculated membership values for specifications to determine the performance score of each robot by executing rules. A performance score combines the individual scores before storage into the fuzzy scores dictionary and returns.

Algorithm 1: Fuzzy Algorithm for Evaluating Robot Performance

Start ()

1. Function: Define Fuzzy Membership Functions ():
2. For each input criterion:
    a. The system requires the definition of membership functions to represent low, medium, and high levels.
    b. Apply membership functions to determine the member value of each incoming input value.
3. Function FuzzyInference(inputs):
4. Initialise fuzzy_scores dictionary.
5. For each robot in the dataset:
    a. Calculate membership values for each specification of the robot.
    b. Apply fuzzy inference rules to determine the performance score of the robot.
    c. Combine the scores for all criteria to obtain the overall performance score.
    d. Store the performance score for the robot in the fuzzy scores dictionary.
6. Return fuzzy_scores.

Stop

4.2 Implementation of a Genetic Algorithm with Fuzzy Logic for Fitness Evaluation

The Genetic Algorithm begins by initialising the population size, number of generations, and weights for criteria, then generating an initial population of chromosomes (Maroua et al., 2025). Each chromosome's fitness is evaluated using fuzzy logic, and the fitness scores are stored. Over the specified number of generations, the algorithm evaluates the fitness of each chromosome, selects the fittest ones as parents, and uses crossover and mutation to produce offspring. These offspring replace the least fit individuals in the current population. This process continues until the specified number of generations is reached, at which point the algorithm returns the fittest chromosome.

Algorithm 2: Implementation of a Genetic Algorithm with Fuzzy Logic for Fitness Evaluation

Start ()

1. Function Initialise Genetic Algorithm ():
   a. Initialise population size, number of generations, and weights for criteria.
   b. Generate an initial population of chromosomes.
2. Function Evaluate Fitness(population):
   a. For each chromosome in the population:
   b. Fitness evaluation needs fuzzy logic algorithms to process the chromosome to get Intent.
   c. Save the fitness score obtained from the chromosome.
3. Function Genetic Algorithm (population, generations):
   a. Perform the operations for the designated number of generations.
   b. Apply fuzzy logic-based evaluation to measure the fitness level of all chromosomes present in the population.
   c. The algorithm will choose suitable chromosomes from the population to become parents for the following generation.
   d. The process includes offspring generation through the application of crossover with mutation functionality.
   e. The least fit individuals from the current population will be replaced by offspring that result from these two groups.
4. Present the most suitable chromosome once the iteration limit reaches its specified value.

Stop

5. PROPOSED HDMP MODEL

5.1 Hybrid Decision Model Process for Robot Selection

A novel approach in optimised robot selection for dynamic cloud environments uses the collective advantages of fuzzy logic and genetic algorithms through the HDMP. The framework combines two separate methods and delivers them as a unified, powerful solution. This system focuses on improving robotic selection within multitasking operations through effective solutions for dynamic, complex events. The HDMP model merges the fuzzy algorithm with the genetic algorithm to obtain their combined superiority. The combination of these two systems produces precise decision outcomes that boost the selection process results for robotic systems. The HDMP develops a reliable robot selection system for dynamic cloud conditions through the unification of fuzzy logic and genetic algorithms. The fuzzy logic judgments about robot performance effectiveness function alongside genetic algorithm optimisation, which improves selection decision-making through multiple iterative steps. The combined model strengthens both decision accuracy and selects robots more effectively and resiliently within the process. The HDMP model stands as a major development for robotic selection because it integrates various technologies to solve complex problems during real-world robot applications.

Algorithm 3: Proposed HDMP

1. FUNCTION obtain_route_set(robot_topology, user_location)
    a. route_set = empty_set
    b. matching_robots = empty_list
    c. FOR EACH robot IN robot_topology DO
        i. IF robot.location EQUALS user_location THEN
            1. ADD robot TO matching_robots
        ii. END IF
    d. END FOR
    e. FOR EACH robot IN matching_robots DO
        i. routes_ros = obtain_routes(robot.location, 'cloud', 'robot_topology')
        ii. FOR EACH route IN routes_ros DO
            1. ADD route TO route_set
        iii. END FOR
    f. END FOR
    g. RETURN route_set
2. END FUNCTION
3. FUNCTION obtain_routes(start_location, end_location, robot_topology)
    a. routes = empty_list
    b. FOR EACH robot IN robot_topology DO
    c. IF robot.location EQUALS start_location AND end_location EQUALS 'cloud' THEN
    d. ADD route FROM robot.location TO 'cloud' TO routes
    e. END IF
    f. END FOR
    g. RETURN routes
4. END FUNCTION

6. RESULTS AND VALIDATION

Table 2 demonstrates TOPSIS Method Validation by showing how the TOPSIS-ARAS hybrid model generates robot alternative performance rankings when applied with the COPRAS and ARAS hybrid model.

For the TOPSIS-ARAS HYBRID MODEL:

- Alternatives are listed from 1 to 12.
- Each alternative in the CCDi column presents its closeness distance (CCDi) performance as a percentage of the total.
- In the % column, the CCDi values are presented using percentage values relative to the total measurement.
- Each alternative receives its placement through CCDi value ranking in the Rank column.

For the HYBRID MODEL OF COPRAS AND ARAS:

- The listed alternatives match the robot options through sequential numbers starting with 1, progressing to 12.
- The QUi column presents quantitative software degree values for each choice through percent-based information.
- The % column demonstrates the values from the QUi scale as a portion of the overall total amount.
- Each alternative receives its rank evaluation directly from the QUi value assessment presented in the Rank column.

These rankings create two separate evaluations of robot alternatives through performance assessments based on each hybrid model. The comparison method determines which new robots' function best according to performance criteria to assist robotic selection through well-informed choices.

6.1. TOPSIS Method Validation

Table 2. Robot ranking the use of COPRAS-ARAS and TOPSIS-ARAS hybrid fashions.

TOPSIS-ARAS HYBRID MODEL				HYBRID MODEL OF COPRAS AND ARAS			
ALTERNATIVES	**CCDi**	**%**	**Rank**	**Alternatives**	**QUi**	**%**	**Rank**
1	0.50707	50.707	5	1	0.57509	57.509	4
2	0.51463	51.463	6	2	0.54150	54.150	5
3	0.54794	54.794	8	3	0.50494	50.494	8
4	0.65795	65.795	12	4	0.33450	33.450	12
5	0.55967	55.967	10	5	0.48578	48.578	10
6	0.57810	57.810	11	6	0.44688	44.688	11

TOPSIS-ARAS HYBRID MODEL				HYBRID MODEL OF COPRAS AND ARAS			
ALTERNATIVES	CCDi	%	Rank	Alternatives	QUi	%	Rank
7	0.49873	49.873	4	7	0.52396	52.396	6
8	0.54569	54.569	7	8	0.51433	51.433	7
9	0.55207	55.207	9	9	0.48618	48.618	9
10	0.44365	44.365	3	10	0.58074	58.074	3
11	0.34017	34.017	2	11	0.62955	62.955	2
12	0.28805	28.805	1	12	0.68203	68.203	1

6.2 Validation Using the COPRAS Approach: TOPSIS Method Validation

Table 3 displays the performance results from the TOPSIS Method Validation, which assesses different robot solutions from multiple parameters. The validation process requires users to compute Positive Ideal Solution (PIS) and Negative Ideal Solution (NIS) as well as Relative Closeness Coefficient (RCCi) for each alternative. PIS represents the quality viable values for every criterion, indicating the most desirable overall performance situation, whilst NIS represents the worst viable values, indicating the least desirable performance situation. RCCi is calculated by evaluating the space of every opportunity from the PIS and NIS, with higher RCCi values indicating nearer proximity to the PIS, suggesting better performance. The rank indicates the rating of every alternative based on its RCCi values, with lower ranks representing better performance.

6.3 Explanation of the Data

TOPSIS bad idea choices the bad ideas in their good ideas, explore 12 of the options, which their Pis the best because there is an alternative. This means that alternative 12 works exceptionally well standard several types and achieves the highest correlation close to the optimal operating conditions and consequently stands out as the top candidate among the alternatives evaluated.

Method 11 follows with a PIS of 0.06195, a NIS of 0.12250, and an RCCi of 0.66415, ranking second. This algorithm performs very well, although slightly less efficiently than method 12. However, method 10, with PIS 0.08230, NIS 0.11152, and RCCi 0.57537, occupies the third place with its high RCCi value indicating a better performance hard, making it another viable candidate. This scheme illustrates the effectiveness of the TOPSIS method in distinguishing between the most efficient methods based on their proximity to the optimal solution.

Alternative 7, with a PIS of 0.09209, NIS of 0.09829, and an RCCi of 0.51629, ranks fourth, demonstrating competitive performance due to its rather high RCCi value. The relaxation of the options complies with a comparable sample, in which those with better RCCi values are ranked higher, indicating better overall performance. For instance, Alternative four, with the lowest RCCi cost of 0.32937, is located at the bottom rank (12), signifying it's farthest from the Positive Ideal Solution and for this reason the least favoured preference. This systematic ranking highlights the overall performance hierarchy among the options, presenting a clean decision-making framework and ensuring that the top-rated robots are selected based on a sturdy quantitative evaluation.

6.4 Validation Using the COPRAS Approach

Table 3. Robot ranking the use of the TOPSIS algorithm.

Possible substitutes	PIS	NIS	RCCi(topsis)	Rank
1	0.09701	0.10230	0.51327	5
2	0.08654	0.08323	0.49024	6
3	0.10345	0.07978	0.43541	8
4	0.13258	0.06511	0.32937	12
5	0.10976	0.07983	0.42107	11
6	0.09303	0.06906	0.42607	10
7	0.09209	0.09829	0.51629	4
8	0.11165	0.08644	0.43638	7
9	0.09733	0.07453	0.43367	9
10	0.08230	0.11152	0.57537	3
11	0.06195	0.12250	0.66415	2
12	0.04653	0.11233	0.70709	1

6.4.1 Validation Using ARAS Method:

To validate the ratings obtained through the COPRAS approach, a further analysis was performed using the ARAS method. The ARAS approach gives a complementary attitude with the aid of rating options primarily based on their utility ratings. This validation process presents a strong evaluation of the overall performance and software of the evaluated options.

6.4.2 COPRAS Method Ranking:

Under the COPRAS approach, options are ranked based on their current overall performance ratings relative to each other. "Ricopras" emerges because the pinnacle performer, securing the great rank, is due to its advanced average overall performance score. "Uicopras" follows intently within the back of, indicating beneficial performance. The final alternatives are ranked, therefore, based on their overall performance rankings.

6.4.3 ARAS Method Ranking:

Contrary to COPRAS, the ARAS approach ranks alternatives based on their utility rankings, in which higher scores correspond to better ranks. "Robot 12" emerges as the top-ranked opportunity, reflecting its superior application compared to different alternatives. The ranking of various options is determined via their software program rankings.

6.5 Case Study: Application of HDMP for Industrial Robot Selection

To prove the practicality of the suggested HDMP, a practical case study was carried out with the help of a representative group of industrial robots, which are commonly used in manufacturing and assembly processes. This case study aims to confirm the efficiency of the HDMP framework in choosing the most appropriate robot where there are several technical and reliability requirements that need to be considered at the same time. In the case study, there were five popular models of industrial robots of notable manufacturers, including ABBA IRB 6700, Fanuc M-20iA, KUKA KR 60, Yaskawa Motoman GP25, and Universal Robots UR10e. The robots are also common in the automobile, electronics, and general manufacturing sectors because they are very precise and flexible. The evaluation criteria were based on both the performance and reliability points of view, such as payload capacity (kg), reach (mm), repeatability (mm), power consumption (kW), cost (USD), and reliability indicators, such as Mean Time Between Failures (MTBF) and maintenance rate. All information used in this research was obtained from official manufacturer brochures, industrial tools specification brochures, and on-field surveys of engineers of the selected model who were well-versed with the specifics of the models. In the case where manufacturer-specific data on the MTBF were not available, the estimates were made based on industry averages reported in the literature on robotics reliability between 2020 and 2024.

Table 4. Comparative Specifications of Industrial and Collaborative Robot Models

Robot Model	Payload (kg)	Reach (mm)	Repeatability (mm)	Power (kW)	Cost (USD)	MTBF (hrs)	Maintenance Freq (per year)
ABB IRB 6700	150	3200	0.05	3	95,000	58,000	1.5
Fanuc M-20iA	20	1811	0.02	1.8	48,000	62,000	1
KUKA KR 60	60	2033	0.06	2.5	70,000	55,000	2
Yaskawa GP25	25	1730	0.03	2.2	52,000	60,000	1.2
UR10e	10	1300	0.05	0.9	42,000	50,000	1.8

The dataset to be used in the comparative analysis is provided in Table 4. Normalisation of the parameters was done through the CRITIC weighting method, and then processed through the HDMP framework, which blends Fuzzy Logic, Genetic Algorithm optimisation and MCDM (TOPSIS, COPRAS and ARAS) techniques.

Once this normalisation and weight computation were complete, the HDMP algorithm was implemented with the assistance of the fuzzy membership function to quantify qualitative attributes (like ease of maintenance) and the genetic algorithm optimising weight adaptation over generations. The final ranking of the robot alternatives was obtained by using the combined MCDM techniques (TOPSIS-ARAS and COPRAS-ARAS hybrids).

Table 5. Comparative Performance Rankings Obtained Through HDMP

Robot Model	Normalised Score (HDMP)	Rank
Fanuc M-20iA	0.875	1
Yaskawa GP25	0.842	2
ABB IRB 6700	0.801	3
KUKA KR 60	0.756	4
UR10e	0.693	5

Table 5, Comparative Performance Rankings Obtained Through HDMP, ranks five robot models based on a Hybrid Distance-based HDMP normalised score. The Fanuc M-20iA achieves the highest score (0.875), followed by the Yaskawa GP25 (0.842), indicating their superior overall performance based on the evaluated criteria.\

The Fanuc M-20iA was ranked number one robot mainly because of the balanced performance, such as low power consumption, high repeatability, and excellent reliability (highest MTBF with minimum maintenance). It was closely followed by the Yaskawa GP25, which provided a good balance of lower costs and coverage performance. The ABB IRB 6700 was powerful but placed third since it is expensive and consumes a lot of energy, and the UR10e was at the bottom of the list due to its low payload and range, despite the low cost. The present case study confirms that the HDMP model is an integrated and data-driven model of selecting an industrial robot that can integrate both performance and reliability dimensions. Reliability parameters like the MTBF that were included played a major role in the ultimate ranking as they demonstrated that the traditional selection techniques of either basing on cost or performance might fail to consider very important parameters of dependability. Therefore, HDMP proves to be practically useful to decision-makers in Industry 4.0 environments that require the optimisation of productivity, precision, and long-term reliability to be done together.

7. COMPARATIVE ANALYSIS

Comparing the numbers generated through the COPRAS and ARAS methods affords valuable insights into the performance and utility of the proposed options. While COPRAS emphasises the performance between alternatives, ARAS focuses on their scoring. Despite differences in style, numbers emphasise superior overall performance: “Ricopras” in Copras and “Robot 12” in ARS. This comparative analysis enables stakeholders to make informed choices by considering all operational and non-operational elements when choosing the most appropriate option for their needs. Unlike traditional methods that consider only cost and performance, HDMP incorporates reliability (MTBF, failure rate), which significantly changes ranking priorities, ensuring long-term operational dependability. The improvement or distinction among the numbers acquired from each method presents a more complete expertise of the overall performance and advantages of the selected techniques and will increase the reliability and consistency of the decision-making method, as given in Table 6

7.1 Validation Using the ARAS Method

Table 6. Robot rating using the ARAS and COPRAS strategies

COPRAS					ARAS			
Possible substitutes	Ri^{copras}	Ui^{copras}	%	Rank	Alternatives	Vi	Ui^{aras}	Rank
1	0.09050	0.83369	83.369	4	0	0.13332 $= V_0$	-	-
2	0.08538	0.78652	78.652	5	1	0.07789	0.58421	4
3	0.08022	0.73899	73.899	8	2	0.07284	0.54635	6
4	0.05338	0.49167	49.167	12	3	0.06805	0.51039	8
5	0.07750	0.71394	71.394	9	4	0.05849	0.43876	12
6	0.07084	0.65256	65.256	11	5	0.06443	0.48325	10
7	0.08276	0.76237	76.237	6	6	0.05990	0.44927	11
8	0.08203	0.75564	75.564	7	7	0.07072	0.53044	7
9	0.07729	0.71201	71.201	10	8	0.07487	0.56162	5
10	0.09173	0.84497	84.497	3	9	0.06477	0.48581	9
11	0.09980	0.91931	91.931	2	10	0.07849	0.58873	3
12	0.10856	1	100	1	11	0.08520	0.63907	2
13	0.10856		Robot 12	0.09365	0.68260	1		

7.2 Validation Process with Cross-Validation and Spearman Rank Correlation:

To ensure the robustness and effectiveness of the hybrid selection-making application, a rigorous validation technique was carried out using pass-validation and Spearman's rank correlation techniques.

7.3 Cross-Validation:

Cross-validation is a broadly used technique for assessing the performance and generalisation capability of predictive models. In our validation procedure, we hired pass-validation to partition the dataset into multiple subsets. Each subset was modified and then used iteratively as both a training set and a testing set. By comparing the overall performance of the hybrid selection-making software program for the duration of splendid subsets of the facts, we ought to investigate its consistency and reliability in predicting rankings.

7.4 Spearman Rank Correlation:

Spearman's rank correlation is a statistical score of the strength and direction of associations in ratings made with specific methods. In our validation procedure, we used Spearman's rank correlation to compare the ratings generated by the hybrid selection-making program with those from traditional methods. A very high Spearman's rank correlation coefficient indicates strong agreement with ratings made in both directions. The interval provides evidence.

7.5 Robustness and Transparency:

Our validation technique was designed to be obvious and sturdy, ensuring the credibility of the effects. By utilising move-validation and Spearman's rank correlation techniques, we addressed issues concerning the reliability and validity of the hybrid desire-making application. This rigorous technique complements confidence in the application's performance and its capability to outperform traditional methods in ranking alternatives correctly and accurately.

By using those strategies, we offer a comprehensive and obvious validation of the hybrid decision-making application, demonstrating its superiority over conventional strategies in generating reliable and steady scores of options. Given in Table 7.

4.4. Validation Using a Method Based on MULTIMOORA and MOORA

Table 7. Robot ranking using the MULTIMOORA and MOORA methods

MOORA			MOORA		
Alternatives	y_i	Rank	Alternatives	$U_i^{multimoora}$	Rank
1	0.04626	5	Robot 1	0.79880	4
2	0.06009	4	Robot 2	0.79836	5
3	0.04145	6	Robot 3	0.70231	8
4	−0.03733	12	Robot 4	0.52626	12
5	0.02456	10	Robot 5	0.65643	10
6	0.00906	11	Robot 6	0.67793	9
7	0.03439	9	Robot 7	0.73838	6

continued on following page

Table 3. Continued

MOORA			MOORA		
Alternatives	y_i	Rank	Alternatives	$U_i^{multimoora}$	Rank
8	0.03551	7	Robot 8	0.64576	11
9	0.03527	8	Robot 9	0.71167	7
10	0.06207	3	Robot 10	0.81755	3
11	0.10233	2	Robot 11	0.92745	2
12	0.11561	1	Robot 12	1.04453	1

8. CONCLUSION AND FUTURE RESEARCH AREA

The integration of robotics cluster matching and the implementation of a hybrid decision-making program represent significant advancements in robotics technology, promising transformative changes in various industries. Graph 1, entitled "multi-robotic localisation and orientation estimation using robotics organisation matching", highlights the pioneering method for progressive conversation and navigation of various robots as defined below in Figure 2. Robots use robotics organisation matching function, so their orientation can be mechanically exactly localised and calculated, enabling them to execute tasks with extraordinary accuracy in quite a few environments. These increased paths hold terrific promise for businesses that rely on complicated automatic solutions, together with manufacturing, logistics, and analytics. Furthermore, the development of a hybrid selection model (HDMP) to pick the maximum fine robots primarily based on numerous criteria is an critical thing of decision support systems for Industry 4.0 utility HDMP riding selections multi-criteria techniques (MCDM) inclusive of TOPSIS, COPRAS, ARAS This sophisticated approach, which affords choice makers with a complete framework for evaluating and deciding on the maximum appropriate robot alternatives, permits companies to make informed decisions in step with their enterprise targets and industry requirements meet.

Figure 2. Multi–Robot Localization and orientation estimation: the use of robotics cluster matching

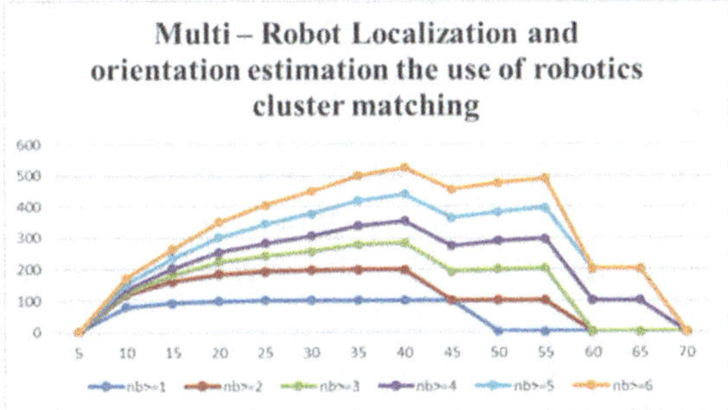

The significance of HDMP is highlighted in the heading "Global Collaborative Robots Market Share, by way of Vertical, 2021 (%)". Collaborative robots, or cobots, have become increasingly more prominent throughout verticals due to their ability to paintings with human beings and increase productivity. HDMP's capability to assess more than one parameter is well aligned with the desires of diverse verticals, and it guarantees that the selected robots are first-rate and appropriate for specific programs in every internal responsibility section.

Furthermore, the validation effects of the HDMP affirm its effectiveness and reliability in generating correct and regular robot choice rankings. Through pass-validation, the use of diverse MCDM methods, and calculating Spearman rank correlation coefficients, the robustness and compatibility of HDMP with enterprise desires are confirmed. This validation system validates the device's potential to provide reliable guidance to empower decision makers, permitting them to make informed decisions that maximise the overall performance of their corporations and force innovation. The study achieved its objective of developing a reliable hybrid model integrating fuzzy logic, GA, and reliability metrics for robot selection. Future work will focus on extending HDMP by adding sustainability indicators and testing in real manufacturing environments.

DECLARATIONS

Competing Interests: On behalf of all authors, the corresponding author states that there is no conflict of interest.

Funding Information: Not Applicable

Author contribution: [Nazim Ali Khan, Ajay Kumar, Naseem Rao] Conceptualisation, Methodology, Software, Investigation, Data curation, Writing – original draft. **[Nazim Ali Khan, Ajay Kumar, Naseem Rao]** Visualisation, Validation, Formal analysis, Writing – review & editing.

Data Availability Statement: Not Applicable

Research Involving Human and /or Animals: Not Applicable

Informed Consent: Not Applicable

REFERENCES

Chen, L., Chen, Z., Zhang, Y., Liu, Y., Osman, A. I., Farghali, M., Hua, J., Al-Fatesh, A., Ihara, I., Rooney, D. W., & Yap, P. S. (2023). Artificial intelligence-based solutions for climate change: A review. *Environmental Chemistry Letters, 21*(5), 2525–2557. DOI: 10.1007/s10311-023-01617-y

Goswami, S. S., & Behera, D. K. (2021). Evaluation of the best smartphone model in the market by integrating fuzzy-AHP and PROMETHEE decision-making approach. *Decision (Washington, D.C.), 48*(1), 71–96. DOI: 10.1007/s40622-020-00260-8

Goswami, S. S., Behera, D. K., Afzal, A., Kaladgi, A. R., Khan, S. A., Rajendran, P., Subbiah, R., & Asif, M. (2021). Analysis of a robot selection problem using two newly developed hybrid MCDM models of TOPSIS-ARAS and COPRAS-ARAS. *Symmetry, 13*(8), 1331. DOI: 10.3390/sym13081331

Khan, N. A., Kumar, A., & Rao, N. (2024). A hybrid robot selection model for efficient decisive support system using fuzzy logic and genetic algorithm. *International Journal of System Assurance Engineering and Management, 15*(6), 2120–2129. DOI: 10.1007/s13198-023-02224-4

Kumar, A., Hussain, S. A. I., & Rai, R. N. (2019). Optimization by AHP-ARAS of EDM process parameters on machining AA7050-10% B 4 C composite. In Advances in Industrial and Production Engineering: Select Proceedings of FLAME 2018 (pp. 285-296). Springer Singapore.

Maroua, B., Laid, Z., Benbouhenni, H., Elbarbary, Z. M. S., Colak, I., & Alammar, M. M. (2025). Z. M. S. Elbarbary, Ilhami Colak, and Mohammed M. Alammer. "Genetic algorithm type 2 fuzzy logic controller of microgrid system with a fractional-order technique.". *Scientific Reports*, *15*(1), 6318. DOI: 10.1038/s41598-025-90239-1

Pytel, K. (2025). Fuzzy logic applied to tunning mutation size in evolutionary algorithms. *Scientific Reports*, *15*(1), 1937. DOI: 10.1038/s41598-025-86349-5

Samsuria, E., Mohd, S. A. M., Wahab, N. A., Romdlony, M. Z., Mohamad, S. Z. A., & Buyamin, S. (2025). An improved adaptive fuzzy-genetic algorithm based on local search for integrated production and mobile robot scheduling in job-shop flexible manufacturing system. *Computers & Industrial Engineering*, *204*, 111093. DOI: 10.1016/j.cie.2025.111093

Stojčić, M., Zavadskas, E. K., Pamučar, D., Stević, Ž., & Mardani, A. (2019). Application of MCDM methods in sustainability engineering: A literature review 2008–2018. *Symmetry*, *11*(3), 350. DOI: 10.3390/sym11030350

Tran, N.-T., Trinh, V.-L., & Chung, C.-K. (2024). An Integrated Approach of Fuzzy AHP-TOPSIS for Multi-Criteria Decision-Making in Industrial Robot Selection. *Processes (Basel, Switzerland)*, *12*(8), 1723. DOI: 10.3390/pr12081723

Vohra, K., Sinha, A. K., & Anand, A. (2024). Development of industry 4.0 based technology selection index using multi criteria decision making. *Operations Research*, *58*(6), 5185–5209. DOI: 10.1051/ro/2024172

Wang, J. J., Miao, Z. H., Cui, F. B., & Liu, H. C. (2018). Robot evaluation and selection with entropy-based combination weighting and cloud TODIM approach. *Entropy (Basel, Switzerland)*, *20*(5), 349. DOI: 10.3390/e20050349

Yoon, K. (1987). A reconciliation among discrete compromise solutions. *The Journal of the Operational Research Society*, *38*(3), 277–286. DOI: 10.1057/jors.1987.44

Chapter 9

Cultural Influences on AI-Driven Decision-Making in Digital Commerce Systems

Chuhui Shao
https://orcid.org/0009-0003-3020-8558
University of Technology Sydney, Australia

Kyeong Kang
https://orcid.org/0000-0003-4252-9802
University of Technology Sydney, Australia

ABSTRACT

This chapter explores how cultural diversity shapes consumer interactions with AI-driven digital commerce systems. While AI technologies revolutionize online shopping through personalized recommendations and intelligent assistants, their effectiveness varies significantly across different cultural backgrounds. This chapter proposes a theoretical framework examining how cultural values influence trust, acceptance, and usage patterns of AI commerce systems. This research introduces the Cultural-Adaptive Decision-Making (CADM) framework to guide the development of culturally-sensitive AI commerce systems. This framework provides theoretical design principles for creating adaptive interfaces that respond to cultural preferences while maintaining system efficiency. The study offers conceptual insights for global commerce companies seeking to optimize AI implementations across multicultural markets, ultimately enhancing user experience and business performance through AI design. This work provides a new perspective and possibilities for AI-driven commerce innovation.

DOI: 10.4018/979-8-3373-5702-7.ch009

INTRODUCTION

The rapid development of artificial intelligence (AI) in digital commerce systems has created unprecedented opportunities for enhanced decision-making processes. Recent studies have demonstrated the significant impact of AI assistants in online shopping environments, with Wang et al. (2025) showing that AI streaming assistants can increase sales by 3.00% while reducing product return rates by 12.55% through intelligent information provision. Similarly, research by Wei et al. (2025) on Virtual Shopping Assistants (VSAs) reveals the complex relationship between co-creation value, experiential value, and consumer purchase intention, Wu et al. (2025) emphasize the importance of human-like attributes and technological capabilities in AI voice assistants for fostering brand loyalty through parasocial interactions.

Drawing from established cultural theories this research examines the intersection of cultural factors, cognitive biases, and human factors in AI-enhanced decision-making environments. The study addresses a significant gap in current literature by exploring how traditional cultural frameworks apply to increasingly AI-mediated engineering practices, particularly in digital commerce backgrounds where technical precision meets commercial viability (Wirtz, 2024).

This research presents a comprehensive analysis of how cultural factors moderate humans' responses to AI commerce systems, automated decision support tools, and intelligent user interfaces in digital commerce platforms. Key findings reveal that high uncertainty avoidance cultures demonstrate stronger preferences for explainable AI systems with detailed algorithmic transparency (Al-Adwan et al., 2022), customers from high power distance cultures exhibit different patterns of trust in hierarchical AI systems compared to their low power distance counterparts, who prefer more egalitarian human-AI interaction models (Lee & Lalwani, 2024).

The research contributes to the emerging field of culturally-adaptive systems by proposing the Cultural-Adaptive Decision-Making (CADM) framework. This theoretical model conceptualizes the dynamic relationships between cultural values, cognitive processes, and AI system design in digital commerce environments. The framework provides practical guidelines for designing AI systems that can effectively support engineering decision-making across diverse cultural backgrounds while maintaining technical accuracy and commercial effectiveness.

Research Aim and Question

Based on the identified gaps, this study aims to investigate how AI-powered features and cultural dimensions influence consumer acceptance and usage preference of AI-powered features in digital commerce platforms. The primary objective is to develop and validate a culturally-adaptive framework for understanding consumer behavior in AI-driven shopping environments.

Research Question:

1. How do AI technology features and cultural dimensions influence consumer behavior in digital commerce?
2. How do cultural dimensions affect preferences for different AI interactions?
3. What cultural factors determine preferences in AI-driven shopping platforms?
4. How can digital commerce platforms adapt AI features to better serve consumers from different cultural backgrounds?

Research Objectives

This research aims to achieve the following specific objectives. Fitstly, develop a comprehensive theoretical framework that integrates cultural theory, technology acceptance models, and decision-making processes in AI-enhanced digital commerce environments. Systematically examine how specific cultural factors influence customers' acceptance and utilization of AI-driven decision support systems.

Secondly, validate the proposed Cultural-Adaptive Decision-Making (CADM) framework across multiple cultural contexts through empirical research involving customers from diverse cultural backgrounds.

Lastly, provide actionable recommendations for AI system designers and developers to create culturally-sensitive interfaces and decision support tools that enhance user experience and decision-making effectiveness across global markets.

LITERATURE REVIEW

Digital Commerce

With the change of technology and consumption trend, the global market has gradually developed from the original E-commerce, mobile commerce and social commerce to comprehensive digital commerce (Wirtz, 2024). Consumers have direct purchasing channels, the ability to shop anytime and anywhere fundamentally changed consumer expectations and behaviors. Developing and revaluating environment of commerce requires now digital marketing methods to adopt consumers' preference, such as digital stores (Rezvi et al., 2025).

The future of shopping is moving from imagination to reality. Artificial intelligence development has had a significant impact on digital commerce adoption, speeding many current trends over time and irreversibly moving customer preferences toward digital channels (Bawack et al., 2022). During recent years, digital commerce has seen the continuous development of new technologies and strategies, such as recommendation systems (Xu et al., 2024), chatbots (Ruan & Mezei, 2022), virtual reality (S. Li et al., 2023), and voice assistants (Lee et al., 2022), providing new ways for consumers to discover and purchase products.

Digital Commerce Platform

Traditional E-commerce Platform. Traditional E-commerce platforms continue to be the foundation of many digital commerce operations, providing the fundamental capabilities needed to present products, process transactions, and manage customer relationships (Mustafa Ayobami et al., 2024). Online shopping platforms, giant companies like Amazon, Alibaba, and eBay, offer commerce functionality to commerce infrastructures, allowing for greater flexibility in creating personalized customer experiences across multiple channels and devices (Warrier et al., 2021). Amazon and Alibaba are global leaders in this market, employing their massive scale and advanced algorithms to match consumers with relevant products from millions of sellers.

Social Platform. Social commerce platforms have become a significant role in digital commerce, blurring the boundaries between content consumption, social interaction, and shopping (Yoesoep Edhie, 2022). This convergence of social media and commerce has created new forms of influence-based marketing, with social media personalities acting an increasingly important role in driving product discovery and purchase decisions (Beichert et al., 2024).

Live Platform. Live shopping, which combines livestreaming with interactive commerce features, shows the latest evolution of this trend, particularly in Asian markets (Tian & Frank, 2024). Live streaming platforms have emerged as dynamic channels within the digital commerce ecosystem. The immediacy and authenticity of live commerce create a sense of presence and urgency that traditional E-commerce often lacks, potentially accelerating purchase decisions through time-limited offers and real-time social proof (Luo et al., 2024). Live platforms represent a significant trend in reducing the experiential gap between physical and online shopping by allowing consumers to witness product demonstrations, ask questions, and make purchases within a single comprehensive environment.

B2C, B2B, C2C and D2C. The digital commerce includes several business models that cater to different market segments and transaction patterns (Rėklaitis & Pilelienė, 2019). Business-to-Consumer (B2C) platforms enable direct transactions between companies and individual consumers, simplifying purchasing processes and providing personalized experiences to increase conversion and loyalty. On the other hand, Business-to-Business (B2B) commerce involves inter-organizational transactions characterized by complex purchasing processes, higher order values, and emphasis on relationship management and customized solutions. Consumer-to-Consumer (C2C) marketplaces enable peer-to-peer exchanges, using trust mechanisms and standardized transaction frameworks to connect individual sellers with buyers, while often monetizing through listing fees or transaction commissions. The relatively newer Direct-to-Consumer (D2C) model represents manufacturers' strategic bypass of traditional retail (Leimstoll & Wölfle, 2021), allowing brands to establish direct relationships with end users, gather first-party data, and exercise greater control over brand presentation and customer experience.

Future in Digital Commerce

The digital commerce environment is fundamentally reshaping consumer behavior across the entire purchase process. Digital commerce has also fostered new expectations for convenience and immediacy (Bernovskis et al., 2024; Chodak, 2024). Consumers increasingly expect seamless, frictionless shopping experiences with features like one-click purchasing, stored payment information, and same-day delivery options. Consumers no longer think in terms of separate channels but expect integrated experiences that use the advantages of both digital and physical environments (Lazić et al., 2024). This convergence represents the next trend in digital commerce, requiring businesses to rethink traditional organizational structures and develop more novel approaches.

Despite a lot of papers conducting digital commerce platforms and business models, there remains a significant lack of research studying the intersection of consumer behavior and cultural differences in new AI-driven digital commerce environments. There is limited empirical investigation into how different cultural groups interpret and respond to various forms of AI-driven shopping experiences. The effectiveness of recommendation systems, chatbots, virtual assistants, and other AI applications may vary substantially across cultures due to differences in technology acceptance, communication preferences, and decision-making processes. This knowledge gap hampers the development of culturally adaptive AI solutions that could enhance the effectiveness of digital commerce across global markets.

Consumer Behavior

Theoretical Foundation of Consumer Behavior

Consumer behavior research has evolved through distinct stages over the past century, reflecting changing academic perspectives and market realities (Rachmad, 2024). Consumer behavioral theory shows in table 1.

Table 1. Consumer Behavioral Theory

Theory	Author	Content
COM-B	Michie et al. (2011)	The COM-B model is a behavior framework with three core elements: capability, opportunity, and motivation. This model shows that behavior change requires appropriate capabilities, suitable opportunities, and sufficient motivation.
TPB	Ajzen (1991)	The TPB model extends the TRA model. It proposes that behavioral intention is the antecedent of behavior, while intention itself is influenced by three factors: attitudes toward the behavior, subjective norms, and perceived behavioral control.
TRA	Fishbein and Ajzen (1975)	The TRA model shows that behavior is primarily determined by behavioral intention, which is influenced by two factors: attitudes toward the behavior and subjective norms.
SOR	Mehrabian and Russell (1974)	The SOR model divides the behavioral process into three sequential stages: external stimuli, internal processing within the individual, and the resulting behavioral response.

Influencing factors of Purchasing Behavior

Digital technology has transformed both consumer behavior and the methods used to study it. Purchasing behavior is formed by a complex interplay of individual, social, and contextual factors that collectively determine what consumers buy, when, where, and how (Toha & Supriyanto, 2023). For example, impulse buying intentions in live streaming are affected by behavior, perceived social pressure and perception of the ease (Li & Kang, 2024a). The digital era has transformed reference group influence through social media platforms, online communities, and influencer marketing, creating new dynamics of social proof and recommendation.

Individual differences represent fundamental variables that shape consistent patterns in digital purchasing behavior across consumers. Demographic factors including age, gender, income, education, and household composition significantly influence purchasing behavior through their effects on needs, preferences, resources, and decision processes (Younus et al., 2015). Age demonstrates measurable variations in technology comfort, advertising receptivity, and loyalty dynamics, The shopping behaviors of people of different ages also have different characteristics. For example, middle-aged people lack online shopping experience and tend to make more interactions before deciding whether to buy (Li & Kang, 2020, 2022c).

Consumer emotions play a key role in shaping purchase behavior throughout the digital commerce journey. Consumers will experience joy and excitement during the purchase process (Razmus et al., 2024), the higher a consumer's brand engagement, the more likely they are to experience joy before making a purchase. If a product's design evokes negative emotions in consumers, for instance, with watches, sharp shapes and discordant color contrasts can worsen consumers' mood, indirectly reducing the product's appeal to them (Yang et al., 2025). In live streaming commerce, according to emotional contagion perspective, when live streamers possess attractive qualities that resonate with consumers, creating a sense of joy and pleasure, the probability of impulse purchasing increases significantly (Li Li et al., 2024).

Socio-cultural factors, including culture, social class, and reference groups, provide the broader background within individual purchasing decisions (Sreen et al., 2018). In live streaming platforms, different social environment can cause different motivations (Kang et al., 2023). Reference groups, which can be different from different cultural groups, significantly impact how consumers make purchase decisions online (Fernandes & Panda, 2018). Family members establish early buying habits and often influence major purchases through shared decision-making and financial considerations. Friends provide trusted recommendations that feel more genuine than marketing messages, making peer suggestions particularly influential when choosing products or platforms. In today's digital environment, social media influencers have become powerful reference figures who shape consumer preferences through content that showcases products in aspirational yet seemingly achievable lifestyles (Li & Pugalia, 2024). These influencers combine the relatability of peers with the reach of traditional media, creating new forms of social proof that guide purchasing decisions. The impact of these reference groups varies dependent on the product type and consumer characteristics, with some purchases being more socially influenced than others. In social commerce, if products have good reputation and information quality, consumers will build trust and be satisfied with the business, and will make repurchases and spread word of mouth (Laradi et al., 2024; Meilatinova, 2021).

Marketing mix elements—product, price, place, and promotion—constitute another set of critical factors influencing purchasing behavior. Product attributes, including functional features, aesthetic qualities, and symbolic associations, are evaluated against consumer needs and preferences, such as website design can influence online consumer purchasing decision-making (Sohaib et al., 2021). Pricing strategies affect not only affordability but also quality perceptions, with price serving as both cost and information (Faith & Agwu, 2018). Distribution channels influence product availability, convenience, and the shopping experience. Promotional activities, including advertising, sales promotion, personal selling, and public relations, shape consumer awareness, attitudes, and purchase intentions (Alhalalmeh et al., 2022). The effectiveness of these marketing mix elements depends on their alignment with consumer characteristics and situational factors, creating complex interaction.

Digital technology has introduced several new factors influencing purchasing behavior, including algorithm-driven recommendations, online reviews and ratings, digital word-of-mouth, and decision tools. Recommendation algorithms shape consideration sets and discovery patterns, often operating invisibly to guide consumer attention toward particular options (Hwangbo et al., 2018). Online reviews serve as prior experiences, allowing consumers to learn from others' experiences before making their own purchases (Liu & Park, 2015). Social media influences word-of-mouth effects, allowing consumption experiences to reach broader audiences and persist longer than traditional person-to-person communication (Kimmel & Kitchen, 2014). New digital decision aids, from augmented reality (AR) applications that virtually place products in consumers' homes to AI assistants that make purchase recommendations, are creating novel influences on purchasing processes (Liu et al., 2025).

Research Gap in Consumer Behavior

While conventional consumer behavior theories have established robust frameworks for understanding purchase decision processes, they were largely developed in pre-digital backgrounds and may inadequately capture the specific dynamics of contemporary digital commerce environments. There exists an important theoretical gap in understanding how established decision-making models require reconceptualization when applied to AI-enhanced shopping environments where algorithmic mediators increasingly shape discovery, evaluation, and choice processes.

Cultural Differences on Consumer Behavior

Cultural Theory

The global market of digital commerce platforms requires a deep understanding of how cultural factors shape consumer behavior, especially multinational companies, which have their own unique ecosystems (Li et al., 2019). Cultural theories provide systematic approaches for examining how deeply culture values, communication norms, and social structures manifest in digital commerce environments and influence consumer responses. Culture theories are shown in table 2.

Table 2. Cultural Theory

Theory	Author	Content
WVS Cultural Map	Haerpfer et al. (2022)	Y-axis: Traditional Values: Emphasis on religion, parent-child ties, deference to authority, traditional family values Secular Rational: Opposite preferences to traditional values X-axis: Survival Values: Emphasis on economic and physical security Self-expression Values: Emphasis on environmental protection, tolerance of diversity, gender equality
GLOBE Project	Bullough et al. (2020)	Power Distance, Uncertainty Avoidance Institutional Collectivism: The degree of encouragement for collective action In-Group Collectivism: Individual's loyalty to their organization or family Gender Egalitarianism: The degree of reducing gender differences Assertiveness: The level of assertive behavior in social relationships Future Orientation, Performance Orientation, Humane Orientation
Cultural Dimensions Theory	Hofstede (2020)	Power Distance Index: High PDI vs. Low PDI Individualism vs. Collectivism Masculinity vs. Femininity Uncertainty Avoidance Index: High UAI vs. Low UAI Long Term Orientation vs. Short Term Orientation Indulgence vs. Restraint
Erin Meyer's Culture Map	Meyer (2014)	Communicating: Low vs. High context Evaluating: Direct vs. Indirect negative feedback Persuading: Principles first vs. Applications first Leading: Egalitarian vs. Hierarchical Deciding: Group decision-making vs. Authoritative decision-making Trusting: Task-based vs. Relationship-based Disagreeing: Confrontational vs. Avoids confrontation Scheduling: Linear time vs. Flexible time
Seven Dimensions of Culture	Trompenaars and Hampden-Turner (2011)	Universalism vs. Particularism Individualism vs. Communitarianism Specific: Separation of work and personal life vs. Diffuse: Overlap between work and personal life Neutral: Emphasis on controlling emotions vs. Emotional: Open expression of feelings Achievement vs. Ascription Sequential vs. Synchronic Time Internal vs. External Control
Cultural Theory	Hall (1976)	High-context vs. low-context cultures Monochronic and polychronic time Proxemics: intimate, personal, social, and public distances

Purchase Behavior in Different Cultures

Cultural dimensions have a profound influence on consumer preferences across product categories, brand characteristics, and shopping experiences. For example, different national cultures can attract different online audiences in live streaming (Kang & Li, 2021; L. Li et al., 2023). The connection between individualism-collectivism and consumer preferences has been extensively documented, with consumers in collectivistic cultures showing greater receptiveness to emphasizing family benefits, in-group acceptance, and social harmony, while placing higher importance on brand reputation and social visibility in purchase decisions (Jin et al., 2008). Individualistic cultures, by contrast, exhibit stronger preferences for products promoting personal identity, self-expression, and individual achievement, responding more positively to unique appeals, personal benefit messaging, and customization options (Kramer et al., 2007). These cultural variations influence not only product selection but also shopping companion preferences, gift-giving practices, and loyalty program structures that resonate with consumers across different cultural backgrounds.

To be more specific, people with different languages and writing systems have specific requirements for platform functions (Li & Kang, 2024b). Not only do platform factors such as website design influence consumer decisions, but cultural factors also have a significant impact on shaping purchasing behavior (Kang & Sohaib, 2016). The interplay between these cultural elements and consumer decision-making processes represents an important dimension that warrants further investigation in the context of online consumer behavior research. For example, different urban and rural areas have different intentions and attitudes towards digital-related platforms and products (Li & Kang, 2022a; Lifu Li et al., 2024)

Website design preferences and user experience expectations vary substantially across cultural contexts, reflecting deeper cultural values and communication patterns (Mushtaha & Troyer, 2009). Social influence operates differently in digital commerce across cultural backgrounds. Muthukrishna and Schaller (2020) believe that the influence of majority opinions is stronger among those individuals who are more sensitive to social influence, spreading more rapidly and becoming more firmly established, even for views that were initially resisted or rejected. These differences extend to influencer marketing effectiveness, for example, online shopping streamers with unique cultures on shopping platforms can enhance consumers' cultural experience and cultural immersion (Li & Kang, 2022b; Li et al., 2022). Technology adoption patterns also reveal important cross-cultural variations in digital environment. Different cultures have different attitudes toward and intention to adopt new technologies in digital commerce (Ashraf et al., 2014). However, trust is an important antecedent of PU and PEOU in most cultures.

Payment preferences in digital commerce reflect deeply held cultural values regarding trust, risk, time orientation, and social relationships (Deufel et al., 2019). In countries with high uncertainty avoidance, consumers tend to prefer more secure payment methods, such as invoices. Payment methods requiring additional steps, such as online bank transfers, are more widely accepted in countries with higher femininity cultural dimensions. Collectivist countries demonstrate greater acceptance of shared payment systems, such as China. In the contract, European countries such as Germany show the highest acceptance rates for high-risk, high-uncertainty payment methods like cryptocurrencies (Busse et al., 2020).

The examination of various factors influencing online purchasing behavior reveals the multidimensional nature of consumer decision-making in digital commerce environments. From individual differences to cultural dimensions, these elements interact in complex ways to shape how consumers discover, evaluate, and ultimately purchase products online. Understanding these influences provides valuable insights for businesses seeking to refine their digital commerce strategies. As technology continues to emerge and consumer expectations shift, recognizing the interplay between these factors becomes increasingly important for creating engaging and effective online shopping experiences.

Research Gap in Cultural Differences and Consumer Behavior

Despite significant advances in cross-cultural consumer behavior research, several critical gaps remain, particularly regarding the nuanced interactions between cultural dimensions and evolving digital shopping environments. While existing literature has established the importance of cultural values in shaping consumer preferences and decision-making processes, most studies have relied on traditional cultural frameworks such as Hofstede's dimensions without adequately capturing the dynamic, evolving nature of cultural values in an increasingly changed world.

AI in Digital Commerce

Theoretical Foundation of Technology Acceptance

The successful adoption of digital commerce platforms fundamentally depends on consumers' willingness to embrace new technologies and integrate them into their purchasing behaviors (Awwad, 2024). Technology acceptance theories provide structured frameworks for understanding the psychological, social, and systemic factors that influence how individuals evaluate and ultimately adopt technological innovations in commercial environments (Musa et al., 2024). The predictive and explanatory power of technology acceptance theories has made them instrumental in guiding platform development, feature prioritization, and implementation strategies across digital commerce. These theories are shown in table 3.

Table 3. Technology Acceptance Theory

Theory	Author	Content
Meta-UTAUT	Dwivedi et al. (2019)	The Meta-UTAUT model extends the UTAUT model by introducing attitude as a new variable, showing that attitudes can influence behavioral intention and use. By combining findings from different studies, this model enhanced predictive ability and provides a more detailed understanding of technology adoption across diverse user groups.
UTAUT2	Venkatesh et al. (2012)	The UTAUT2 model extends the UTAUT model by adding three key factors: hedonic motivation, price value, and habit. This model retains age, gender, and experience as moderators.
TAM3	Venkatesh and Bala (2008)	TAM3 maintains the core TAM relationships while introducing detailed antecedents to perceived usefulness and perceived ease of use. TAM3 specifically emphasizes that the determinants of perceived usefulness and perceived ease of use do not cross-influence each other, providing a more specific understanding of the psychological processes.
UTAUT	Venkatesh et al. (2003)	The UTAUT model identifies four key factors of technology acceptance and use: performance expectancy, effort expectancy, social influence, and facilitating conditions. UTAUT has four moderating variables—gender, age, experience, and voluntariness of use—that influence the impact of the four main determinants on behavioral intention and use behavior.
TAM2	Venkatesh and Davis (2000)	TAM2 expands the TAM by adding additional constructs, including social influence processes and cognitive processes. The social influences include subjective norm, voluntariness, and image, acknowledging the impact of social pressure on technology acceptance. This extended model provides a more comprehensive understanding of the factors influencing perceived usefulness and the conditions under which social influences.
TAM	Davis (1989)	TAM identifies two primary determinants of technology acceptance: perceived usefulness (the degree to which a person believes using a system would enhance job performance) and perceived ease of use (the degree to which a person believes using a system would be free of effort). The model shows that these factors influence behavioral intentions and actual use.

AI Applications in Digital Commerce

Artificial Intelligence is changing everything, including redefine the relationship of digital commerce and their customers (Dwivedi et al., 2021; Saleh & Zeebaree, 2025). The goal of Artificial Intelligence is to understand human and build intelligent system (Ertel, 2024). The application of artificial intelligence in consumer behavior analysis has undergone a remarkable transformation since its early implementations, evolving from basic rule-based systems to sophisticated neural networks capable of identifying complex behavioral patterns (Boukrouh & Azmani, 2024).

Artificial intelligence has transformed numerous aspects of digital commerce, with several key technologies now playing vital roles in enhancing customer experience. Among these innovations, visual search allows consumers to discover products through images rather than keywords, enabling more intuitive discovery for visually complex items, Ara et al. (2025) found that AI-powered visual search methods can improve user satisfaction. Intelligent search and recommendation systems explore user behavior and preferences to surface the most relevant products, helping both consumers and merchants by creating more efficient paths to discovery. Voice shopping assistants represent another significant advancement, providing conversational interfaces that enable hands-free shopping experiences and open new accessibility options, moreover, it can also help platforms grow customers (Sun et al., 2025). These complementary AI applications collectively address different facets of the consumer journey, reducing work and creating more engaging shopping experiences across diverse platforms.

Intelligence searching and recommendation help small companies have higher competition (Chandra Reddy et al., 2025). Their research demonstrated that AI-powered search and recommendation engines create opportunities for smaller businesses to compete more effectively against larger, established companies. By analyzing consumer behavior patterns and personalizing product suggestions in real-time, these systems enable small companies to maximize their limited product assortments and create more relevant customer experiences.

Among the diverse AI applications transforming digital commerce, intelligent chatbots have emerged as the most widely implemented technology across the sector. AI chatbots can provide product selections and recommendations directly based on consumer requirements (Rahevar & Darji, 2024). Chatbots utilizing artificial intelligence technology have significantly improved the accuracy of product recommendations, helping consumers save time and enhancing user satisfaction. Recent advancements in natural language processing have significantly enhanced these systems' ability to understand detailed customer requests and respond with increasingly human-like interactions.

AI in Enhancing Customer Experience. The customer experience dimension of AI applications in digital commerce has drawn significant interest in recent research. Singh et al. (2024) provide valuable insights into the acceptance of Online Shopping Assistants (OSAs) in digital commerce interactions, highlighting how these AI-powered tools compensate for the absence of personalized human assistance in online shopping environments. Their research reveals that anthropomorphism, attitude, ease of use, enjoyment, privacy concerns, trust, and perceived usefulness significantly influence technology acceptance. Interestingly, their findings emphasize that user demographics, particularly gender and education, moderate acceptance intentions, suggesting the importance of tailored OSA implementation strategies for different customer segments.

Building on this understanding of AI-driven customer assistance, Mahapatra et al. (2022) explore AI's potential in shaping the customer journey through cognitive marketing in e-commerce websites. Their study of consumers in Odisha, India reveals a positive relationship between AI implementation and customer experience, particularly in influencing consciousness and intention to purchase. The research demonstrates how AI-enhanced elements—such as interactive product images, videos, and personalized content—contribute to both cognitive and affective dimensions of the purchase decision process. This work importantly bridges theoretical understanding with practical applications in regional markets, showing how AI can be effectively deployed to enhance customer engagement in specific cultural contexts.

Taking a more comprehensive perspective, Nalla and Reddy (2024) examine AI-driven big data analytics for enhanced customer experience in digital commerce. Their research illuminates how real-time insights, predictions, and personalized recommendations enabled by AI technologies reform customer interactions at every aspect. By analyzing techniques such as machine learning, natural language processing, and neural networks, they demonstrate how these technologies can be deployed to understand customer behavior patterns, preferences, and purchasing habits.

The research by Khrais (2020) adds another dimension by focusing on AI's role in forming consumer demand in digital commerce. The study emphasizes how personalization and product customization through AI influence customer behavior toward specific products and brands. Moreover, Khrais raises important ethical considerations regarding AI systems, particularly concerning explainability. The research advances the concept of explainable artificial intelligence as a solution to the “black box” problem in AI decision-making, suggesting that making AI systems more interpretable and comprehensible is crucial for user trust and ethical implementation.

AI-Driven Marketing and Advertising. The transformation of marketing strategies through AI applications represents another significant theme in research. Sabharwal et al. (2022) examine the relationship between AI, digital advertising, and marketing strategy formulation. Their research reveals how AI has fundamentally changed communication with consumers, creating new opportunities for storytelling and marketing engagement. The study emphasizes the growing importance of AI and machine learning in digital marketing, showing that businesses must adopt these technologies to remain competitive. Their findings highlight how AI-driven marketing not only saves time and money for marketers and customers but also effectively engages consumers' attention without human intervention.

Extending this perspective, Alqurashi et al. (2023) focus specifically on AI's impact on personalized content marketing. Their large-scale study confirms that organizations implementing AI experience significant enhancements in personalized content marketing efforts. The research demonstrates how AI's capabilities in analyzing complex data and tailoring content to individual preferences drive engagement and customer satisfaction. They particularly emphasize the strategic integration of

AI channels as crucial for optimizing marketing strategies and developing deeper connections with target audiences.

Hicham et al. (2023) provide a strategic structure for leveraging AI in marketing decision-making. Their analysis examines recent AI developments in marketing, including predictive analytics for customer behavior analysis, chatbot integration for enhanced customer support, and AI-driven content personalization strategies. The research explores AI applications across various marketing and their influence on different marketing sectors, offering insights into how businesses can strategically implement AI to change marketing operations and decision-making procedures.

AI Impact on Business. Beyond customer experience and marketing, research has also examined how AI influences broader business performance and operations in digital commerce. With diverse inputs through sophisticated neural architectures, AI techniques are widely used in decision making (Duan et al., 2019). Chen et al. (2022) apply the resource-based view to investigate the relationship between AI capability, management, driven decision-making, and firm performance in digital commerce. They identify important moderating factors, demonstrating how innovation culture positively moderates the link between firm creativity and AI-driven decision-making, as well as between decision-making and business performance.

Hindarto (2023) examines e-commerce's role in increasing sales, highlighting how modern technology and internet proliferation have created significant opportunities for businesses to expand their sales reach. While not only focused on AI, but this research also provides context for understanding the technological environment in which AI operates. The study emphasizes how digital commerce enables businesses to engage in global sales without physical location or time constraints, offering consumers more convenient and efficient shopping experiences.

Tang et al. (2023) focus on a specific operational application of AI in E-commerce: data-intensive inventory forecasting for international E-commerce service automation. Their case study demonstrates how AI approaches can optimize inventory management systems to provide reliable supply, enhance flow among supply chain members, fulfill changing customer requirements, and enable digital commerce service automation. Their evaluation of different AI methods reveals that the method performs the best performance in terms of accuracy and reasonable calculation time.

AI Technology Acceptance. The acceptance and implementation of AI technologies in digital commerce form another significant research direction. Young consumers, such as Generation Z, are more likely to use technical, user-friendly AI tools to enhance their shopping experience (Pu et al., 2025). The utility and user-friendly of AI systems can improve consumers' attitudes toward AI (Kang & Alnefaie, 2024). Wang et al. (2023) provide an empirical evaluation of the TAM for AI in digital commerce. Their study examines how AI can be made more effective and profitable in digital commerce and how entrepreneurs can use AI technology to achieve business goals. Singh (2021) provides a broader perspective on the AI and E-commerce ecosystem from a customer's view. This research discusses how the E-commerce ecosystem has grown exponentially, even during global crises like the COVID-19 pandemic. It highlights the critical role of data-driven approaches and AI/ML systems in exploring vast customer-generated data. However, the study identifies significant challenges related to cybersecurity, with AI/ML systems making E-commerce vulnerable to data theft, privacy invasion, data manipulation, and fraud.

Challenges and Ethical Considerations. Several studies have addressed the challenges and ethical considerations related with AI implementation in digital commerce. Singh (2021) emphasizes cybersecurity and privacy concerns, highlighting how AI systems have made e-commerce vulnerable to various threats. Khrais (2020) focuses specifically on ethical considerations related to AI explainability. The research argues that despite AI's contributions to digital commerce, its ethical soundness remains contentious, particularly regarding the concept of explainability.

Future Directions and Emerging Trends. Several studies point to future directions and emerging trends in AI applications for digital commerce. Hicham et al. (2023) discuss how AI represents the most recent example of disruptive technology with the potential to revolutionize marketing completely. They identify exciting developments suggesting these areas will continue to evolve and transform digital commerce.

Nalla and Reddy (2024) highlight opportunities related with AI adoption in big data environments, discussing how these technologies can transform customer experiences and boost business growth in competitive digital commerce landscapes. Their research suggests that AI-driven analytics will play an increasingly important role in predicting market trends, enhancing supply chain management, and raising customer satisfaction.

Theoretical and Methodological Approaches. The reviewed literature employs diverse theoretical and methodological approaches to study AI applications in digital commerce. Chen et al. (2022) apply the resource-based view to understand how AI capabilities contribute to firm performance, while Wang et al. (2023) and Singh et al. (2024) utilize the TAM to examine AI adoption factors. These theoretical frameworks provide valuable views for understanding different aspects of AI implementation and impact in digital commerce backgrounds.

Methodologically, the studies employ various approaches, including quantitative surveys (Alqurashi et al., 2023; Singh et al., 2024), case studies (Tang et al., 2023) and mixed methods (Khrais, 2020). Many researches employ advanced analytical methods such as partial least squares structural equation modeling to examine relationships among variables (Chen et al., 2022; Singh et al., 2024; Wang et al., 2023), reflecting the methodological comprehensiveness of research in this field.

Integration Analysis. Integrating findings across these studies reveals several key topics and insights. First, AI applications in digital commerce cross the whole customer journey, from initial engagement and product discovery to post-purchase support and relationship management. These technologies enable new levels of personalization, prediction, and automation, transforming how businesses interact with customers and manage operations.

Second, while AI offers significant benefits for enhancing customer experiences, improving marketing effectiveness, and optimizing business operations, its implementation also raises important challenges related to privacy, security, and ethics. The tension between data utilization and privacy protection emerges as a critical consideration for businesses implementing AI in digital commerce contexts.

Third, the successful implementation of AI in digital commerce depends on multiple factors, including technological capabilities, organizational readiness, customer acceptance, and strategic alignment. Studies emphasize the importance of considering these factors deeply when creating AI strategies for digital commerce.

Fourth, cultural and demographic factors significantly influence AI acceptance and effectiveness in digital commerce contexts. Several studies highlight how factors such as gender, education, and cultural background moderate the impact of AI implementations, suggesting the need for contextually sensitive approaches to AI application.

Cultural Difference in AI

The adoption of AI technologies demonstrates significant cross-cultural variation, reflecting deeper cultural values and psychological differences that shape technology acceptance patterns. Recent research has begun to illuminate these cultural dependencies, though substantial gaps remain in understanding how to design culturally-adaptive AI systems for global digital commerce.

Barnes et al. (2024) demonstrate that cultural identity fundamentally shapes how individuals perceive and interact with AI systems. Their research reveals that individualists tend to view AI as external to the self and interpret AI features as potential infringements upon their uniqueness, autonomy, and privacy. In contrast, collectivists are more prone to view AI as an extension of the self and interpret AI features as facilitating conformity to consensus and environmental responsiveness. Empirical evidence from Tubadji et al. (2021) provides quantitative support for these cultural differences, cultural bias appears to be associated with social capital rather than fear of novelty, suggesting that cultural acceptance of AI is driven by relational and trust-based factors rather than mere technological anxiety.

Cross-cultural studies reveal stark regional differences in AI acceptance. Mantello et al. (2023) found that East Asian respondents are significantly more likely to express trust toward AI-enabled automated management systems compared to Western subjects. This cross-cultural discrepancy extends beyond simple acceptance rates to encompass different concerns about AI deployment. Their research indicates that disadvantaged ethnicities, gender groups, and income classes may experience heightened stress and anxiety from AI monitoring tools, with these effects varying by cultural background. The challenge of achieving cross-cultural cooperation in AI adoption is compounded by barriers identified by ÓhÉigeartaigh et al. (2020). Their research on AI ethics and governance reveals that misunderstandings between cultures play a more important role in undermining cross-cultural trust than fundamental disagreements. This finding suggests that many apparent cultural conflicts regarding AI adoption may stem from miscommunication rather than irreconcilable value differences, indicating opportunities for culturally-sensitive design interventions.

Despite growing recognition of cultural factors in AI adoption, significant research gaps persist. Mogaji et al. (2024) critically evaluate the Technology Acceptance Model (TAM) in the context of generative AI and identify fundamental limitations in its cultural applicability. Their analysis reveals that existing technology acceptance frameworks require substantial reconceptualization to accommodate cultural nuances in AI adoption, particularly in service-intensive sectors like hospitality and digital commerce.

Arif (2025) found the contrast between implementations in technologically advanced nations like Finland and Japan versus developing regions like Sub-Saharan Africa and South Asia illustrates that infrastructure and digital literacy create foundational constraints interacting with cultural factors. The emphasis on linguistic diversity and governmental policies indicates successful AI deployment requires understanding not only individual cultural preferences but also institutional and infrastructural ecosystems within which AI operates. Kim and Yoo (2021) demonstrate that cultural variations manifest even at interface design levels, with preferences for 2D versus 3D AI agents correlating with religious and philosophical traditions. The finding that male preferences for female AI agents remain consistent cross-culturally while female preferences diverge significantly, South Korean, Chinese, and Russian women preferring female agents versus American and Brazilian women preferring male agents, suggests that gender dynamics within cultures shape expectations for AI interaction partners, indicating culturally-adaptive design must account for intersectional cultural-gender factors.

Wang (2025) provides critical empirical evidence through multilevel analysis across 20 countries, revealing that countries with higher power distance and individualism demonstrate lower favorable perceptions of AI. The negative correlation between indulgence and favorable AI views indicates that cultures emphasizing gratification may perceive AI as threatening to human-centric values. Prabhakaran et al. (2022) expose a fundamental structural problem: geographic concentration of

AI development in a handful of countries has resulted in systems embedding specific cultural values and interpretive frameworks. The analysis reveals that training data systematically underrepresents global cultural diversity, meaning AI systems are not culturally neutral tools but fundamentally shaped by their development environments. Liu (2024) reveals that while AI offers technical solutions to communication barriers such as linguistic limitations, effectiveness depends critically on cultural alignment. The examination of AI applications in content creation, translation, and cultural preservation demonstrates that AI technologies can either bridge or exacerbate cultural divides depending on design and deployment, suggesting cultural adaptation is not merely optimization but necessity for avoiding friction in global marketplaces.

These studies collectively demonstrate that AI adoption cannot be understood through culturally-neutral frameworks. Cultural dimensions influence not only whether individuals adopt AI technologies, but also how they perceive AI capabilities, what concerns they prioritize, how they establish trust in AI systems, and what design features they find acceptable or problematic. However, current research remains limited in providing actionable frameworks for designing culturally-adaptive AI systems that can effectively serve diverse global markets while respecting cultural values and preferences.

Research Gap in AI Applications in Digital Commerce

Research examining the intersection of cultural factors, consumer behavior, and artificial intelligence technologies remains in its beginning stages, creating a substantial knowledge gap as these technologies rapidly transform global commerce. While general research on technology acceptance across cultures is well-established, specific investigation into how cultural dimensions influence consumer responses to AI-driven personalization, recommendation systems, conversational agents, and automated decision support tools is still limited. The question of how to design culturally adaptive AI systems that can effectively respond to diverse cultural preferences while avoiding stereotypical assumptions represents a crucial frontier for both theoretical development and practical implementation.

Research Gap

Current research in decision-making has primarily focused on technical optimization, algorithmic efficiency, and system performance metrics, with limited consideration of cultural factors that influence user acceptance and effectiveness. While extensive research exists on cultural influences in general technology acceptance

(Kang & Sohaib, 2016) and on AI implementations in digital commerce (Chen et al., 2022), there remains a significant gap in understanding their intersection.

The existing literature on technology acceptance across cultures provides valuable insights through frameworks, however, these models were developed primarily for general technology adoption and may not adequately capture the unique requirements.

Similarly, while AI research has made significant advances in developing sophisticated recommendation systems (Xu et al., 2024), natural language processing interfaces (Zhou et al., 2020), most of these developments focus that assumes universal user preferences and interaction patterns. This approach fails to account for the substantial cultural variations in communication styles, decision-making processes, risk tolerance, and authority relationships that characterize different cultural backgrounds.

Building upon the analysis of existing theories and research gaps, we can propose an integrated theoretical framework that conceptualizes cross-cultural digital commerce as a dynamic socio-technical ecosystem with continuous interactions between technological systems, human behavior, cultural contexts, and AI applications. This framework, which can term the Cultural-Adaptive Decision-Making (CADM) framework, acknowledges the complex, dynamic relationships between these aspects while providing specific theoretical mechanisms to explain their interactions.

CADM FRAMEWORK

Drawing from cultural theory, technology acceptance models, and human-AI interaction research, the CADM framework conceptualizes the dynamic relationships between cultural values, AI system design, and consumer responses in digital commerce environments.

The CADM framework comprises five interconnected components. Cultural dimensions serve as moderating variables, including power distance, uncertainty avoidance, individualism-collectivism, and communication context preferences. AI system characteristics represent technological features such as recommendation algorithms, conversational interfaces, personalization mechanisms, and decision automation levels. Cognitive processes mediate relationships between cultural values and behaviors, encompassing perceived usefulness, perceived ease of use, trust formation, and risk assessment. User responses capture attitudes toward AI features, intention to use AI-driven tools, and engagement levels. Behavioral outcomes represent ultimate consequences including purchase decisions, platform loyalty, and continued AI usage.

Cultural dimensions moderate the relationships between AI system characteristics and user responses through cognitive processes. High power distance cultures respond more positively to AI systems with authority signals and expert endorsements, while low power distance consumers prefer egalitarian AI interactions. Uncertainty avoidance influences requirements for AI explainability, with high uncertainty avoidance cultures demanding detailed algorithmic transparency and low uncertainty avoidance cultures accepting streamlined recommendations. Individualism-collectivism affects the effectiveness of personalization versus social proof, where individualistic consumers favor unique recommendations while collectivistic consumers engage more with popularity indicators. These cultural moderations operate through cognitive processes that shape perceptions of usefulness, trust development, and risk assessment, creating feedback loops that reinforce or diminish AI adoption based on cultural alignment of system design.

METHODOLOGY

This research employs a mixed-methods research approach designed to provide comprehensive understanding of cultural influences on AI-driven decision-making. The methodology combines quantitative assessment of cultural factors and technology acceptance with qualitative exploration of user experiences and decision-making processes.

A structured survey instrument will be administered to customers across multiple cultural backgrounds. The survey incorporates validated scales for measuring cultural factors, technology acceptance variables, and AI system evaluation criteria specific to commerce platforms.

Semi-structured interviews and focus group discussions will be conducted to explore their experiences with AI-driven digital commerce systems. These qualitative methods will provide deeper insights into how cultural values manifest in actual decision-making processes and system interactions. Cultural analysis will examine communication patterns, trust development mechanisms, risk assessment approaches, and collaborative decision-making preferences. Statistical analysis will employ multi-group structural equation modeling to test the moderating effects of cultural factors on relationships between AI system characteristics and user acceptance outcomes.

EXPECTED CONTRIBUTIONS

Theoretical Implications

This research makes significant theoretical contributions to the intersection of cultural studies, technology acceptance, and decision-making research by addressing critical gaps identified in the introduction.

The primary contribution involves the development of the Cultural-Adaptive Decision-Making (CADM) framework that integrates cultural theory with AI system design principles and consumer behavior models. This framework directly addresses the theoretical gap regarding the lack of comprehensive models explaining how traditional cultural frameworks apply to AI-mediated digital commerce environments. While existing literature has examined cultural influences on general technology acceptance and AI implementations in isolation, the CADM framework uniquely synthesizes these streams to explain the dynamic interplay between cultural values, cognitive processes, and AI system characteristics in shaping consumer responses.

The framework extends existing technology acceptance models by incorporating cultural dimensions as critical moderating variables. Conventional consumer behavior theories were largely developed in pre-digital contexts and inadequately capture the specific dynamics of AI-enhanced shopping environments where algorithmic mediators increasingly shape discovery, evaluation, and choice processes. The CADM framework reconceptualizes these established models for AI-driven contexts by specifying how cultural factors moderate the relationships between AI system features and acceptance outcomes, providing theoretical mechanisms that explain cross-cultural variance in AI adoption patterns.

The framework also contributes to cultural theory application in digital contexts by demonstrating how established cultural dimensions manifest in novel technological environments. Cultural frameworks have been extensively validated in traditional organizational and consumer contexts, and their application to AI-mediated interactions represents a theoretical extension that enriches understanding of how enduring cultural values shape responses to emerging technologies.

Managerial Implications

The practical implications of this research are substantial for engineering organizations operating in global markets. The study provides comprehensive design guidelines for creating culturally-adaptive AI interfaces and decision support systems specifically tailored for commerce platforms, enabling developers to create more effective and user-friendly systems. Additionally, the research offers detailed implementation strategies for global engineering companies seeking to deploy AI-driven digital commerce systems across diverse cultural markets, helping organizations navigate the complexities of cross-cultural technology deployment. The CADM framework provides actionable recommendations for optimizing AI system performance across different cultural contexts, including guidance on interface design patterns, information presentation strategies, and interaction models that align with specific cultural values and preferences.

CONCLUSION

This research addresses a critical gap in understanding how cultural factors influence the adoption and effectiveness of AI-driven decision-making systems in digital commerce environments. By proposing the Cultural-Adaptive Decision-Making (CADM) framework, this research offers theoretical advancement and conceptual guidance for creating more effective, culturally-sensitive AI systems for global commerce platforms.

The research contributes to the broader objectives of innovative decision-making in commerce by theorizing how consideration of cultural factors could potentially enhance the effectiveness of AI-driven systems while supporting better human-computer interaction across diverse global contexts. As engineering practice becomes increasingly dependent on AI technologies and global collaboration, understanding and accommodating cultural differences emerges as essential for achieving both technical excellence and commercial success.

For platform designers and developers, this research recommends prioritizing modular AI architectures that allow flexible cultural adaptation without requiring complete system redesigns, including adjustable communication styles, configurable information density controls, and adaptive decision support modes. International businesses expanding digital commerce operations should adopt systematic market entry strategies that begin with cultural assessment, invest in cross-cultural user research, and establish regional AI governance teams including local cultural experts. Success metrics should be culturally contextualized rather than applying universal benchmarks. Policymakers and regulators should develop frameworks that acknowledge cultural diversity while maintaining consumer protections, avoiding one-size-fits-all AI regulations and instead allowing cultural variation in areas such as data privacy expectations and algorithmic transparency requirements. Governments should support research initiatives advancing understanding of cultural factors in AI adoption and develop educational programs to improve AI literacy across diverse cultural populations.

This work provides new ways for future empirical investigation and provides a foundation for developing culturally-sensitive AI commerce systems that can serve global markets effectively while respecting cultural diversity.

REFERENCES

Ajzen, I. (1991). The theory of planned behavior. *Organizational Behavior and Human Decision Processes*, *50*(2), 179–211. DOI: 10.1016/0749-5978(91)90020-T

Al-Adwan, A. S., Al-Debei, M. M., & Dwivedi, Y. K. (2022). E-commerce in high uncertainty avoidance cultures: The driving forces of repurchase and word-of-mouth intentions. *Technology in Society*, *71*, 102083. Advance online publication. DOI: 10.1016/j.techsoc.2022.102083

Alhalalmeh, M., Alkhawaldah, R. A., Mohammad, A., Al-Quran, A., Hijjawi, G., & Al-Hawary, S. (2022). The Effect of Selected Marketing Activities and Promotions on the Consumers Buying Behavior. *Business: Theory and Practice*, *23*(1), 79–87. DOI: 10.3846/btp.2022.13929

Alqurashi, D. R., Alkhaffaf, M., Daoud, M. K., & Ahmad, J., Al-Gasawneh, & Alghizzawi, M. (2023). Exploring the Impact of Artificial Intelligence in Personalized Content Marketing: A Contemporary Digital Marketing. *Migration Letters : An International Journal of Migration Studies*, *20*(S8), 548–560.

Ara, J., Ghodke, S., Akter, J., & Roy, A. (2025). Optimizing E-Commerce Platforms with AI-Enabled Visual Search: Assessing User Behavior, Interaction Metrics, and System Accuracy. *Journal of Economics. Finance and Accounting Studies*, *7*(3), 9–17. DOI: 10.32996/jefas

Arif, S. (2025). Cross-Cultural perspectives on AI in education: Case studies from global classrooms. *AI EDIFY Journal*, *2*(1), 12–20.

Ashraf, A. R., Thongpapanl, N. T., & Auh, S. (2014). The Application of the Technology Acceptance Model under Different Cultural Contexts: The Case of Online Shopping Adoption. *Journal of International Marketing*, *22*(3), 68–93. DOI: 10.1509/jim.14.0065

Awwad, B. (2024). *The AI Revolution: Driving Business Innovation and Research*. Springer.

Barnes, A. J., Zhang, Y., & Valenzuela, A. (2024). AI and culture: Culturally dependent responses to AI systems. *Current Opinion in Psychology*, *58*, 101838. DOI: 10.1016/j.copsyc.2024.101838

Bawack, R. E., Wamba, S. F., Carillo, K. D. A., & Akter, S. (2022). Artificial intelligence in E-Commerce: A bibliometric study and literature review. *Electronic Markets*, *32*(1), 297–338. DOI: 10.1007/s12525-022-00537-z

Beichert, M., Bayerl, A., Goldenberg, J., & Lanz, A. (2024). Revenue Generation Through Influencer Marketing. *Journal of Marketing, 88*(4), 40–63. DOI: 10.1177/00222429231217471

Bernovskis, A., Sceulovs, D., & Stibe, A. (2024). Society 5.0: Shaping the future of e-commerce. *Journal of Open Innovation, 10*(4), 100391. Advance online publication. DOI: 10.1016/j.joitmc.2024.100391

Boukrouh, I., & Azmani, A. (2024). Artificial Intelligence Applications in E-Commerce: A Bibliometric Study from 1995 to 2023 Using Merged Data Sources. *International Journal of Professional Business Review, 9*(4), e4537. Advance online publication. DOI: 10.26668/businessreview/2024.v9i4.4537

Bullough, A., Cotton, R., Dastmalchian, A., Dorfman, P., & Egri, C. (2020). GLOBE 2020: The Latest Findings on Cultural Practices, Culture Change, and Leadership Ideals. *Proceedings - Academy of Management, 17374*(1), 17374. Advance online publication. DOI: 10.5465/AMBPP.2020.17374symposium

Busse, K., Tahaei, M., Krombholz, K., Zezschwitz, E. v., Smith, M., & Tian, J. (2020). Cash, Cards or Cryptocurrencies? A Study of Payment Culture in Four Countries. *2020 IEEE European Symposium on Security and Privacy Workshops*, 200-209. https://doi.org/DOI: 10.1109/EuroSPW51379.2020.00035

Chandra Reddy, B. R. V. N., Konakalla, L., Kumar, G. P., Ganapathi, K. S., Santhosh, M., & Rao, M. N. (2025). AI-Powered E-commerce Platform for Small Businesses. *International Journal of Innovative Science and Research Technology, 10*(4), 495–501. DOI: 10.38124/ijisrt/25apr420

Chen, D., Esperanca, J. P., & Wang, S. (2022). The Impact of Artificial Intelligence on Firm Performance: An Application of the Resource-Based View to e-Commerce Firms. *Frontiers in Psychology, 13*, 884830. DOI: 10.3389/fpsyg.2022.884830

Chodak, G. (2024). *Future of E-commerce*. Springer. DOI: 10.1007/978-3-031-55225-0

Davis, F. D. (1989). Perceived usefulness, perceived ease of use, and user acceptance of information technology. *Management Information Systems Quarterly, 13*(3), 319–340. DOI: 10.2307/249008

Deufel, P., Kemper, J., & Brettel, M. (2019). Pay now or pay later: A cross-cultural perspective on online payments. *Journal of Electronic Commerce Research, 20*(3), 141–154.

Duan, Y., Edwards, J. S., & Dwivedi, Y. K. (2019). Artificial intelligence for decision making in the era of Big Data – evolution, challenges and research agenda. *International Journal of Information Management*, *48*, 63–71. DOI: 10.1016/j.ijinfomgt.2019.01.021

Dwivedi, Y. K., Hughes, L., Ismagilova, E., Aarts, G., Coombs, C., Crick, T., Duan, Y., Dwivedi, R., Edwards, J., Eirug, A., Galanos, V., Ilavarasan, P. V., Janssen, M., Jones, P., Kar, A. K., Kizgin, H., Kronemann, B., Lal, B., Lucini, B., & Williams, M. D. (2021). Artificial Intelligence (AI): Multidisciplinary perspectives on emerging challenges, opportunities, and agenda for research, practice and policy. *International Journal of Information Management*, *57*, 101994. Advance online publication. DOI: 10.1016/j.ijinfomgt.2019.08.002

Dwivedi, Y. K., Rana, N. P., Jeyaraj, A., Clement, M., & Williams, M. D. (2019). Re-examining the Unified Theory of Acceptance and Use of Technology (UTAUT): Towards a Revised Theoretical Model. *Information Systems Frontiers*, *21*(3), 719–734. DOI: 10.1007/s10796-017-9774-y

Ertel, W. (2024). *Introduction to Artificial Intelligence*. Springer Nature., DOI: 10.1007/978-3-658-43102-0

Faith, D. O., & Agwu, E. (2018). A Review of the Effect of Pricing Strategies on the Purchase of Consumer Goods. *International Journal of Research in Management. Science & Technology*, *2*(2), 88–102.

Fernandes, S. F., & Panda, R. (2018). Social reference group influence on women buying behaviour: A review. *Journal of Commerce and Management Thought*, *9*(2), 273–291. DOI: 10.5958/0976-478X.2018.00019.8

Fishbein, M., & Ajzen, I. (1975). *Belief, Attitude, Intention, and Behavior: An Introduction to Theory and Research*. Addison-Wesley Publishing Company., https://books.google.com.au/books?id=8o0QAQAAIAAJ

Haerpfer, C., Inglehart, R., Moreno, A., Welzel, C., Kizilova, K., Diez-Medrano, J., Lagos, M., Norris, P., Ponarin, E., & Puranen, B. (2022). *World Values Survey Trend File (1981-2022) Cross-National Data-Set*. JD Systems Institute & WVSA Secretariat. https://doi.org/DOI: 10.4232/1.14021

Hall, E. T. (1976). *Beyond culture*. Anchor.

Hicham, N., Nassera, H., & Karim, S. (2023). Strategic Framework for Leveraging Artificial Intelligence in Future Marketing Decision-Making. *Journal of Intelligent Management Decision*, *2*(3), 139–150. DOI: 10.56578/jimd020304

Hindarto, D. (2023). The Role of E-Commerce in Increasing Sales Using Unified Modeling Language. [IJSECS]. *International Journal Software Engineering and Computer Science*, *3*(2), 120–129. DOI: 10.35870/ijsecs.v3i2.1503

Hofstede, G. (2020). The 6-D model of national culture. *Geert Hofstede. http:// geerthofstede. com/culture-geert-hofstede-gert-jan-hofstede/6d-model-of-national-culture/(accessed: 19.02. 2020.).*

Hwangbo, H., Kim, Y. S., & Cha, K. J. (2018). Recommendation system development for fashion retail e-commerce. *Electronic Commerce Research and Applications*, *28*, 94–101. DOI: 10.1016/j.elerap.2018.01.012

Jin, B., Yong Park, J., & Kim, J. (2008). Cross-cultural examination of the relationships among firm reputation, e-satisfaction, e-trust, and e-loyalty. *International Marketing Review*, *25*(3), 324–337. DOI: 10.1108/02651330810877243

Kang, K., & Alnefaie, A. E. A. (2024). Cultural Transformations of E-commerce Consumer Behavior and Intention Toward Using Artificial Intelligence (AI) Assistants. *The 57th Hawaii International Conference on System Sciences.*

Kang, K., & Li, L. (2021). *Exploring the Relationships between Cultural Content and Viewers' Watching Interest: A Study of Tiktok Videos Produced by Chinese Ethnic Minority Groups* Proceedings of the 18th International Conference on e-Business, Kang, K., Li, L., & Namisango, F. (2023). Dissimilar Social Settings Impact on User Motivations and Activities on Live-Streaming Digital Platforms. In *E-service Digital Innovation*. IntechOpen. https://doi.org/DOI: 10.5772/intechopen.112787

Kang, K., & Sohaib, O. (2016). Individualists vs. Collectivists in B2C E-Business Purchase Intention. *Journal of Internet and e-business Studies*, 1-11. https://doi.org/ DOI: 10.5171/2016.948644

Khrais, L. T. (2020). Role of Artificial Intelligence in Shaping Consumer Demand in E-Commerce. *Future Internet*, *12*(12), 226. Advance online publication. DOI: 10.3390/fi12120226

Kim, Y., & Yoo, H. (2021). Cross-cultural comparison of preferences for the external appearance of artificial intelligence agents. *Social Behavior and Personality*, *49*(11), 1–23. DOI: 10.2224/sbp.10824

Kimmel, A. J., & Kitchen, P. J. (2014). Word of mouth and social media. *Journal of Marketing Communications*, *20*(1-2), 2–4. DOI: 10.1080/13527266.2013.865868

Kramer, T., Spolter-Weisfeld, S., & Thakkar, M. (2007). The Effect of Cultural Orientation on Consumer Responses to Personalization. *Marketing Science*, *26*(2), 246–258. DOI: 10.1287/mksc.1060.0223

Laradi, S., Alrawad, M., Lutfi, A., & Agag, G. (2024). Understanding factors affecting social commerce purchase behavior: A longitudinal perspective. *Journal of Retailing and Consumer Services, 78*, 103751. Advance online publication. DOI: 10.1016/j.jretconser.2024.103751

Lazić, A., Milić, S., & Vukmirović, D. (2024). The Future of Electronic Commerce in the IoT Environment. *Journal of Theoretical and Applied Electronic Commerce Research, 19*(1), 172–187. DOI: 10.3390/jtaer19010010

Lee, H., & Lalwani, A. K. (2024). Power Distance Belief and Consumer Purchase Avoidance: Exploring the Role of Cultural Factors in Retail Dynamics. *JMR, Journal of Marketing Research, 61*(2), 349–367. DOI: 10.1177/00222437231182600

Lee, S., Oh, J., & Moon, W.-K. (2022). Adopting Voice Assistants in Online Shopping: Examining the Role of Social Presence, Performance Risk, and Machine Heuristic. *International Journal of Human-Computer Interaction, 39*(14), 2978–2992. DOI: 10.1080/10447318.2022.2089813

Leimstoll, U., & Wölfle, R. (2021). Direct to Consumer (D2C) E-Commerce: Goals and Strategies of Brand Manufacturers. In *New Trends in Business Information Systems and Technology* (pp. 237-250). https://doi.org/DOI: 10.1007/978-3-030-48332-6_16

Li, J., Chen, L., Yi, J., Mao, J., & Liao, J. (2019). Ecosystem-specific advantages in international digital commerce. *Journal of International Business Studies, 50*(9), 1448–1463. DOI: 10.1057/s41267-019-00263-3

Li, L., Chen, X., & Zhu, P. (2024). How do e-commerce anchors' characteristics influence consumers' impulse buying? An emotional contagion perspective. *Journal of Retailing and Consumer Services, 76*, 103587. Advance online publication. DOI: 10.1016/j.jretconser.2023.103587

Li, L., & Kang, K. (2020). Analyzing shopping behavior of the middle-aged users in tiktok live streaming platform. *AMCIS 2020 Proceedings.*

Li, L., & Kang, K. (2022a). Impact of opportunity and capability on e-entrepreneurial motivation: A comparison of urban and rural perspectives. *Journal of Entrepreneurship in Emerging Economies, 16*(4), 932–953. DOI: 10.1108/JEEE-06-2022-0178

Li, L., & Kang, K. (2022b). The Role of Cultural Attractors in Live Streaming Content: Regional Cultural Perspective Using Multi-Group Analysis. *PACIS 2022 Proceedings, 49.*

Li, L., & Kang, K. (2022c). Understanding the real-time interaction between middle-aged consumers and online experts based on the COM-B model. *Journal of Marketing Analytics, 12*(3), 570–582. DOI: 10.1057/s41270-022-00196-1

Li, L., & Kang, K. (2024a). Discovering online Chinese consumers' impulse buying in live streaming by the theory of planned behavior. *Journal of Economic Analysis*, *3*(2), 60. Advance online publication. DOI: 10.58567/jea03020008

Li, L., & Kang, K. (2024b). Discussing Social Media Using Affordance of Young Chinese Ethnic Minority Group Users. *Cultural Arts Research and Development*, *4*(1), 12–23. DOI: 10.55121/card.v4i1.64

Li, L., Kang, K., Feng, Y., & Zhao, A. (2022). Factors affecting online consumers' cultural presence and cultural immersion experiences in live streaming shopping. *Journal of Marketing Analytics*, *12*(2), 250–263. DOI: 10.1057/s41270-022-00192-5

Li, L., Kang, K., & Namisango, F. (2024). Online younger viewers' motivation to purchase virtual gifts: Multi-group analysis of urban-rural backgrounds. *Aslib Journal of Information Management*. Advance online publication. DOI: 10.1108/AJIM-03-2024-0201

Li, L., Kang, K., & Sohaib, O. (2023). An Analysis of Cultural Content on Short Video Platforms: Cultural Protection Perspective. In *E-Business and Telecommunications* (pp. 29-50). https://doi.org/DOI: 10.1007/978-3-031-36840-0_2

Li, L., & Pugalia, S. (2024). How Wanghong Live Streamers Affect Online Users' Purchasing Behavior on Live Streaming Platforms: A PLS-SEM Approach. *Journal of Global Information Technology Management*, *27*(4), 277–292. DOI: 10.1080/1097198X.2024.2410675

Li, S., Zhu, B., & Yu, Z. (2023). The Impact of Cue-Interaction Stimulation on Impulse Buying Intention on Virtual Reality Tourism E-commerce Platforms. *Journal of Travel Research*, *63*(5), 1256–1279. DOI: 10.1177/00472875231183163

Liu, H., Zhang, P., Cheng, H., Hasan, N., & Chiong, R. (2025). Impact of AI-generated virtual streamer interaction on consumer purchase intention: A focus on social presence and perceived value. *Journal of Retailing and Consumer Services*, *85*, 104290. Advance online publication. DOI: 10.1016/j.jretconser.2025.104290

Liu, Z. (2024). New Dynamics in Cross-Cultural Exchange: A Study on the Role and Impact of Artificial Intelligence Technology in the Global Cultural Market. *Academic Journal of Business & Management*, *6*(12), 137–143.

Liu, Z., & Park, S. (2015). What makes a useful online review? Implication for travel product websites. *Tourism Management*, *47*, 140–151. DOI: 10.1016/j.tourman.2014.09.020

Luo, X., Cheah, J.-H., Hollebeek, L. D., & Lim, X.-J. (2024). Boosting customers' impulsive buying tendency in live-streaming commerce: The role of customer engagement and deal proneness. *Journal of Retailing and Consumer Services*, *77*, 103644. Advance online publication. DOI: 10.1016/j.jretconser.2023.103644

Mahapatra, D. M., Patra, S. K., & Baral, S. K. (2022). Unleashing the Potential of Artificial Intelligence (AI) in Customer Journey of Cognitive Marketing and Consciousness Intention in E-Commerce Websites. *Proceedings of the 2nd International Conference on Sustainability and Equity (ICSE-2021)*, 44-51. https://doi.org/DOI: 10.2991/ahsseh.k.220105.006

Mantello, P., Ho, M.-T., Nguyen, M.-H., & Vuong, Q.-H. (2023). Bosses without a heart: Socio-demographic and cross-cultural determinants of attitude toward Emotional AI in the workplace. *AI & Society*, *38*(1), 97–119. DOI: 10.1007/s00146-021-01290-1

Mehrabian, A., & Russell, J. A. (1974). *An approach to environmental psychology*. The MIT Press.

Meilatinova, N. (2021). Social commerce: Factors affecting customer repurchase and word-of-mouth intentions. *International Journal of Information Management*, *57*, 102300. Advance online publication. DOI: 10.1016/j.ijinfomgt.2020.102300

Meyer, E. (2014). *The culture map: Breaking through the invisible boundaries of global business*. Public Affairs Books.

Michie, S., Stralen, M. M., & West, R. (2011). The behaviour change wheel: A new method for characterising and designing behaviour change interventions. *Implementation Science : IS*, *6*(1), 1–12. DOI: 10.1186/1748-5908-6-42

Mogaji, E., Viglia, G., Srivastava, P., & Dwivedi, Y. K. (2024). Is it the end of the technology acceptance model in the era of generative artificial intelligence? *International Journal of Contemporary Hospitality Management*, *36*(10), 3324–3339. DOI: 10.1108/IJCHM-08-2023-1271

Musa, H. G., Fatmawati, I., Nuryakin, N., & Suyanto, M. (2024). Marketing research trends using technology acceptance model (TAM): A comprehensive review of researches (2002–2022). *Cogent Business & Management*, *11*(1), 2329375. Advance online publication. DOI: 10.1080/23311975.2024.2329375

Mushtaha, A., & Troyer, O. D. (2009). Cross-Culture and Website Design: Cultural Movements and Settled Cultural Variables. *Internationalization, Design and Global Development: Third International Conference, IDGD 2009, Held as Part of HCI International 2009*, 69–78. https://doi.org/DOI: 10.1007/978-3-642-02767-3_8

Mustafa Ayobami, R., Hameedat Bukola, O., Timothy Tolulope, O., Wilhelmina Afua, A., Onyeka Chrisanctus, O., & Adedoyin Tolulope, O. (2024). E-commerce and consumer behavior: A review of AI-powered personalization and market trends. *GSC Advanced Research and Reviews, 18*(3), 066-077. https://doi.org/DOI: 10.30574/gscarr.2024.18.3.0090

Muthukrishna, M., & Schaller, M. (2020). Are Collectivistic Cultures More Prone to Rapid Transformation? Computational Models of Cross-Cultural Differences, Social Network Structure, Dynamic Social Influence, and Cultural Change. *Personality and Social Psychology Review*, *24*(2), 103–120. DOI: 10.1177/1088868319855783

Nalla, L. N., & Reddy, V. M. (2024). AI-Driven Big Data Analytics for Enhanced Customer Journeys: A New Paradigm in E-Commerce. *2*(1), 719-740.

ÓhÉigeartaigh, S. S., Whittlestone, J., Liu, Y., Zeng, Y., & Liu, Z.ÓhÉigeartaigh. (2020). Overcoming barriers to cross-cultural cooperation in AI ethics and governance. *Philosophy & Technology*, *33*(4), 571–593. DOI: 10.1007/s13347-020-00402-x

Prabhakaran, V., Qadri, R., & Hutchinson, B. (2022). Cultural incongruencies in artificial intelligence. arXiv preprint arXiv:2211.13069.

Pu, L., Radics, R., Umar, M., Jeremiah, F., & Quan, Z. (2025). The potential of AI tools in shaping digital consumers' behavior: Investigating e-commerce engagement of Chinese Generation Z. *Asia Pacific Journal of Marketing and Logistics*, *37*(9), 2720–2737. Advance online publication. DOI: 10.1108/APJML-08-2024-1048

Rachmad, Y. E. (2024). *The Evolution of Consumer Behavior: Theories of Engagement, Influence, and Digital Interaction*. PT. Sonpedia Publishing Indonesia.

Rahevar, M., & Darji, C. S. (2024). The adoption of AI-driven Chatbots into a recommendation for e-commerce systems to targeted customer in the selection of product. *International Journal of Management. Economics and Commerce*, *1*(2), 128–137. DOI: 10.62737/m1vpdq75

Razmus, W., Razmus, M., & Grabner-Kräuter, S. (2024). Joy and excitement in the purchase process: The role of materialism and brand engagement. *Journal of Product and Brand Management*, *33*(3), 330–344. DOI: 10.1108/JPBM-04-2022-3937

Rėklaitis, K., & Pilelienė, L. (2019). Principle Differences between B2B and B2C Marketing Communication Processes. *Management of Organizations: Systematic Research*, *81*(1), 73–86. DOI: 10.1515/mosr-2019-0005

Rezvi, R. I., Rahman, K. O., Hasan, M. A., Nasrullah, F., Nusrat, N., Jishan, S. S., & Ahmed, S. (2025). How Digital Marketing Affiliates the Digital Stores: A Deep Dive into Shopify, Amazon, Walmart, and Other E-Commerce Giants. *Journal of Computer Science and Technology Studies*, *7*(2), 95–101. DOI: 10.32996/jcsts.2025.7.2.8

Ruan, Y., & Mezei, J. (2022). When do AI chatbots lead to higher customer satisfaction than human frontline employees in online shopping assistance? Considering product attribute type. *Journal of Retailing and Consumer Services*, *68*, 103059. Advance online publication. DOI: 10.1016/j.jretconser.2022.103059

Sabharwal, D., Sood, R. S., & Verma, M. (2022). Studying the relationship between artificial intelligence and digital advertising in marketing strategy. *Journal of Content. Community and Communication*, *16*(8), 118–126. DOI: 10.31620/JCCC.12.22/10

Saleh, R. A., & Zeebaree, S. R. M. (2025). Artificial Intelligence in E-commerce and Digital Marketing: A Systematic Review of Opportunities, Challenges, and Ethical Implications. *Asian Journal of Research in Computer Science*, *18*(3), 395–410. DOI: 10.9734/ajrcos/2025/v18i3601

Singh, C., Dash, M. K., Sahu, R., & Kumar, A. (2024). Investigating the acceptance intentions of online shopping assistants in E-commerce interactions: Mediating role of trust and effects of consumer demographics. *Heliyon*, *10*(3), e25031. DOI: 10.1016/j.heliyon.2024.e25031

Singh, R. (2021). A Study of Artificial Intelligence and E-Commerce Ecosystem – A Customer's Perspective. *International journal of research in engineering, science and management, 4*(2), 78-87.

Sohaib, O., Kang, K., & Miliszewska, I. (2021). Uncertainty Avoidance and Consumer Cognitive Innovativeness in E-Commerce. *Journal of Global Information Management*, *27*(2), 59–77. DOI: 10.4018/JGIM.2019040104

Sreen, N., Purbey, S., & Sadarangani, P. (2018). Impact of culture, behavior and gender on green purchase intention. *Journal of Retailing and Consumer Services*, *41*, 177–189. DOI: 10.1016/j.jretconser.2017.12.002

Sun, C., Shi, Z., Liu, X., & Ghose, A. (2025). The Effect of Voice AI on Digital Commerce. *Information Systems Research*, *36*(2), 1147–1166. Advance online publication. DOI: 10.1287/isre.2022.0140

Tang, Y. M., Chau, K. Y., Lau, Y.-y., & Zheng, Z. (2023). Data-Intensive Inventory Forecasting with Artificial Intelligence Models for Cross-Border E-Commerce Service Automation. *Applied Sciences (Basel, Switzerland)*, *13*(5), 3051. Advance online publication. DOI: 10.3390/app13053051

Tian, Y., & Frank, B. (2024). Optimizing live streaming features to enhance customer immersion and engagement: A comparative study of live streaming genres in China. *Journal of Retailing and Consumer Services*, *81*, 103974. Advance online publication. DOI: 10.1016/j.jretconser.2024.103974

Toha, M., & Supriyanto, S. (2023). Factors influencing the consumer research process: Market target, purchasing behavior and market demand (literature review of consumer behavior). *Danadyaksa: Post Modern Economy Journal*, *1*(1), 1–17. DOI: 10.69965/danadyaksa.v1i1.5

Trompenaars, F., & Hampden-Turner, C. (2011). *Riding the waves of culture: Understanding diversity in global business*. Nicholas Brealey International.

Tubadji, A., Denney, T., & Webber, D. J. (2021). Cultural relativity in consumers' rates of adoption of artificial intelligence. *Economic Inquiry*, *59*(3), 1234–1251. DOI: 10.1111/ecin.12978

Venkatesh, V., & Bala, H. (2008). Technology Acceptance Model 3 and a Research Agenda on Interventions. *Decision Sciences*, *39*(2), 273–315. DOI: 10.1111/j.1540-5915.2008.00192.x

Venkatesh, V., & Davis, F. D. (2000). A Theoretical Extension of the Technology Acceptance Model: Four Longitudinal Field Studies. *Management Science*, *46*(2), 186–204. DOI: 10.1287/mnsc.46.2.186.11926

Venkatesh, V., Morris, M. G., Davis, G. B., & Davis, F. D. (2003). User Acceptance of Information Technology: Toward a Unified View. *Management Information Systems Quarterly*, *27*(3), 425–478. DOI: 10.2307/30036540

Venkatesh, V., Thong, J. Y. L., & Xu, X. (2012). Consumer Acceptance and Use of Information Technology: Extending the Unified Theory of Acceptance and Use of Technology. *Management Information Systems Quarterly*, *36*(1), 157–178. DOI: 10.2307/41410412

Wang, C., Ahmad, S. F., Bani Ahmad Ayassrah, A. Y. A., Awwad, E. M., Irshad, M., Ali, Y. A., Al-Razgan, M., Khan, Y., & Han, H. (2023). An empirical evaluation of technology acceptance model for Artificial Intelligence in E-commerce. *Heliyon*, *9*(8), e18349. DOI: 10.1016/j.heliyon.2023.e18349

Wang, L., Huang, N., He, Y., Liu, D., Guo, X., Sun, Y., & Chen, G. (2025). Artificial Intelligence (AI) Assistant in Online Shopping: A Randomized Field Experiment on a Livestream Selling Platform. *Information Systems Research*, isre.2023.0103. Advance online publication. DOI: 10.1287/isre.2023.0103

Wang, S. (2025). Public Perceptions of Artificial Intelligence in 20 Countries: Assessing Individual-and Country-Level Factors. *Cross-Cultural Research*, •••, 10693971251336803. DOI: 10.1177/10693971251336803

Warrier, U., Singh, P., Jien, C. W., Kee, D. M. H., Yi, G. Z., Jiann, T. W., Liang, T. Y., Sb, G., Nair, S., Nair, R. K., Lokhande, S. D., & Ganatra, V. (2021). Factors that Lead Amazon.com to A Successful Online Shopping Platform. *International Journal of Tourism and Hospitality in Asia Pasific*, *4*(1), 7–17. DOI: 10.32535/ijthap.v4i1.1017

Wei, B., Huang, J., Ren, R., & Hua, M. (2025). Virtual Shopping Assistant, Real User Experience: The Impact of Co-creation and Experiential Value on Consumer Purchase Intention. *Lecture Notes in Computer Science*, *15769*, 214–233. Advance online publication. DOI: 10.1007/978-3-031-93861-0_14

Wirtz, B. W. (2024). *Digital business and electronic commerce: Strategy, business models and technology*. Springer Nature., DOI: 10.1007/978-3-031-50289-7

Wu, Z., Aw, E. C. X., Tan, G. W. H., & Ooi, K. B. (2025). Speak and shop! Transforming retail experience with AI voice assistants. *Internet Research*. Advance online publication. DOI: 10.1108/INTR-02-2024-0149

Xu, K., Zhou, H., Zheng, H., Zhu, M., & Xin, Q. (2024). Intelligent Classification and Personalized Recommendation of E-commerce Products Based on Machine Learning. *arXiv preprint arXiv:2403.19345*.

Yang, C. Y., Wang, C. Y., & Pao, C. L. (2025). The impact of negative emotions on product purchase decisions: A study of watch design. *BMC Psychology*, *13*(1), 447. DOI: 10.1186/s40359-025-02775-0

Yoesoep Edhie, R.Yoesoep Edhie Rachmad. (2022). Social Media Marketing Mediated Changes In Consumer Behavior From E-Commerce To Social Commerce. *International Journal of Economics and Management Research*, *1*(3), 227–242. DOI: 10.55606/ijemr.v1i3.152

Younus, S., Rasheed, F., & Zia, A. (2015). Identifying the Factors Affecting Customer Purchase Intention. *Global Journal of Management and Business Research*, *15*(2), 8–13.

Zhou, M., Duan, N., Liu, S., & Shum, H.-Y. (2020). Progress in Neural NLP: Modeling, Learning, and Reasoning. *Engineering (Beijing)*, *6*(3), 275–290. DOI: 10.1016/j.eng.2019.12.014

Chapter 10
Synergizing Unconventional Approaches for Hybrid Power Generation From Emerging Renewable Technologies

Md. Ziaul Islam
https://orcid.org/0009-0009-9160-2173
West Zone Power Distribution PLC, Bangladesh

ABSTRACT

The global energy transition needs solutions that go beyond traditional solar and wind sources, which have issues with reliability and location. This chapter focuses on hybrid power generation systems that utilize unconventional renewable resources such as footstep power generators, thermoelectric harvesters, noise-to-power converters and other unorthodox approaches to energy storage. Each of these technologies has limitations in terms of location and output. But when these are combined into hybrid systems, they become more resilient, stable and scalable. Strategies include mapping regional energy needs, designing modular systems with smart controllers and using AI management to optimize different energy flows. Future energy independence will be shaped by IoT, advanced materials and community driven manufacturing enabling flexible systems. By focusing on collaboration across fields and human centered design, the chapter argues that hybridizing unconventional renewables is not just a supplementary option but a crucial strategy for sustainable and decentralized power systems.

DOI: 10.4018/979-8-3373-5702-7.ch010

INTRODUCTION

The global energy sector is changing rapidly due to climate change, resource depletion, and increasing energy demand. Nations are searching for sustainable options to replace fossil fuels. The push for clean energy is heightened by worries about energy security, geopolitical issues, and the social and environmental effects of carbon-heavy power generation. While solar and wind energy have become central to renewable energy plans around the world, they face limitations due to their intermittency, reliance on weather conditions, and site-specific restrictions. Solar energy production varies with cloud cover, seasonal changes, and day-night cycles. Wind energy relies heavily on local wind patterns, making these sources unreliable for continuous power supply. As a result, even areas with large investments in renewable energy often need extra energy storage or backup systems, which increase costs and complicate energy infrastructure. This situation highlights the need for new methods that go beyond typical renewable technologies to ensure a stable, reliable, and decentralized power supply that can meet the diverse needs of today's societies (International Energy Agency, 2024).

In light of these challenges, unconventional renewable energy sources have emerged as promising alternatives that can complement traditional energy sources. These technologies capture energy from non-standard sources like human movement, temperature differences, microbial activity, and ambient noise. For example, footstep power generators collect kinetic energy from people walking in urban areas and convert it into electrical power to supply lighting or small electronic devices. Thermoelectric harvesters use temperature differences between surfaces or fluids to generate electricity without moving parts. This method allows for energy recovery from industrial waste heat, natural thermal gradients, or even body heat. Noise-to-power converters, still being developed, aim to turn sound energy into electricity, potentially providing low-cost energy recovery in busy urban areas. Additionally, bio-electrochemical systems, like microbial fuel cells, use microbial processes to produce electricity from organic materials, creating sustainable energy options in wastewater treatment facilities, agricultural waste, and other organic-rich environments. While each of these technologies has its limitations regarding output, efficiency, and infrastructure needs, combining them into hybrid systems can create synergies that address individual weaknesses, improving overall energy reliability and scalability.

The hybrid power system concept is compelling because it allows multiple unconventional energy sources to work together in a single, coordinated setup. By thoughtfully integrating complementary energy harvesters, hybrid systems can respond to changing environmental conditions and varying energy demand. This ensures a steady power supply even when some sources perform poorly. For instance, a system that merges footstep power generators and thermoelectric harvesters in a bustling transit center can gather both kinetic and thermal energy, providing reliable electricity for lighting, signage, and small devices. Similarly, adding microbial fuel cells to thermoelectric harvesters in industrial or wastewater treatment settings can create dual-function systems that treat waste while generating electricity. This kind of hybridization not only improves system efficiency but also supports energy solutions tailored to specific locations, reducing reliance on centralized grids and increasing energy resilience in remote or underserved areas (González & Moya Rodríguez, 2025)

In addition to technical benefits, hybrid systems using unconventional renewables also connect with broader social, economic, and environmental goals. Generating energy locally can empower communities by decreasing dependency on national grids, reducing transmission losses, and offering energy access to areas where traditional infrastructure may not be feasible. The modular nature of hybrid systems allows for scaling and customization, enabling designs that fit local energy sources, usage patterns, and community needs. Advances in sensor networks, Internet of Things (IoT) devices, and smart controllers help these systems optimize themselves in real time, adjusting energy flow based on conditions and user demand. These improvements not only boost system performance but also lower operational costs and extend the lifespan of equipment, making hybrid power solutions financially appealing for both urban and rural settings.

Finally, incorporating unconventional renewable sources into hybrid systems represents more than just a technological breakthrough. It signifies a shift in energy thinking, focusing on adaptability, sustainability, and user-centered design. By merging technical creativity with social insight and supportive policies, hybrid systems can lead to energy designs that are resilient to environmental challenges, responsive to local needs, and aligned with global decarbonization targets. This chapter highlights hybrid power generation from unconventional renewable resources as a transformative approach to distributed energy systems. It sets the stage for discussions on design strategies, implementation challenges, and future opportunities within this broader context. Through this analysis, the chapter shows that hybridization is not just a small improvement over traditional renewables; it is a crucial step toward creating the next generation of sustainable, resilient, and decentralized power infrastructures (Amin & Lee, 2022).

Figure 1. Hybrid Technology

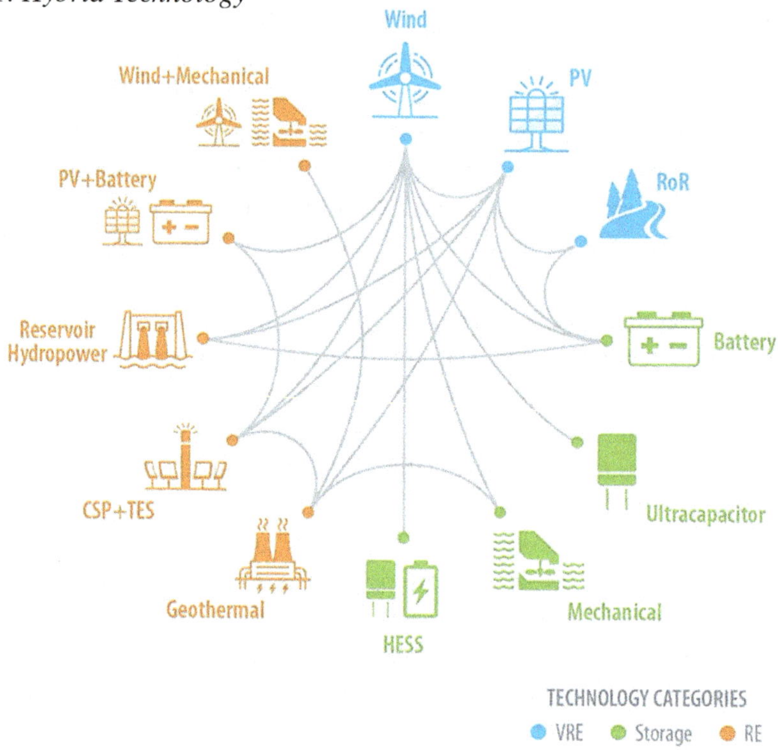

OVERVIEW OF UNCONVENTIONAL RENEWABLE TECHNOLOGIES

As the limits of conventional renewables become clearer, unconventional renewable technologies have drawn interest for their potential to diversify and stabilize energy systems. Unlike solar and wind, these sources use energy from underutilized environmental phenomena or human activities, often turning everyday processes into electricity. Their outputs may be modest, but when combined in hybrid systems, they can greatly improve energy resilience, lessen dependence on centralized grids, and offer localized solutions suited to specific areas and applications. This section looks at some of the most promising unconventional renewable technologies, focusing on how they work, their practical uses, and their potential for hybrid integration (Golubev et al. 2021).

1. Footstep Power Generators

Footstep power generators, also known as piezoelectric floor systems, convert mechanical pressure from human movement into electrical energy using piezoelectric materials. When pressure is applied, these materials produce a small electric charge depending on the mechanical stress. In high-traffic areas like subway stations, airports, shopping malls, and stadiums, this energy can be collected on a meaningful scale. For example, a 1-square-meter piezoelectric floor mat can generate between 5 to 10 watts per hour, based on footfall density and mat efficiency. Besides generating electricity, these systems can also act as smart sensors for counting pedestrians, detecting occupancy, or monitoring structural health. While the energy density is lower than that of solar or wind, placing multiple mats in busy areas or combining them with other unconventional sources can provide steady and reliable power for local lighting, signage, and small electronic devices. Additionally, the modular design of these systems allows for easy scaling. Ongoing research into advanced piezoelectric polymers and nanomaterials promises to boost energy conversion efficiency significantly (Zhang & Wu, 2023).

Another innovative technology, Pavegen tiles are kinetic energy harvesting floor tiles that convert the mechanical energy of footsteps into electricity. Each tile contains piezoelectric or electromagnetic mechanisms that generate power when compressed by foot traffic. The energy produced can be used to power low-energy devices, such as streetlights, signage, or sensors, or stored in batteries for later use. Pavegen tiles are often installed in high-footfall areas like train stations, shopping centers, and urban plazas to maximize energy capture. Beyond energy generation, these tiles also provide real-time data on pedestrian traffic, supporting smart city planning and sustainable infrastructure initiatives (Pavegen, 2025). Here's a clear outline for a Footstep-to-Power Generator System which converts human walking energy into electrical power.

Table 1. Outline for a Footstep-to-Power Generator System

Step	Process	Description
1. Pressure Application	Human footsteps press the top plate.	Converts body weight into mechanical force.
2. Mechanical Transmission	Springs, levers, or rack-and-pinion system.	Amplifies and transfers motion to the generator.
3. Power Generation	Micro-dynamo or piezoelectric module.	Converts mechanical energy to electrical.
4. Rectification & Regulation	Diode bridge and voltage regulator.	Converts AC to stable DC output.

continued on following page

Table 1. Continued

Step	Process	Description
5. Energy Storage	Battery or super capacitor.	Stores energy for continuous use.
6. Output Utilization	LED lights, sensors, chargers.	Demonstrates the system's practical value.

2. Thermoelectric Harvesters

Thermoelectric harvesters use the Seebeck effect, which creates a voltage when there is a temperature difference across a conductive material. This technology can capture waste heat from industrial processes, building HVAC systems, vehicle engines, or even human body heat and turn it into usable electricity. Thermoelectric generators (TEGs) have no moving parts, making them highly durable and maintenance-free. Even though their current efficiency is low—typically 5 to 8% for small-scale devices—their integration into hybrid systems allows for continuous energy capture from sources that would otherwise be wasted. Applications include powering sensors, monitoring devices, or supporting larger power systems. New materials like nanostructured semiconductors, organic thermoelectric polymers, and flexible thermoelectric films are expected to significantly improve conversion efficiency, making TEGs an appealing part of hybrid energy systems in both urban and rural areas (Enescu, 2019).

3. Noise-to-Power Conversion Systems

Acoustic energy harvesting, or noise-to-power conversion, captures energy from ambient sound vibrations. While the energy density of typical environmental noise is low, advancements in electromechanical transducers and piezoelectric membranes have made it possible to extract micro-watt to milli-watt levels, enough to power low-energy devices such as sensors, IoT nodes, or LED lighting. Urban areas, especially near transportation hubs, construction sites, or industrial zones, present significant opportunities for noise-to-power systems. Although still in development, combining acoustic harvesters with other energy sources in a hybrid system can provide extra energy, particularly in places where other renewable resources are scarce or unreliable. Research continues to enhance energy capture efficiency through resonant structures, metamaterials, and adaptive acoustic filters (Hossain et al. 2021).

4. Microbial Fuel Cells (MFCs)

Microbial fuel cells use the metabolic activities of microorganisms to produce electricity from organic materials like wastewater, agricultural residues, or food waste. In an MFC, bacteria break down organic matter, releasing electrons that are collected at the anode and flow to the cathode to generate an electric current. This technology offers two benefits: it treats wastewater while generating electricity. MFCs are especially suitable for decentralized applications in rural areas, wastewater treatment plants, and agricultural processing facilities. Although the power output per unit volume is relatively low, modular designs and hybrid integration with thermoelectric generators or small-scale solar panels can increase total energy production. Advances in electrode materials, optimizing microbial consortia, and biofilm engineering are steadily enhancing energy yields and operational longevity, positioning MFCs as a viable part of hybrid renewable systems (Tan et al. 2021).

5. Other Unorthodox Energy Storage and Harvesting Approaches

In addition to the technologies mentioned above, several unconventional methods focus on capturing energy from overlooked environmental phenomena. Examples include vibration energy harvesters from infrastructure motion, piezoelectric energy from road traffic, and energy recovery from temperature changes in water pipelines or buildings. Innovations in energy storage, such as supercapacitors, flow batteries, and new chemical storage systems, complement these harvesters by storing inconsistent or low-output energy for future use. Hybrid systems that combine multiple unconventional harvesters with advanced storage methods can provide continuous power, even under changing environmental conditions.

6. Advantages and Challenges

Unconventional renewable technologies offer distinct advantages: they use readily available and often untapped energy sources, provide decentralized and localized power solutions, and can be designed modularly for flexibility and scalability. However, each technology faces its own challenges. Output per unit is usually low, efficiency can be limited, and deployment often needs specialized infrastructure or integration with complementary sources. Hybridization addresses many of these issues by combining different sources, enabling energy storage, and optimizing management through smart controllers and algorithms. Furthermore, new materials and production techniques, including nanomaterials, bio-compatible polymers, and

3D-printed modular components, are expected to improve efficiency, lower costs, and simplify deployment in various environments.

7. Integration Potential

Integrating these unconventional technologies into hybrid systems allows for complementary energy generation, where the limitations of one source are balanced by another. For example, footstep generators can provide intermittent energy in urban settings, while thermoelectric harvesters capture otherwise wasted thermal energy, and microbial fuel cells work continuously in organic-rich waste environments. By strategically combining sources based on environmental availability and usage needs, hybrid systems enhance reliability, optimize for specific locations, and build resilience against fluctuations in individual energy outputs. This integration also allows for greater energy independence for both urban and rural communities, contributing to decentralized, sustainable power infrastructures.

Figure 2. Unconventional Renewable Technologies

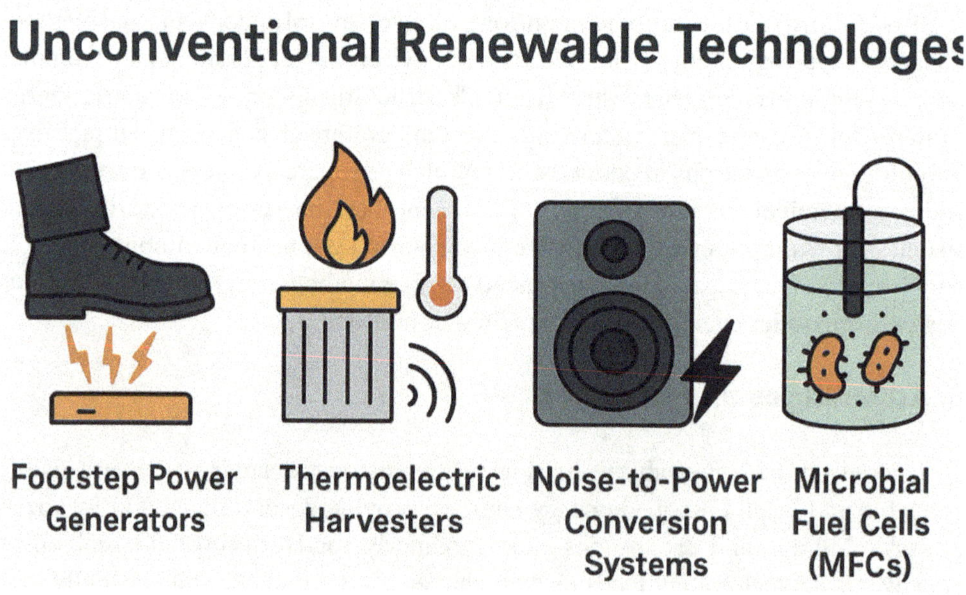

HYBRID POWER SYSTEM DESIGN AND STRATEGIES

The integration of alternative renewable energy technologies into hybrid power systems represents a significant shift in energy generation. It focuses on reliability, scalability, and solutions tailored to specific locations. Unlike traditional renewable sources like solar and wind, which rely heavily on consistent environmental conditions, hybrid systems take advantage of varying energy flows from different sources to create resilient, self-sustaining networks. Designing and deploying these systems requires a multi-layered strategy that includes technical, operational, and socio-economic factors. This section examines these strategies, emphasizing system layout, modular integration, smart control methods, flexible management, and practical deployment.

1. System Architecture and Topology

A hybrid power system is primarily defined by its layout, which dictates how multiple energy sources connect, synchronize, and operate together. The system's topology is crucial because it directly impacts efficiency, stability, and scalability. Common designs include series, parallel, and mixed configurations. In a series setup, the output from one energy source can feed into or aid another. For example, thermoelectric harvesters can provide low-voltage input to a larger energy storage unit linked to microbial fuel cells. Parallel configurations allow each energy source to work independently while a central management unit optimizes the overall system's output. Mixed architectures blend elements of both approaches, allowing for localized energy harvesting along with centralized optimization.

Modular design is another vital aspect of hybrid architecture. Modular systems enable individual parts—such as footstep power mats, thermoelectric units, or MFC modules—to be added or removed based on energy needs, environmental availability, or maintenance. This flexibility improves scalability and lowers the risk of system failure. In urban settings, modular hybrid systems can be set up in subway stations, commercial buildings, or university campuses. In rural regions, modules can be tailored to meet community-specific needs such as irrigation pumps, micro-grids, or off-grid lighting systems. This modular approach also supports gradual upgrades, making it easier to incorporate new technologies or more efficient components without needing to replace the entire system (Yalla et al. 2022).

2. Energy Flow Optimization

A key design challenge is optimizing energy flow between different sources with changing outputs. Unlike traditional systems, hybrid architectures must manage intermittent and low-output sources to provide a steady power supply. This is done through smart controllers that monitor input from each source, manage energy storage, and prioritize loads based on demand. Advanced systems use adaptive algorithms to predict changes in source output—like foot traffic for footstep generators or temperature fluctuations for thermoelectric harvesters—and adjust energy distribution accordingly.

Incorporating energy storage is essential for smoothing out intermittent generation. Technologies like super capacitors, lithium-ion batteries, and new flow batteries provide temporary storage, allowing hybrid systems to maintain a stable supply even when some sources are underperforming. The choice of storage depends on factors like output voltage, energy density, cost, and lifespan. For instance, supercapacitors work well for quick energy bursts from footstep or noise-to-power harvesters, while lithium-ion or flow batteries are better suited for larger, slower energy flows from microbial fuel cells or thermoelectric units. By coordinating storage options with source variability, hybrid systems achieve resilience and consistent availability for critical applications (Darney, 2022).

3. Adaptive Management and AI Integration

To improve efficiency and dependability, hybrid power systems increasingly use AI-driven management strategies. Machine learning algorithms can forecast energy availability, predict demand patterns, and optimize distribution in real-time. For example, AI can analyze previous foot traffic data in a metro station to forecast when footstep mats will produce peak energy and adjust high-priority loads accordingly. Similarly, temperature sensors linked to thermoelectric units can make predictive adjustments to energy routing to storage or usage points. This intelligent management ensures that energy flows match supply conditions and user demands, minimizing waste and boosting overall system performance.

Furthermore, AI can aid in predictive maintenance by identifying components likely to fail or degrade before they drop below acceptable performance levels. This is especially important for hybrid systems located in remote or decentralized areas, where maintenance access may be challenging. By combining real-time monitoring, predictive analysis, and automated control, hybrid systems become self-optimizing, able to adjust continuously to environmental and operational changes (O'Dwyer et al. 2020).

4. Integration with IoT and Sensor Networks

Incorporating IoT devices and sensor networks improves hybrid system performance through detailed monitoring, data collection, and remote management.

Sensors can track environmental factors like temperature, vibration, foot traffic, and organic substrate levels, providing real-time data to controllers. IoT connectivity enables various hybrid systems across a city, campus, or rural community to connect, share energy, and optimize performance together. This creates a distributed energy network where hybrid nodes not only offer localized power but also play a role in a larger smart grid.

IoT-enabled hybrid systems also support user-focused energy management. This allows consumers or operators to prioritize energy use, schedule critical loads, and engage in demand-response programs. This level of flexibility boosts energy efficiency, lowers operational costs, and encourages sustainable energy practices (Saleem et al. 2017).

5. Deployment Strategies and Practical Considerations

Successfully deploying hybrid systems requires careful regional energy profiling to identify available alternative energy sources, usage trends, and environmental limitations. Urban areas benefit from kinetic, acoustic, and thermal energy sources due to high levels of human activity and infrastructure. Rural and agricultural regions can take advantage of microbial fuel cells, low-speed wind, and thermal gradients from water or soil. Mapping these opportunities ensures hybrid systems are optimized for the location to maximize energy capture and impact.

Engaging with the community is another important factor. Involving local stakeholders in system design, assembly, and maintenance enhances social acceptance, ensures operational continuity, and promotes energy independence. Community-based manufacturing using 3D-printed modules or locally sourced materials cuts costs and empowers users to take charge of their energy infrastructure. Training programs for operators and technicians can further enhance sustainability, especially in remote or resource-limited areas.

6. Resilience and Sustainability Considerations

Hybrid systems provide increased resilience by diversifying energy inputs and buffering against environmental changes. In times of climate unpredictability or grid failures, multi-source hybrid architectures continue to supply electricity for essential services such as lighting, communication, and water pumping. Sustainability is further strengthened by the minimal environmental impact of unconventional renewables, modular design, and local energy production. By strategically combining complementary sources, hybrid systems reduce reliance on large-scale energy transmission and support fair access to clean energy.

Figure 3. Hybrid Power Production Strategies

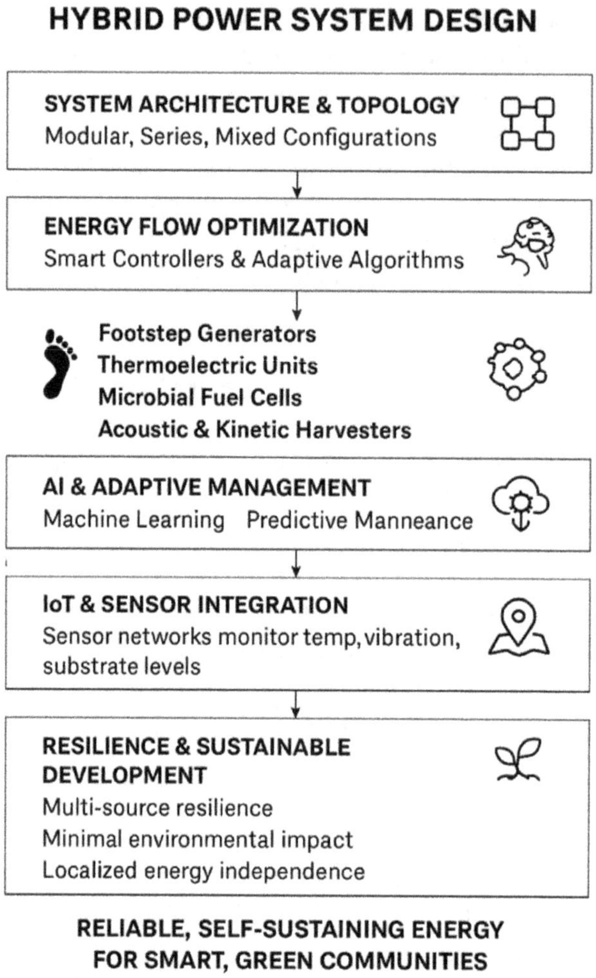

IMPLEMENTATION AND CASE STUDIES

The practical use of hybrid power systems based on unconventional renewable resources connects theoretical design with real-world energy solutions. While hybrid architectures promise improved reliability, resilience, and location-specific optimization, their deployment requires careful planning, adjustment to environmental and socio-economic contexts, and strategic integration with existing infrastructure. This section looks at urban, rural, and community-based implementation approaches, showcasing successful case studies and lessons learned from earlier deployments.

1. Urban Implementation

Urban environments offer both challenges and opportunities for unconventional hybrid systems. High population density, constant human activity, and extensive infrastructure provide ample sources of kinetic, thermal, and acoustic energy. Footstep power generators, for example, are well-suited for metro stations, shopping malls, airports, and university campuses. In a busy subway station, footstep mats installed along corridors or stairs can capture energy from thousands of daily commuters. When combined with thermoelectric harvesters that capture heat from HVAC systems and acoustic energy recovery in noisy areas, these hybrid systems can supply sustainable electricity for lighting, digital signage, or environmental sensors (Unegbu et al. 2025).

A notable case is the deployment of piezoelectric floor systems in European cities like Amsterdam and London. In these projects, kinetic energy from pedestrian traffic is harvested to illuminate public spaces, showing the feasibility of small-scale urban energy generation. When combined with other unconventional sources, these systems highlight the potential for distributed urban power networks that support traditional electricity grids. IoT integration further improves operational efficiency by providing real-time monitoring of energy flows, predicting peak generation times, and optimizing energy distribution to prioritized loads.

2. Rural and Off-Grid Applications

Rural areas, especially those distant from centralized grids, benefit from hybrid systems designed to take advantage of locally available unconventional resources. Microbial fuel cells (MFCs) are an excellent example, using organic waste from agriculture or livestock to generate electricity while also treating waste. In small pilot projects in India and Sub-Saharan Africa, MFC-based hybrid systems have powered LED lighting, small pumps, and sensor networks in remote villages. When paired with thermoelectric generators that capture temperature differences in soil or water bodies, these hybrid systems provide a continuous, reliable energy supply, even in places where solar and wind availability may be sporadic.

Another rural implementation includes low-speed wind turbines combined with footstep or vibration energy harvesters in communal areas such as schools or marketplaces. These setups allow small communities to operate essential infrastructure, including water pumps, refrigeration units, and communication devices, without relying on national grids. Modular and scalable system designs ensure that as energy demand increases, additional units can be added without major retrofitting. This flexibility is vital in rural areas where resource availability and community needs change seasonally and regionally(Li et al. 2020).

3. Community-Based Manufacturing and Energy Independence

Hybrid systems are even more relevant when designed for community-based assembly and management. Using locally manufactured components, 3D-printed modules, and easily replaceable parts allows communities to take responsibility for their energy infrastructure, lowering maintenance costs and reliance on outside service providers. In pilot projects in Southeast Asia, communities have successfully installed hybrid systems using footstep mats, micro-hydro generators, and MFC modules made from locally sourced materials. Training programs for technicians and local operators have promoted sustainability, enabling communities to gradually expand systems while adjusting to changing energy needs.

Community-driven implementations also bring social and economic benefits beyond electricity generation. Local manufacturing and maintenance create job opportunities, promote technical skill development, and encourage innovation in energy solutions. Furthermore, community ownership ensures that hybrid systems are maintained responsibly, reducing the chances of neglect or misuse. These factors together contribute to energy resilience, social empowerment, and fair access to decentralized power (Chire Saire et al, 2021).

4. Industrial and Institutional Applications

Hybrid systems are not limited to urban or rural community settings; they are also valuable in industrial and institutional contexts. Wastewater treatment plants, food processing units, and agricultural facilities generate significant organic and thermal energy, which can be captured using MFCs, thermoelectric harvesters, and vibration-based energy capture. For example, a hybrid system combining MFCs and TEGs in a medium-scale wastewater facility can treat effluent while generating enough electricity to power sensors, pumps, and lighting within the plant.

Institutions such as universities and research campuses provide another excellent testbed for hybrid systems. High footfall, laboratory waste heat, and controlled acoustic environments offer predictable energy inputs. Hybrid systems deployed in these contexts not only generate electricity but also serve as teaching tools, demonstrating sustainable energy principles to students and researchers while providing real operational data for ongoing improvement (Shezan et al. 2021).

5. Challenges and Mitigation Strategies

Despite the promising applications, implementing hybrid systems faces several challenges. The low power density of unconventional sources requires integration with multiple complementary technologies to produce significant output. Additionally, environmental variability—such as changes in pedestrian traffic, temperature gradients, or organic waste availability—necessitates adaptive energy management and storage solutions to ensure continuous operation.

Financial and policy barriers also affect adoption. High upfront costs, lack of regulatory incentives, and limited technical expertise in some regions can slow progress. Overcoming these challenges involves smart design, policy support, and community engagement. Grants, subsidies, and training programs can help with early adoption, while modular and scalable designs lower risk and enable gradual investment. Pilot projects with thorough monitoring and performance evaluation provide evidence to support larger-scale implementation.

CASE STUDY EXAMPLES

1. Urban Metro Station Hybrid Floor System: In Amsterdam, a pilot installation of piezoelectric floor mats combined with thermoelectric panels and vibration harvesters powered corridor lighting, meeting up to 30% of energy needs for localized lighting. Real-time monitoring via IoT devices optimized energy capture based on pedestrian density.
2. Rural Microbial Fuel Cell Hybrid System in India: A village-level deployment combined MFCs with thermoelectric harvesters for powering LED lighting and water pumps. The system used locally made modules, trained community operators, and modular expansion to meet seasonal energy needs.
3. Industrial Wastewater Hybrid System: A medium-scale treatment facility used MFCs and TEGs to manage effluent and provide electricity for plant operations. AI-based controllers improved energy distribution between storage batteries and operational loads, increasing overall energy efficiency by 15%.
4. University Campus Energy Harvesting Network: Footstep generators mixed with vibration and thermal harvesters in high-traffic campus areas supplied energy for lighting, environmental sensors, and IoT devices. The system acted as both a functional power source and a research platform for students studying sustainable energy systems.

OPTIMIZATION OF SMART HYBRID POWER HARVESTING SYSTEMS INTEGRATING CONVENTIONAL AND UNCONVENTIONAL RENEWABLE SOURCES

The design and implementation of smart hybrid power harvesting systems require a complete engineering approach that combines both traditional and new renewable energy sources. This integration aims for efficient, reliable, and improved energy output. A hybrid system is designed to use complementary energy sources, managing variability and space limitations while providing consistent power to meet changing load needs. The design starts with an assessment of possible energy sources and specific site conditions. Traditional sources, like solar photovoltaic (PV) panels and wind turbines, form the foundation of the system due to their higher energy density and established technology. Factors like irradiance profiles, tilt angles, and temperature coefficients help determine the selection of solar PV arrays. Engineering also considers shading, array orientation, and series-parallel configurations, which affect voltage and current outputs. Wind turbines are chosen based on local wind speed patterns, turbulence, hub height, and power characteristics. Proper turbine sizing, blade design, and generator choice are necessary to enhance conversion efficiency across various wind conditions. Both renewable sources can be variable; therefore, the hybrid setup must limit energy shortfalls by engineering complementary generation profiles. For example, solar energy peaks during the day, while wind often picks up at night or during windy seasons. These complementary traits help engineers create hybrid systems that lessen the need for storage or backup generation while ensuring reliable supply.

Adding unconventional energy sources brings extra engineering challenges and opportunities. Kinetic energy harvesting through piezoelectric or electromagnetic power tiles, such as Pavegen systems, captures mechanical energy from people walking. The engineering design focuses on optimizing the conversion from mechanical to electrical energy, improving energy efficiency, and matching load impedance to maximize power extraction. Thermoelectric generators turn temperature differences into electricity using the Seebeck effect and require careful thermal management to maximize the temperature difference across the modules. Engineers choose materials with high efficiency, improve heat sink designs, and implement thermal coupling to enhance energy conversion. Systems that convert noise to electricity use piezoelectric or resonant acoustic devices to gather vibrational energy from sound sources. They require careful adjustment of resonance frequencies and placement in noisy environments for maximum power collection. Microbial fuel cells (MFCs) utilize electrochemical reactions from microorganisms in organic materials. Choices in electrode materials, biofilm management, and fluid dynamics are key to optimizing energy output. Each unconventional source usually generates lower power density than traditional sources, but their inclusion offers extra energy, boosts system reliability, and allows use of local energy resources that would generally go untapped.

Power electronics play a central role in engineering smart hybrid systems, allowing for energy conversion, regulation, and the integration of multiple sources into a single output. Maximum power point tracking (MPPT) controllers are used for each renewable source to change operating conditions for peak efficiency. For solar PV modules, MPPT algorithms like Perturb and Observe (P&O) and Incremental Conductance (IncCond) continuously monitor voltage and current to ensure operation near peak power points, compensating for variations due to changes in irradiance and temperature. Wind turbine outputs are managed through power converters that adapt variable AC or DC voltages to the hybrid system bus, often using full-bridge or half-bridge converters with synchronous rectification for high efficiency. For unconventional sources, low-power converters are designed, such as boost converters and buck-boost topologies, to fit the low-voltage, high-impedance outputs to the system bus, ensuring smooth integration. Bidirectional converters help manage energy flow to and from storage units, allowing for charging during extra generation and careful discharge during high demand. The integration of various converters and MPPT units requires careful planning to avoid conflicts and power losses, often handled through centralized or decentralized energy management systems.

Energy storage is vital in hybrid system engineering, acting as a buffer for variable generation and stabilizing output. Lithium-ion batteries are preferred for their high energy density and quick charging cycles, while flow batteries offer scalability for larger setups, and supercapacitors provide high power density for short-term support. Engineers use battery management systems (BMS) to monitor state-of-charge (SOC), state-of-health (SOH), temperature, and voltage across cells. They implement balancing methods to avoid overcharging, undercharging, or overheating. Proper sizing of storage units relies on careful load profiling, forecasting generation, and reliability needs. For instance, storage capacity is designed to serve critical loads during low generation periods while keeping operational efficiency by reducing deep cycles that can harm battery life. Smart energy management algorithms enhance the interaction between various storage types, favoring supercapacitors for high-power, short-duration loads and batteries for sustained energy needs, which improves system longevity and efficiency.

The control strategy of a smart hybrid system is a complex engineering challenge that incorporates real-time monitoring, predictive analytics, and adaptable decision-making to optimize energy flows. Supervisory control units or microcontrollers gather data from sensors that track irradiance, wind speed, vibration levels, temperature differences, load demand, and SOC of storage units. Predictive algorithms, often based on machine learning, forecast energy availability and load changes, allowing for proactive adjustments in generation priorities, storage use, and load management strategies. For example, if solar irradiance is predicted to drop due to cloud cover, wind or unconventional sources can be prioritized early on, while storage units are managed to ensure an uninterrupted supply. Engineers create layered control systems. Primary controllers handle individual source optimization, secondary controllers manage the integration of multiple sources, and tertiary controllers oversee predictive load balancing, grid interfacing, and fault detection. Communication protocols like Modbus, CAN, or IoT-enabled wireless networks ensure reliable, real-time data exchange. Redundancy and fail-safe systems help maintain reliability during component failures.

Optimizing the output in smart hybrid systems involves considering multiple objectives like efficiency, reliability, cost, and environmental impact. System modeling and simulations using tools such as MATLAB/Simulink, HOMER, or PSS/E allow engineers to assess energy generation profiles, power losses, voltage stability, and component sizing in various scenarios. Efficiency optimization aims to reduce conversion losses, match source and storage features, and ensure MPPT and converter systems work in high-efficiency areas. Reliability optimization includes having backups for crucial parts, fault-tolerant control strategies, and strong mechanical designs to withstand environmental stresses like wind and temperature extremes. Cost optimization looks at overall spending, operational costs, and lifecycle expenses, guiding the choice of materials, parts, and strategies that provide the best return on investment. Environmental considerations drive material choices, emissions reduction, and sustainable disposal or recycling of components. Engineers use multi-objective optimization algorithms like genetic algorithms, particle swarm optimization, or linear programming to meet these needs at the same time, creating a balanced and practical hybrid power system.

Managing loads is another important engineering aspect in optimizing hybrid system output. Smart load controllers or demand-side management methods allow for dynamic adjustments in energy consumption to match available generation, reducing the need for excessive storage and lowering dependence on the grid. Loads are categorized into critical and non-critical, and scheduling algorithms help shift energy-heavy operations to high-generation times, such as during sunny days for solar or windy periods for wind turbines. In industrial settings, production can be synchronized with energy availability. In urban or residential areas, IoT-enabled smart devices adjust operations based on real-time energy pricing and generation patterns. Grid interfaces are designed to meet local utility standards by managing voltage, reactive power, and harmonics to ensure safe and reliable connections with the main grid. Engineers often use bidirectional inverters and smart meters to allow energy export during surplus generation, which boosts economic benefit and supports distributed generation projects.

Thermal and mechanical factors are crucial in engineering hybrid systems. Solar PV modules need thermal management to prevent efficiency drops at high temperatures, using passive heat sinks or active cooling systems. Wind turbines require structural analysis to withstand changing wind loads, fatigue, and vibration. Kinetic energy tiles must have strong mechanical designs to handle repeated impact while maximizing conversion efficiency. Thermoelectric and microbial systems need optimized thermal and fluidic designs to maintain conditions that support high energy conversion. Environmental protection features, such as encapsulation and corrosion-resistant materials, extend system lifespan and reliability. Engineers conduct finite element analysis, thermal simulations, and mechanical stress tests to ensure all components work effectively and safely under expected conditions.

Case studies of engineering implementations illustrate the practicality of smart hybrid power harvesting systems. Urban microgrids that combine solar PV, small wind turbines, footstep power tiles, and thermoelectric generators show how local power supply can continuously support lighting, sensors, and low-power electronics. Industrial plants using hybrid systems with solar, wind, thermoelectric recovery, and microbial fuel cells have significantly reduced grid reliance and costs. In these cases, engineers carefully size components, design converters, use advanced MPPT and energy management methods, and incorporate IoT-enabled monitoring to improve performance. Performance metrics such as energy yield, system efficiency, storage use, and reliability are continuously assessed to optimize system operation and guide maintenance. Optimization remains an evolving engineering task requiring adaptive control, predictive maintenance, and system adjustments to address changes in environment, load patterns, and technology. Smart hybrid power harvesting systems entails integrating both traditional and unconventional energy sources, advanced power electronics, effective energy storage management, predictive control strategies, and thorough optimization methods. By leveraging complementary generation capabilities, adaptive MPPT, smart energy management, and solid mechanical and thermal designs, hybrid systems can provide reliable, efficient, and sustainable energy. Load management, grid interfacing, and IoT-enabled monitoring enhance overall performance, supporting proactive decision-making, real-time optimization, and reduced energy waste. Economic, environmental, and reliability factors inform choices in components, system sizing, and operational methods, ensuring that engineering solutions are practical, scalable, and aligned with sustainability goals. Combining solar, wind, kinetic, thermoelectric, noise, and microbial energy sources in a unified, intelligently controlled framework showcases the future of hybrid power systems. These engineering-focused designs tackle energy variability and efficiency issues while promoting decentralized, resilient, and optimized energy solutions suitable for urban, industrial, and rural settings. Ongoing innovation in materials, control algorithms, and predictive analytics will further improve the effectiveness of smart hybrid systems, making them key players in the global shift toward sustainable, reliable, and intelligent energy networks. Through careful engineering design, precise component integration, and adaptive optimization, smart hybrid power harvesting systems offer a feasible, cost-effective, and environmentally friendly approach to meet increasing energy demands using both traditional and unconventional renewable resources efficiently.

Figure 4. Hybrid Power Harvesting Scheme

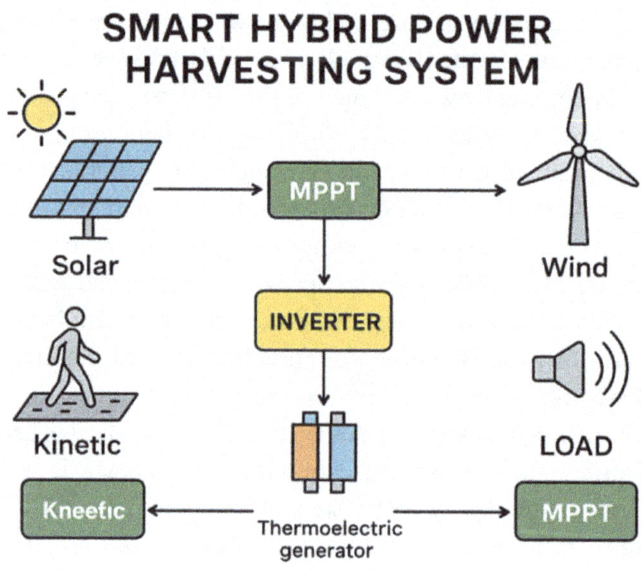

FUTURE PROSPECTS AND EMERGING TRENDS

The future of hybrid power systems based on unconventional renewable resources has significant, largely untapped potential. This potential is driven by technological innovation, societal needs, and the need for sustainable energy. Current hybrid implementations show their feasibility and benefits. However, trends in IoT, artificial intelligence, new materials, and community-based manufacturing are set to change the capabilities, scalability, and impact of these systems. This section looks at these trends and how they will change hybrid systems from localized setups to self-optimizing, adaptive networks that contribute meaningfully to decentralized energy infrastructures (Sarker et al.2023).

Integrating the Internet of Things (IoT) into hybrid energy systems is set to transform energy management and system efficiency. IoT devices such as sensors, communication modules, and cloud-based monitoring platforms collect real-time data on environmental factors, energy flows, and user demand patterns. For example, footstep power generators in urban transit hubs can use IoT sensors to monitor pedestrian traffic density. This provides predictive insights into energy availability and optimizes allocation to important needs like lighting or emergency systems. Similarly, thermoelectric harvesters in industrial settings can send temperature gradient data to cloud-based controllers. This allows for predictive energy routing and storage decisions. IoT networks also support distributed coordination among multiple hybrid nodes in urban, rural, or campus environments. Hybrid systems can monitor performance collectively, share excess energy, and respond dynamically to changes in energy availability or demand. This interconnected framework enables smart microgrids, where unconventional hybrid systems not only meet local energy needs but also contribute to larger community or city-scale energy resilience. Additionally, IoT-driven monitoring helps with predictive maintenance. This identifies potential component failures before they occur, reducing downtime and operational costs (Orie et al. 2025).

Artificial intelligence (AI) is becoming crucial in optimizing hybrid power systems, especially given the variability and low output of many unconventional sources. Machine learning algorithms can forecast energy availability, prioritize loads in real time, and manage storage systems for maximum efficiency. For instance, AI can analyze historical environmental and usage data to predict when high footstep or vibration energy generation will happen. This allows for the efficient allocation of stored energy to critical systems. Moreover, AI creates self-optimizing systems that can adapt to changes in the environment, user behavior, and unexpected shifts. In hybrid setups, this means that various energy sources like microbial fuel cells, thermoelectric harvesters, and footstep generators can work well together without manual input. Adaptive algorithms can balance inputs, redirect energy flows to storage or consumption nodes, and adjust operational parameters to increase output and reduce energy waste. In remote or decentralized locations, AI management cuts down the need for continuous human oversight, allowing for autonomous operation over long periods (Rashid et al. 2024).

The rise of fabrication technologies such as 3D printing and modular assembly kits allows communities to build, deploy, and maintain hybrid energy systems locally. This approach encourages energy independence by reducing reliance on centralized manufacturing and distribution, while also promoting local innovation and economic growth. In rural or resource-limited areas, locally fabricated modules such as thermoelectric units, microbial fuel cell arrays, or piezoelectric mats can be assembled gradually to meet changing energy needs.

Community-driven production also improves system sustainability. Local operators trained in assembly, monitoring, and maintenance can lower operational costs and extend system lifespan, ensuring that hybrid installations remain functional over time. Additionally, community involvement fosters ownership, which increases the likelihood of responsible use and long-term engagement. When combined with IoT and AI management, locally produced hybrid systems create adaptive, resilient microgrids that can operate autonomously while contributing to larger energy networks.

Future hybrid systems are expected to become self-optimizing networks that respond to real-time changes in environmental conditions, energy demand, and infrastructure availability. By integrating AI, IoT, and advanced materials, these networks can constantly adjust energy capture, storage, and distribution to maintain peak performance. For instance, a hybrid system in a smart city could redirect footstep-generated energy during busy pedestrian hours while also tapping into thermoelectric energy from HVAC systems and outputs from nearby organic waste facilities. These adaptive networks not only maximize energy efficiency but also improve resilience to environmental challenges. In uncertain climates, instances of grid failure, or resource shifts, multi-source hybrid systems maintain energy continuity by utilizing complementary inputs. This flexibility is especially valuable in decentralized or off-grid areas where energy reliability is crucial for lighting, communication, water pumping, and other essential services (Elias et al. 2025).

The future of hybrid energy systems goes beyond technology to include collaboration among engineers, cognitive scientists, urban planners, and policy-makers. Designing systems that focus on people, are economically viable, and are environmentally sustainable requires insights from various fields. Studies on cognition help with understanding energy usage patterns and behavioral interventions, making sure hybrid systems fit with human activity and societal needs. Urban planning perspectives optimize system placement, accessibility, and integration with existing infrastructure, while policy frameworks support adoption through incentives, standards, and regulations. Collaboration across different disciplines also promotes innovation in design, operation, and scaling. For example, engineers can produce modular hardware, AI experts can optimize energy flows, and social scientists can create engagement programs that encourage community involvement. By merging these perspectives, hybrid energy systems can grow into robust, adaptive, and socially integrated solutions instead of isolated technical setups (Babatunde et al. 2025)

Deploying unconventional hybrid systems widely has important environmental and societal effects. By diversifying energy inputs and decreasing reliance on fossil fuels, hybrid systems support decarbonization and climate change mitigation. Decentralized systems cut down on transmission losses, lower infrastructure costs, and improve energy access in underserved areas. In cities, hybrid systems aid smart city initiatives, reduce peak demand on central grids, and contribute to sustainable public infrastructure. In rural and off-grid regions, they provide reliable electricity for education, healthcare, irrigation, and small businesses, fostering economic growth and social empowerment.

Figure 5. Future Opportunities

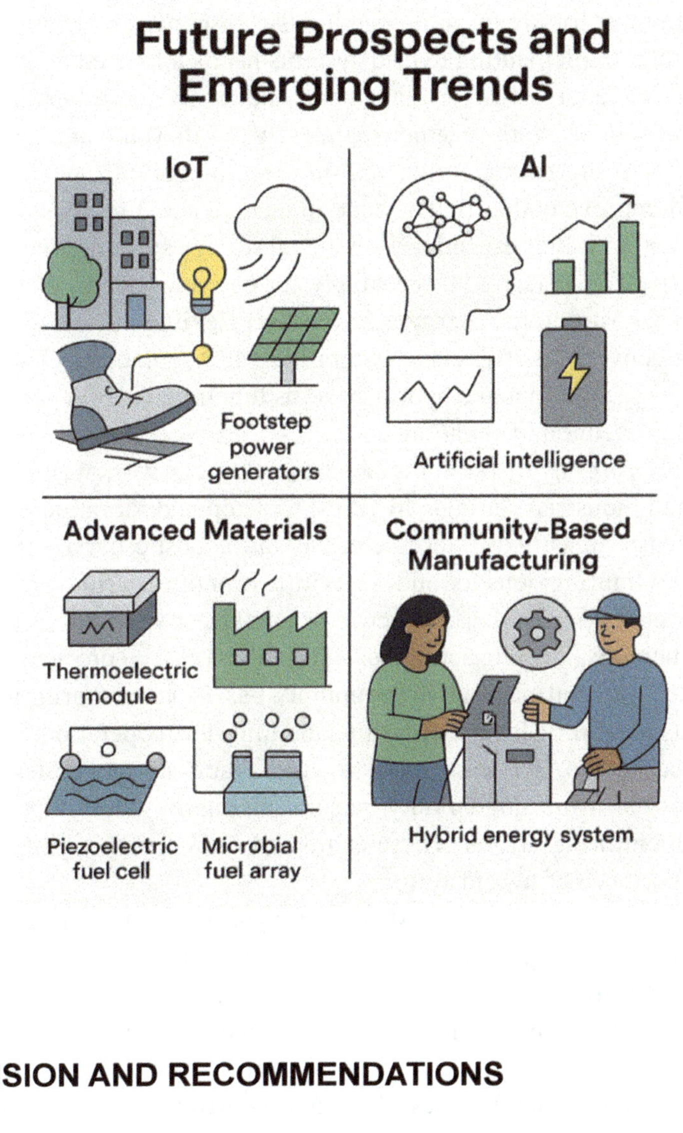

CONCLUSION AND RECOMMENDATIONS

Hybrid power generation systems that use unconventional renewable resources offer a valuable way to tackle the growing challenges of global energy demand, environmental sustainability, and decentralized electricity access. This chapter has looked at the limits of traditional renewables like solar and wind, focusing on their intermittency, dependency on specific locations, and the need for additional solutions. By adding technologies such as footstep power generators, thermoelectric harvesters, noise-to-power converters, and microbial fuel cells to hybrid setups, energy systems become more resilient, reliable, and efficient based on location. Hybridization allows for the combined use of different energy sources, turning individually limited options into integrated networks that can provide consistent and adaptable power for urban, rural, and industrial scenarios.

Designing and implementing hybrid systems needs a layered approach. Modular design, flexible energy flow management, and smart controllers help different unconventional sources work together seamlessly. AI-driven management and IoT integration offer energy forecasting, real-time optimization, and self-regulating control, which improve both efficiency and independence. Energy storage options, like supercapacitors, batteries, and new chemical storage systems, manage irregular outputs, allowing for a steady power supply for essential needs. Using advanced materials like nanostructured thermoelectrics and flexible piezoelectric polymers boosts energy conversion efficiency, durability, and adaptability. These technical strategies work together to make hybrid systems functional across various environmental and socio-economic situations.

The practical use of hybrid systems has shown they can work well in urban, rural, community, and industrial settings. In cities, footstep and vibration energy can be harnessed in busy areas, with thermoelectric and acoustic harvesters supporting energy needs for lighting, sensors, and smart infrastructure. In rural areas, microbial fuel cells can team up with thermoelectric or low-speed wind harvesters to power off-grid communities, farming activities, and small industrial processes. Using 3D-printed modules and local materials in community-based manufacturing and assembly encourages sustainability, local ownership, and long-term operation. For industrial and institutional setups, like wastewater treatment facilities and college campuses, hybrid systems can work double duty, generating energy while supporting waste treatment or monitoring efforts. These examples show the flexibility, scalability, and social importance of hybrid systems.

Current trends suggest that the next wave of hybrid energy systems will be more adaptable, self-optimizing, and focused on human needs. IoT networks and AI tools will promote real-time energy management, preventive maintenance, and coordinated actions among multiple hybrid nodes. New materials and manufacturing methods will increase conversion efficiency, lessen system weight and size, and facilitate installation in unconventional spaces. Community-led deployment models will give local people more control, leading to greater energy independence and grassroots innovation. Collaboration among engineers, urban planners, cognitive scientists, and policymakers will help ensure that hybrid systems not only perform well technically but also meet societal needs, economic realities, and principles of fairness.

In summary, hybrid power systems that rely on unconventional renewable resources are essential for a sustainable and decentralized energy future—not just an addition to traditional renewables. By combining various energy sources, smart management approaches, flexible designs, and community involvement, hybrid systems can effectively tackle the challenges of energy reliability and environmental sustainability. Policymakers, researchers, and practitioners should prioritize collaboration across fields, invest in adaptable technologies, and support pilot projects that showcase practical applications. Ongoing research into advanced materials, AI optimization, and blending social and technical aspects will improve performance, making hybrid systems a key part of resilient, fair, and environmentally friendly energy infrastructures globally.

In the end, embracing hybrid power generation from unconventional renewable sources provides a route to flexible, resilient, and human-focused energy systems. These systems can respond to the changing demands of the 21st century while pushing forward global sustainability initiatives. By viewing hybridization as a critical strategy rather than a mere option, stakeholders can ensure that energy transition efforts are inclusive, innovative, and successful in tackling both present and future challenges.

REFERENCES

Amin, M., & Lee, C. (2022). Hydrogen production through renewable and non-renewable energy processes and their impact on climate change. *International Journal of Hydrogen Energy*, *47*(48), 27174–27189. DOI: 10.1016/j.ijhydene.2022.07.172

Babatunde, O., Akintayo, B., Ighravwe, D., & Olanrewaju, O. (2025). Transdisciplinary approach to accelerate the adoption of hybrid renewable energy systems through sustainable design. *Frontiers in Built Environment*, *1*, 1520883. Advance online publication. DOI: 10.3389/fbuil.2025.1520883

Darney, P. E.P. Ebby Darney. (2022). Power flow optimization of a hybrid energy system with salp swarm algorithm. *Journal of Electronics and Informatics*, *4*(4), 266–274. DOI: 10.36548/jei.2022.4.006

Elias, F., Ekpo, S., Alabi, S., Uko, M., Enahoro, S., Ijaz, M., Ji, H., Unnikrishnan, R., & Olasunkanmi, N. (2025). Design of multi-sourced MIMO multiband hybrid wireless RF-perovskite photovoltaic energy harvesting subsystems for IoTs applications in smart cities. *Technologies*, *13*(3), 92. Advance online publication. DOI: 10.3390/technologies13030092

Enescu, D. (2019). Thermoelectric energy harvesting: Basic principles and applications. In *Green Energy Advances*. IntechOpen., DOI: 10.5772/intechopen.83495

Golubev, V. A., Verbnikova, V. A., Lopyrev, I. A., Voznesenskaya, D. D., Alimov, R. N., Novikova, O. V., & Konnikov, E. A. (2021). Energy evolution: Forecasting the development of non-conventional renewable energy sources and their impact on the conventional electricity system. *Sustainability (Basel)*, *13*(22), 12919. DOI: 10.3390/su132212919

González Cusa, Y., Moya Rodríguez, J. L., Vieira de Melo, S. A. B., & Andrade Torres, E. (2025). Optimization of hybrid renewable energy systems for power generation: A bibliometric review. *International Journal for Innovation Education and Research*, *9*(3), 369–380. DOI: 10.31686/ijier.vol9.iss3.3015

Hossain, J., Sadad Ovi, N., & Monirujjaman Khan, M. (2021). Design and investigation of energy harvesting system from noise. *Energy and Power Engineering*, *13*(8), 307–321. DOI: 10.4236/epe.2021.138021

International Energy Agency. (2024).. . *World Energy Outlook, 2024*. Advance online publication. DOI: 10.1787/9789266829873-en

Li, J., Liu, P., & Li, Z. (2020). Optimal design and techno-economic analysis of a solar-wind-biomass off-grid hybrid power system for remote rural electrification: A case study of west China. *Energy*, *208*, 118387. Advance online publication. DOI: 10.1016/j.energy.2020.118387

O'Dwyer, E., Pan, I., Charlesworth, R., Butler, S., & Shah, N. (2020). Integration of an energy management tool and digital twin for coordination and control of multi-vector smart energy systems. *Sustainable Cities and Society*, *62*, 102412.

Orie, R., Otabil, L., Agorua, J., & Iqbal, M. T. (2025). Enhanced IoT-based optimization for a hybrid power system in Cartwright, Labrador. *Energies*, *18*(7), 1566. Advance online publication. DOI: 10.3390/en18071566

Pavegen. (2025). What can Pavegen power? https://www.pavegen.com/what-can-pavegen-power

Rashid, A., Biswas, P., Biswas, A., Al Nasim, M. A., Gupta, K. D., & George, R. (2024). Present and future of AI in renewable energy domain: A comprehensive survey. arXiv preprint arXiv:2406.16965. DOI: 10.48550

Saire, C. J. E., Gastelo Roque, J. A., & Canziani, F. (2021). Study of a hybrid photovoltaic-wind smart microgrid using data science approach. arXiv preprint arXiv:2105.08510. DOI: 10.48550

Saleem, Y., Crespi, N., Rehmani, M. H., & Copeland, R. (2017). Internet of Things-aided smart grid: Technologies, architectures, applications, prototypes, and future research directions. arXiv preprint arXiv:1704.08977. /arXiv.1704.08977DOI: 10.48550

Sarker, A. K., Azad, A. K., Rasul, M. G., & Doppalapudi, A. T. (2023). Prospect of green hydrogen generation from hybrid renewable energy sources: A review. *Energies*, *16*(3), 1556. Advance online publication. DOI: 10.3390/en16031556

Shezan, S. K. A., Rawdah, S., Ali, S., & Rahman, Z. (2021). Design and implementation of an islanded hybrid microgrid system for a large resort center for Penang Island with the proper application of excess energy. arXiv preprint arXiv:2107.01032. /arXiv.2107.01032DOI: 10.48550

Tan, W. H., Chong, S., Fang, H.-W., Pan, K.-L., Mohamad, M., Lim, J. W., Tiong, T. J., Chan, Y. J., Huang, C.-M., & Yang, T. C.-K. (2021). Microbial Fuel Cell Technology—A Critical Review on Scale-Up Issues. *Processes (Basel, Switzerland)*, *9*(6), 985. DOI: 10.3390/pr9060985

Unegbu, H. C. O., Yawas, D. S., Dan-asabe, B., & Alabi, A. A. (2025). Integrating renewable energy solutions in sustainable building projects: A case study of Nigerian urban centers. *Discover Civil Engineering*, *2*(1), 67. Advance online publication. DOI: 10.1007/s44290-025-00226-8

Yalla, S. P., Subudhi, P. S., & Ramachandaramurthy, V. K. (2022). Topological review of hybrid RES-based multi-port converters. *IET Renewable Power Generation*, *16*(6), 1087–1106. DOI: 10.1049/rpg2.12356

Zhang, F., Li, X., & Wu, Y. (2023). Footstep-powered floor tile: Design and evaluation of an electromagnetic frequency up-converted energy harvesting system enhanced by a cylindrical Halbach array. *Sustainable Energy Technologies and Assessments*, *60*, 103571. Advance online publication. DOI: 10.1016/j.seta.2023.103571

Chapter 11
EEGNet-Based Emotion Recognition Using Multimodal EEG and ECG Signals:
A Hybrid AI-Driven Model

T. Nithya
https://orcid.org/0000-0001-9738-9910
Velalar College of Engineering and Technology, India

V. Chandrasekaran
Velalar College of Engineering and Technology, India

S. Mahendrakumar
https://orcid.org/0000-0001-6966-2545
Velalar College of Engineering and Technology, India

N. V. Keerthana
https://orcid.org/0000-0001-8878-2083
Velalar College of Engineering and Technology, India

C. Vasuki
https://orcid.org/0009-0003-7926-2045
Nandha Engineering College, India

R. Loganathan
https://orcid.org/0009-0004-3659-846X
Paavai Engineering College, India

P. Sathyasutha
Gnanamani College of Technology, India

R. Maheshwari
https://orcid.org/0009-0001-5935-8475
Velalar College of Engineering and Technology, India

ABSTRACT

Physiological signal-based emotion recognition has emerged as a significant field of study with applications in interactive system interfaces, affective computing and healthcare. The DREAMER dataset to explore the effectiveness of EEGNet, a compact

DOI: 10.4018/979-8-3373-5702-7.ch011

convolutional neural network designed for EEG-based interfaces between brain and computers in emotion classification. EEGNet can attain superior emotion recognition accuracy by learning directly from raw EEG data, thus outperforming traditional feature-based methods. The experimental results demonstrate that EEGNet attains accuracy for valence classification and arousal classification, significantly surpassing conventional machine learning approaches. This paper demonstrates the usefulness of EEGNet for creating EEG-based emotion identification systems as a versatile and effective solution for real-world application. The possible future concepts under this research include multimodal data integration and subject independent classification, which enhance deep learning in physiological signal processing.

1. INTRODUCTION

Human emotions or affect are a result of psychological and physiological changes due to human consciousness. It plays a major role in interpersonal communication, not only among humans but also between humans and machines. The time spent on social media, online purchasing platforms, e-games, and other digital activities keeps escalating due to the ease with which it is available to people. However, human-computer interfaces are still lacking since they don't possess the emotional intelligence required to analyze and understand the emotional data needed to leverage it. In order to tackle this, systems can be equipped with the ability to decipher the emotional data. The human brain is not a blank canvas but stores all past experiences, which aids in everyday decision-making. Brain activity leads to emotion generation or change in emotional state, and could be represented by using physiological signals like Electro-Encephalo-Gram (EEG) for brain activity and Electro-CardioGram (ECG) for signals from the heart (Jafari et al., 2023). Human behaviour is dependent on emotional state, which in turn is a manifestation of behaviours and a reflection of thoughts caused by psychological conditioning. Emotions can be acquired through numerous ways, such as facial expression analysis, animated features from texts, EEG, Galvanic Skin Response (GSR), and other physiological signals (Gursesli et al., 2024). Physiological signals are difficult to manipulate or feign and accurately mirror an individual's state of thoughts and emotions(X. Wang et al., 2023).

The primary manner in which emotion recognition is used in medical settings is when a healthcare provider reacts to a patient's emotional state and modifies the course of treatment if needed (Hasnul et al., 2021). To do this, the following steps must be followed: induction signal acquisition, preprocessing, feature extraction, and classification (Raheel et al., 2020). Another example of emotion driving people's behaviour could be seen in shopping during the COVID-19 pandemic, where people showed an increase in arousal and a decrease in pleasure in shopping(Yu & Wang,

2022) . The customers showed an urgency in shopping due to their risk awareness and were wary of not getting their health negatively affected. Affect and emotions also play a part in sustainable behaviour change in numerous aspects (Rolls, 2023). The effect reaction imparted a significant role in the forecasting of perceived climate risks, predisposition towards mitigation measures, adaptation behaviour, acceptance of policies, willingness to integrate technologies, and more.

In spite of recent progress in multimodal emotion recognition, there are still some gaps in current research. Firstly, most studies only use EEG or ECG signals, which reduces the classification robustness because there are no complementary physiological signals. Secondly, conventional machine learning methods are greatly based on hand-crafted features, which can miss sophisticated spatiotemporal patterns in data. Thirdly, though deep learning models like CNNs and LSTMs are promising, they tend to be dataset-specific and do not generalize well across subjects.

The contribution of this chapter is filling the above gaps by combining EEG and ECG signals via a hybrid deep learning model (EEGNet with dynamic aggregation). The method enhances robustness, improves explain ability, and preserves temporal and spatial relationships.

The key findings of this research are:

- A multimodal emotion recognition system superior to standard ML, CNN, and LSTM baselines.
- Evidence of high classification accuracy on various emotional states using the DREAMER dataset.
- Practical insights in actual health care applications, HCI, and mental health surveillance.

This effort thus makes an effort to close the gap between traditional emotion recognition methods and contemporary multimodal deep learning methodologies, and emphasize the need for interpretable and scalable models.

2. LITERATURE REVIEW

Analysing spatiotemporal patterns in the data, through the employment of Convolutional Neural Network (CNN),to classify emotions based on EEG signals with a significant level of accuracy (Kumawat & Jain, 2021). The framework mainly integrates three basic emotions of sadness, happiness, and anger. Similarly, a hybrid model using a combination of CNNs with Long Short-Term Memory (LSTM) networks is proposed(Raheel et al., 2020). In the multimodal dataset, the model showed better recognition rates. To discover emotions, Raheel et al. (2021) introduced the use of multiple physiological markers, such as GSR, Heart Rate Variability (HRV), and EEG (Yu & Wang, 2022). These signals reflect a person's thoughts and feelings are hard to fake or influence. Although this study highlighted how multi-modal data might improve accuracy, limitations like computing complexity and integrating signals from different sources nevertheless have to be tackled. The CNN was used to discern emotions according to EEG signals, and reached higher accuracy rates of 90.65% for arousal, together with 90.01% for valence (Kumawat & Jain, 2021). This model works very efficiently for a considerable volume of EEG data by showing improved feature extraction and reduced noise impact. A cyclic CNN that used both CNN and Recurrent Neural Networks (RNN) models with an attention mechanism was proposed by Tao et al. (2020). By integrating the spatial and temporal features from EEG signals, the model showed better performance with the DEAP and DREAMER datasets (Katsigiannis & Ramzan, 2017). Exploration of the usage of different machine learning algorithms for face feature-based emotion recognition demonstrated the effectiveness of using multimodal data for emotion identification tasks and demonstrated how combining speech and facial expressions can increase accuracy (Gursesli et al., 2024).

Applications in human-computer interaction and mental health monitoring need emotion recognition from EEG signals. The traditional approach involves manual feature extraction from EEG recordings, which is often inaccurate and prone to missing complex spatiotemporal patterns of brain activity. This weakness underlines the need for more advanced, automated systems that are capable of making accurate emotion classification from raw EEG data. Challenges, including data redundancy and the complexity in extracting required features from full-channel EEG signals, reduce the efficiency of traditional machine learning methods.

Techniques based on EEG alone may not be sufficient to fully capture the intricacy of human emotions, given that they are influenced by numerous physiological and extrinsic factors, although deep learning frameworks like EEGNet have shown promise in automatically extracting features from raw EEG data (Pan et al., 2023). This shows there is a need for multicomponent approaches combining EEG with other data for better accuracy and more robust performance. The main issues that arise are individual differences in emotion expression and ensuring that a model can guarantee consistent performance across different datasets. To close this gap between people and technology, our work focuses on the EEGNet model while exploring multimodal methods to boost the accuracy and reliability of systems employed for emotion identification.

It pursues developing high-order techniques to precisely detect and analyse human emotions from physiological signals like EEG and ECG. By employing deep learning architectures such as EEGNet, the accuracy and reliability of the emotion recognition systems can be increased. The basic goal is to use the features in EEG and ECG signals to improve the classification of emotions. Discussions on the techniques used, findings, and conclusions on the method used in data preprocessing are provided. Recommendations and directions for future study and development in the area of emotion recognition are also provided. The paper seeks to fill research gaps and is structured as follows: the data collection and pre-processing, methodologies used, findings, and conclusion. The primary purpose of this study is to assist in the development of effective and scalable emotion recognition systems for practical uses, including mental health monitoring and human-computer interaction. The incorporation of emotional intelligence into technology by making such systems more accurate, robust, and interpretable can enhance user experience across various applications. Such advancements will make the system more responsive and adaptive to individuals, especially those with communication difficulties.

Although several emotion theories stating the emotion invocation proposed by neuropsychologists bear similarities in the psychological and physiological aspects, their order of occurrence and the intensity of the inherent processes vary. These psychological and psychological elements comprise the emotional experience. Emotion representation frameworks are proposed to promote emotion modelling, observation, and recognition. These frameworks are one-dimensional or multidimensional. Recent technology advancements like Moderate-to-Vigorous Physical Activity (MVPA) have led to the leveraging of new tools that can be used to understand how neural representations accord with psychological and physiological compositions and their relationship (Ding et al., 2018)(Toraman & Dursun, 2021).

Previous works on emotion recognition can be classified into three categories:

- Conventional Machine Learning Techniques: SVM, Random Forest, and Logistic Regression with hand-engineered EEG features like PSD and wavelets. These are light but not generalizable.
- Deep Learning Techniques: CNNs extract spatial features, and LSTMs represent temporal dependencies. Single-modality dependency and overfitting restrict their scalability.
- Multimodal Fusion Strategies: EEG fusion with ECG, GSR, or facial cues has experienced gains in performance, while current fusion approaches tend to have high complexity and low interpretability.

Herein, the current directions are furthered with a scalable EEGNet-based multimodal system that fills the gap between single-modality limitations and promises enhanced interpretability.

3. DATA COLLECTION

EEG and ECG are two multimodal physiological signals that are obtained during controlled emotional elicitation experiments and are stored in the DREAMER database, an open-source resource for emotion identification research (Katsigiannis & Ramzan, 2017). 23 volunteers were chosen, made to watch 18 film snippets to elicit particular emotional reactions, which made up the dataset. To ensure that they were successful in eliciting the intended emotional reactions, these videos underwent a thorough evaluation and validation procedure.

During the experiment, EEG signals were recorded using a wireless EEG headset, creating an Mx14 matrix, where M represents the quantity of samples and each column represents one of the EEG channels. A Mx2 matrix, with each column denoting one of the ECG channels, was used simultaneously to collect ECG data using an ECG sensor. A detailed depiction of both brain and cardiac activity is offered by this dual model method during emotional simulation. Following each film clip, participants completed a self-assessment rating on a five-point scale to evaluate their emotional states along three dimensions: valence, arousal, and dominance. This was done to supplement the physiological data. An in-depth examination of the connection between physiological signals and self-reported emotional states is made possible by these subjective ratings. This data provided insightful information about the participants' emotional experiences. The dataset is methodically arranged into structured variables with different categories for stimulus clips and baseline clips before each stimulus. The architecture improves the dataset's efficiency for emotion recognition studies and enables reliable comparison analysis.

3.1 Data Pre-Processing

Raw EEG and ECG data can be affected by high-frequency noise, power line interference, and baseline drift (Subasi et al., 2021). The preprocessing pipeline includes filtering to remove noise, standardization for consistent scaling between participants and channels. Baseline correction is performed to reduce individual variations. Additionally, feature extraction decreases dimensionality by identifying valuable data from the signals. To ensure the dataset is complete and functional, missing data is handled. These steps transform the raw physiological data into organized data. Figure 1 represents the sample EEG and ECG signal representation.

Figure 1. Raw EEG and ECG Signal Representation

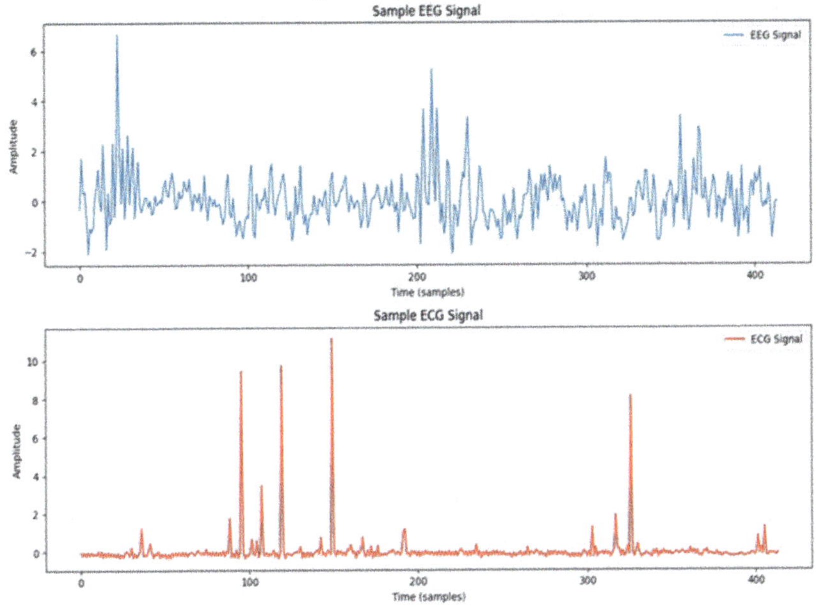

3.2 Exploratory Data Analysis

Data exploration can better depict the time and frequency-domain properties of these EEG and ECG data (Huang et al., 2020). Although some channels, such as SampEn, are invariant concerning the observation time, others, like RNSSD and MeanNN, are sensitive amplitude variables and periodic functions for the EEG data. The rolling mean of RMSSD smooths over transient noise, showing underlying dynamics and transient fluctuations. A peak in the frequency spectrum, situated between 10 and 20 Hz, shows significant alpha or beta brain activity. In this regard, periodic patterns of the ECG data show heart rate variability and the presence of varied amplitudes in channels such as psdalpha_1 and psdbeta_1. While the frequency spectrum has peaks in lower frequency ranges, such as 0,04-0,15 Hz, corresponding to heart rate variability, attention is drawn to the irregularities of the pattern with the application of standard deviation and rolling mean. These findings will give a strong foundation for additional preprocessing and modelling, by highlighting the importance of investigating time-domain as well as frequency-domain variables for EEG and ECG data. Figure 2 represents the rolling Mean is a pivotal technique for analysing time series data, particularly for rolling statistics.

The formula represented in (1) is employed to level out the brief fluctuations and feature longer-term patterns within the data.

$$RM_n = \frac{1}{Ws} \sum_{l=n-Ws+1}^{n} Ac_l \tag{1}$$

Where,

RM_n is the moving average at time n
Ws is the window size
Acl is the actual data concerning time 1
$l=n-Ws+1$ is the summation upto l=n

Figure 2. Rolling Mean for EEG and ECG data

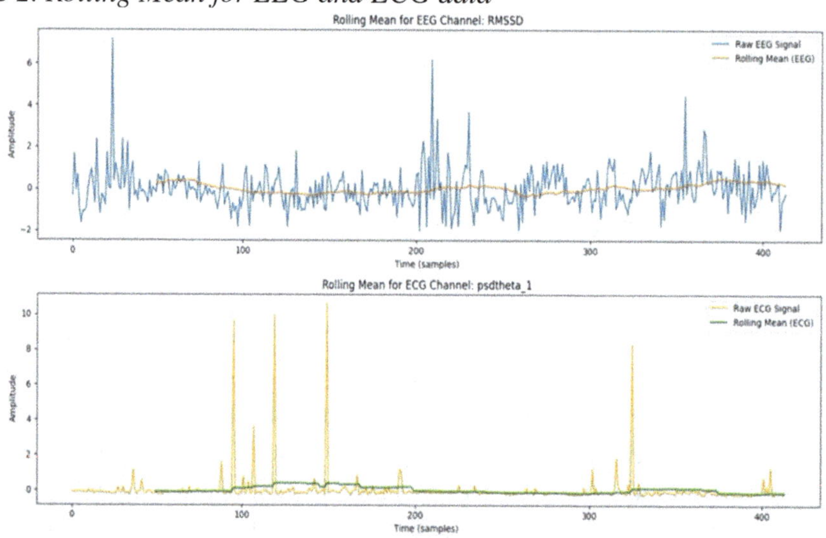

3.3 Baseline Correction

Biomedical time series data, for instance, those collected by heart and brain activity sensors, could exhibit a Direct Current offset. Ambient noise or sensor drift is one of the reasons why it could be nonzero. Such an offset would lead to biased assessments in the frequency and time domains. Baseline correction subtracts the mean value MAn from each point in the signal a_k so that the signal is centred around zero (Kumawat & Jain, 2021).Taking this precaution ensures that real signal dynamics are employed rather than artifacts of any subsequent calculations, like variance or spectral power analysis. Removal of baseline offset aids in making signals more interpretable and suitable for feature extraction. These adjustments are represented in (2),

$$Mx\ adjuste\,d^{(t)} = x(t) - \frac{1}{M}\sum x(k) \tag{2}$$

Where,

$x_{adjusted}(t)$ is the adjusted signal after removing the baseline offset.
(t) is the original signal at time t
x(k) is the overall count of the signal's samples.
(k) denotes the value of the initial signal at the index of discrete time k.

3.4 Filtering

Noise from unrelated sources, such as motion artifacts, low-frequency drift, or powerline interference, often accompanies raw biomedical signals (Xingyi Wang et al., 2023). This is mitigated by filtering, which eliminates the unwanted parts of the signals and retains only the important frequency ranges. In most cases, a band-pass filter is employed that allows the passing of signals in a specified range. The frequencies between 0.5 and 50 Hz are maintained for the brain activity signals. For cardiac data, a frequency range of 0.04 to 0.15 Hz is used to highlight heart rate variability. Thecomparison of Raw and Filtered EEG signal distribution is represented in Figure 3. Filtering sharpens the data and makes it easier to identify physiological patterns that are important for feature extraction and downstream analysis that as represented in (3),

$$xfiltered(t) = \mathscr{F} - 1[H(f) \cdot \mathscr{F}x(t)] \tag{3}$$

Where,

$x_{filtered}(t)$ is the time-domain filtered signal.
$\mathscr{F}$ denotes the Fourier transform.
(f) is the bandpass filter permits frequencies within a given range.

Figure 3. Comparison of Raw and Filtered EEG signal distribution

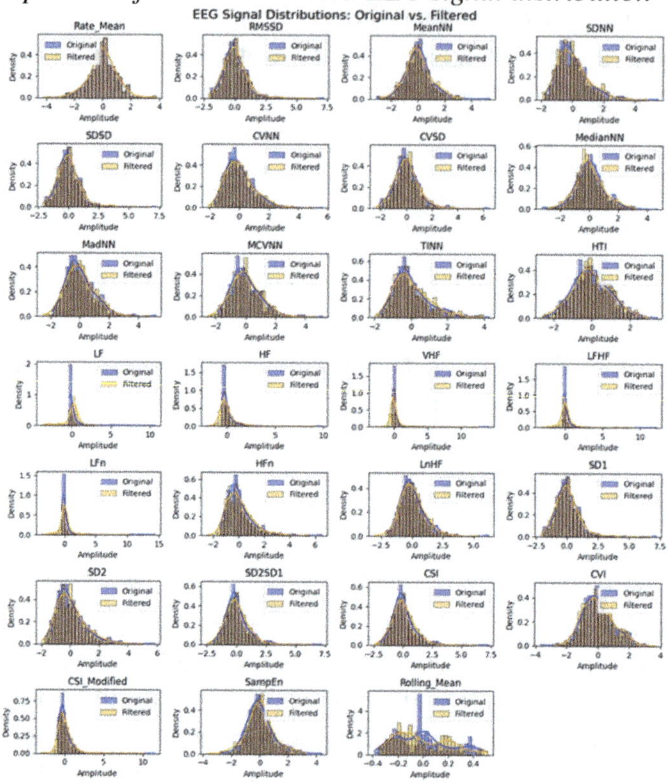

3.5 Normalization

Features extracted from biomedical time series data often have heterogeneous scales. For example, time domain measures like heart rate variability could have much smaller ranges than power spectral densities. In the training process, larger-scale features might be more common, and this could affect machine learning models. Normalization standardizes these features to a common range. Normalization ensures uniformity across features and helps machine learning algorithms treat each variable equally is represented in (4). This gets rid of biases stemming from differing scales and strengthens the model's potential to recognize profound patterns in data.

$$X_{norm} = \frac{X - \mu}{\sigma} \tag{4}$$

Where,

X_{norm} is the normalized EEG signal.
X represents the initial EEG signal.
μ is the average of the signal values.
σ is the signal reading's standard deviation.

These preprocessing steps, like baseline correction, filtering, and normalization, work together to prepare the time-series data for analysis and predictive modelling. The data becomes cleaner, more consistent, and more interpretable for downstream applications, including emotion recognition, by removing noise, reducing biases, and standardizing feature scales.

4. METHODOLOGY

To classify emotions based onEEGNet, this proposed hybrid system of deep learning utilizes EEG as well as ECG (Farokhah et al., 2023; García-Martínez et al., 2019). The model processes the modality-specific features and puts them together using an adaptive procedure of fusion (Deng et al., 2021). The architecture used for categorizing the emotions features valence, arousal, and dominance. Figure 4 represents the EEGNet architecture.

Figure 4. EEGNetArchitecture (Katsigiannis & Ramzan, 2017)

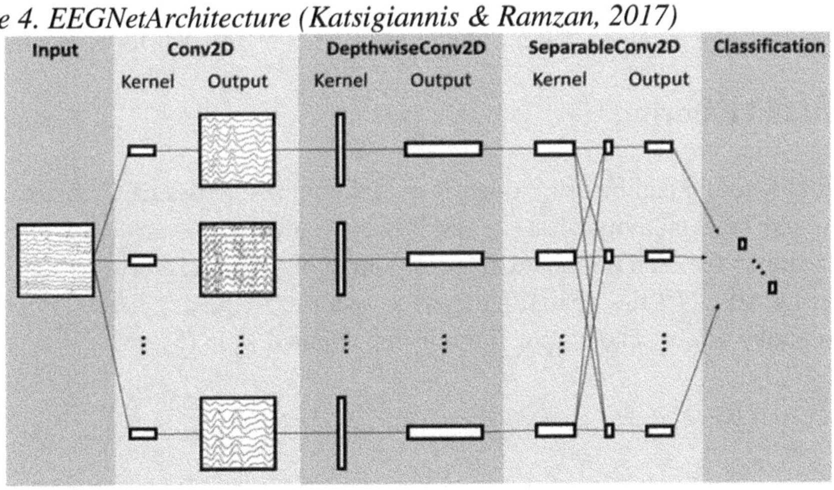

The lines in Figure 4 illustrate the connection between the inputs and outputs, which are also referred to as feature maps. The network first uses a temporal convolution (second column) followed by a depth-wise convolution (middle column), separately linked to every feature map, to learn spatial filters peculiar to a certain frequency. Separable convolution (fourth column) is made up of the depth-wise convolution that learn temporal data summary for every feature map individually and the pointwise convolution, which learns the best way to put together the feature (Katsigiannis & Ramzan, 2017).

4.1temporal Filtering

Frequency bands relevant to the tasks are extracted through temporal convolutions. This technique can extract frequency-specific information from raw EEG signals. EEGNet can record frequency information over 2 Hz by using a temporal convolution with a kernel size equal to half the sampling rate. The resulting signal is as follows in (5),

$$z_{freq}(t) = (s(t)*k(t))*f_{band}(t) \quad (5)$$

Where,

$z_{\mathrm{freq}}(t)$ is the resulting signal that contains the isolated frequency-specific information.
(t) represents the original EEG signal that is being processed.
(t) is the convolution kernel designed to capture relevant frequency characteristics.
$f_{\mathrm{band}}(t)$ is the bandpass filter applied to isolate the desired frequency range.

4.2 Spatial Filtering

Spatial filtering can be used to learn spatial patterns across EEG channels that differentiate brain activity classes. Depth-wise convolutions are used to extract spatial features for each frequency band. By applying spatial filtering after temporal filtering, EEGNet ties spatial patterns to specific frequency bands, improving discriminative power. The output signal is represented as in (6),

$$(t) = \left(\left(x(t)*h_t\right)(t)*h_m\right) \quad (6)$$

Where,

(t) is the output signal for the m-th spatial filter after applying depth-wise convolution.
(t) is the raw EEG signal across multiple channels.
$h(t)$ is the temporal convolution kernel used for temporal filtering.

h_m is the depthwise convolution kernel for the m-th frequency band.

4.3 Separable Convolutions

Separable convolutions are used to reduce model complexity by decoupling temporal and spatial feature extraction. Separable convolutions consist of a depth-wise convolution for temporal feature extraction followed by a pointwise convolution for spatial mixing. This approach minimizes the number of trainable parameters while retaining robust spatiotemporal representations in (7),

$$(t) = (((t) * g_d)(t) * g_p) \quad (7)$$

Where,

(t) is the output signal after feature extraction.
(t) is the input signal.
(t) is the depthwise convolution kernel used for temporal feature extraction.
g_p is the pointwise convolution kernel used for spatial mixing.

4.4 Feature Fusion Using Dynamic Aggregation Module

The Dynamic Aggregation Module fuses features extracted from the EEG and ECG processing branches (Can et al., 2023; Sepúlveda et al. 2021; Sharafi et al., 2022). EEGNet extracts spatial as well as temporal attributes from EEG signals, while the ECG branch captures dynamic heart variations and morphological patterns (Hasnul et al., 2021). The DAM adaptively assigns weights to these features based on their relevance to emotion recognition, enabling the model to integrate fine-grained spatial features from EEG and contextual physiological information from ECG. The fused feature vector F_{combined} is defined as follows in (8),

$$F_{combined} = \alpha EEG \cdot FEEG + \alpha ECG \cdot FECG \quad (8)$$

Where,

F_{fusion} is the combined feature vector,
F_{EEG} and F_{ECG} are the feature vectors from the EEG and ECG branches, respectively, α_{EEG} and α_{ECG} are the trainable weights for the EEG and ECG features.

This fusion mechanism enhances the model's ability to leverage complementary features from both modalities, providing a more comprehensive representation for emotion recognition.

4.5 Fully Connected Layers

The aggregated feature vector from the DAM passes through one or more dense layers. These layers enhance the feature augmentation by identifying complex mappings between the fused features and the target emotion dimensions. It allows the model to project the fused characteristics into a place that best reflects the emotional categories. Each fully connected layer transforms the input feature vector F_{combined} into a high-level representation y, as follows in (9),

$$y = M \cdot F_{combined} + c \tag{9}$$

Where,

y is the output vector.
M denotes the weight matrix.
c represents the offset vector.

4.6 Softmax Activation and Final Prediction

The last layer of the model generates class-specific probabilities for the valence, arousal, and dominance dimensions. It does this by applying a SoftMax activation mechanism to the fully linked levels' output. SoftMax normalizes the logits, computing the probability of each class as in (10),

$$C_i = \frac{e^{zi-zj}}{\sum_{j=1}^{C} e^{c}} \tag{10}$$

Where,

$z_i$and$z_j$is the logit for class.
C represents the overall number of classes
(c_i) signifies the probability associated with class i

The outcome is interpretable and probabilistic since the projected class aligns with the one that has the highest probability. This procedure is the conceptual inference mechanism of the model.

5. MODEL TRAINING

The training step is crucial for the model to perform at its best on emotion recognition tests. A structured training approach is adopted to improve feature learning, avoid overfitting, and increase generalization across various subjects due to the complexity of EEG and ECG signals.

5.1 Data Augmentation

The generality of the model is improved using data augmentation, particularly in the DREAMER dataset, with subject-to-subject variations in the physiological signals (Salama et al., 2021)(Nakisa et al., 2018). For EEG and ECG signals, standard augmentation methods include noise injection, scaling, and time-shifting. These changes preserve the emotional information encoded in the signals and simulate variances that occur in reality.

- Time-shifting represents a realistic, time-lagged perspective by randomly shifting data points along the time axis.
- Scaling increases or decreases the amplitude of data by applying some random scaling factor to the amplitudes.
- To replicate sensor noise or crosstalk, random Gaussian noise will be added.

These methods became widely used for time-series data analysis of EEG and ECG signals because of their ease and effectiveness. The augmented signal is represented which integrates of Gaussian noise, is represented in (11),

$$[z' = z + (0,\ \sigma^2)] \tag{11}$$

Where,

z is the original signal
z'is the augmented signal
$(0, \sigma^2)$ represents Gaussian noise characterized by a mean and variance of 0 and $\sigma^2 respectively.$

5.2 Loss Function

For emotion recognition in the three dimensions of valence, arousal, and dominance, binary cross-entropy loss is utilized for each dimension. This method treats each dimension as a separate binary classification task, which fits well with the multi-label nature of the problem. Binary cross-entropy provides a clear probabilistic interpretation of the model's predictions (Nakisa et al., 2018). For a single space sample i, with $x_i \in [0,1]$, which is said to be the true label (For instance, low/high valence). $\hat{x}_i \in [0,1]$ is said as predicted probability (from the model, for instance, sigmoid output). The loss function with binary cross-entropy is defined as in (12),

$$Loss_i = -[\, x_i \cdot log\,(\hat{x}_i) + (1 - x_i)\,{}^\circ \log(1 - \widehat{x_i})\,] \tag{12}$$

Where,

$Loss_i$ represents the binary cross-entropy loss.
x_i indicates the true label for the i-th sample.

The loss function for one dimension over all the samples by considering the set $x = [x_1, x_2, \ldots, x_n], \hat{x}_i = [\hat{x}_1, \hat{x}_2, \ldots \hat{x}_n]$. Then the total loss is defined as in (13),

$$Loss_{dim} = -\frac{1}{m}\sum_{i=1}^{m}[\, x_i\,{}^\circ \log(\widehat{x_i}) + (1 - x_i)^\circ \log(1 - \hat{x}_i)] \tag{13}$$

Where,

dim can be valence, arousal, or sominance.

The weighted total loss is represented as in (14),

$$L_{total} = \alpha^\circ L_{valence} + L_{arousal} + L_{dominance} \tag{14}$$

Where,

$\mathscr{L}$ total, $\mathscr{L}v$, $\mathscr{L}a$, $\mathscr{L}$ represent total loss, valence loss, arousal loss, and dominance loss, respectively.

5.3 Optimizer

Adam optimizer is used for training because of its ability to adjust the learning rate, which helps in handling the high variability of EEG and ECG readings. Flexible learning rates combined with the benefits of momentum-based optimization make it suitable for deep learning models in high-variance, noisy data such as EEG and ECG (Y. Wang et al., 2023). The Adam optimizer is defined as in (15),

$$\partial_{t+1} = \partial_t - \mu \cdot \frac{\widehat{n}_t}{\sqrt{\widehat{\beta}_t} + \epsilon} \tag{15}$$

Where,

$\widehat{n}_t$ is the bias-corrected estimation of the mean gradient.
$\widehat{\beta}_t$ is the bias-corrected estimation of the squared gradient.

5.4 Activation Function

The network uses the Rectified Linear Unit (ReLU), which enables the model to capture complicated patterns in the EEG and ECG data with higher efficiency without suffering from the vanishing gradient problem, thus saving computations. The output produced is as follows in (16),

$$[(t) = \max(0,\ t)] \tag{16}$$

Where,

(t) represents the output produced by the activation function.
t is the input to the activation function

5.5 Dropout and Regularization

To prevent overfitting, Dropout and L2 regularization are applied, and a dropout rate of 30% is applied after each fully connected layer (Mai et al., 2021). During training, neurons are randomly deactivated with probability p = 0.3. L2 regularization is used to prevent large weights. The regularization term is added to the loss function as follows in (17),

$$Loss_{tot} = Loss_{data} + \gamma \circ R(\partial) \tag{17}$$

Where,

$Loss_{tot}$ represents the complete loss function, including the regularization component.
$Loss_{data}$ is the primary loss function without regularization.
γ denotes the regularization coefficient.
$R(\partial)$ indicates the term of regularization.
∂ is the weight.

To minimize the prediction error, while also preventing the model from becoming too complex and overfitting, an effective way to enhance this is by using Mean Squared Error (MSE) loss integrated with L2 regularization to penalize large weights. The main goal is to minimize the loss, so predictions get closer to the true labels. The MSE measures the average squared difference between the true value and the predicted value, as represented in (18).

$$Loss_{MSE} = \frac{1}{m}\sum_{i=1}^{m}(x_i - \hat{x}_i)^2 \tag{18}$$

Where,

m is the number of samples.
x_i is the true output obtained for the ith sample.
$\hat{x}_i$is the predicted output obtained for the ith sample.

The L2 regularization penalizes the sum of the squares of the model parameters, typically represented for the weights is represented in (19),

$$R_{L2}(\partial) = \sum_{j=1}^{p}\partial_j^2 \tag{19}$$

Where,

∂_j^2 is the jth weight of the model.
p is the total number of weights.

The combined loss function is also called as ridge regression loss function, which helps in preventing the weights from becoming too large, which helps prevent overfitting. It is represented in (20),

$$Loss_{tot} = \frac{1}{m}\sum_{i=1}^{m}\widehat{(x_i - x_i)}^2 + \delta\sum_{j=1}^{p}\partial_j^2 \tag{20}$$

Where,

δ is the regularization parameter that controls the strength of the penalty.
δ= 0 means no regularization.
δ= ∞ means heavy regularization.

5.6 Learning Rate Scheduling

To improve training efficiency, a step-based learning rate scheduler is used (Liu et al., 2021; Mai et al., 2023). It decreases the learning rate by 0.1 factors at every 20 epochs. The learning rate for the k-th epoch can be computed as in (21),

$$[\eta_k = \eta_0 \times \gamma^{\cdot k \cdot}] \tag{21}$$

Where,

η is the learning rate at k-th epoch,
η_0 is the learning rate the initial epoch.
γ^k is the decay factor.

5.7 Hyperparameter Rationale

The hyperparameters in this study were selected based on a combination of literature, experimental test and validation optimization. A dropout rate of 0.3 was chosen to reduce over-fitting based on previously reported EEG-based emotion recognition studies (Mai et al., 2021). This dropout rate was determined to provide model robustness while avoiding over-fitting during pilot studies. The Adam algorithm was utilized since it supports adaptive learning rate which works well for noisy and high variance physiological data like the EEG and ECG signals (Wang et al., 2023). An empirically verified step learning rate scheduling, that decays by 0.1 every 20 epochs, is found to help both the stability of training and the convergence speed. ReLU was used to prevent the vanishing gradients and accelerate the convergence, referring to prior practice in deep learning implementations for EEG. These hyperparameters were optimized, based on an unbiased validation subset of the DREAMER dataset, to balance between overfitting and underfittig.

6. RESULTS AND DISCUSSION

The performance evaluation of the model depends on a number of measures used in multiclass classification. Among those, in addition to F1-score for balancing precision and recall, particularly for imbalanced datasets, and accuracy, there exists AUC-ROC. The area under the Receiver Operating Characteristic curve measures the model's ability to differentiate between classes at different thresholds. The confusion matrix also provides a comprehensive view of categorization performance by giving information about true positives, false negatives, true negatives, and false positives.

These measures give a comprehensive evaluation of the model's performance in emotion recognition tasks. The performance analysis of the EEGNet model in the detection of emotional states is executed. The model's effectiveness in distinguishing between various emotions is discussed and analysed. Challenges faced in emotion recognition and their real-world implications are discussed.

During the performance, the EEGNetmodel's efficiency is calculated using precision, recall. F1score, accuracy. These are vital for models reliably implementing emotional states to prevent false classifications, especially for subtle emotional categories like happiness, sadness, and anger. The parameters expansion is represented as T_rP = True Positive, F_aN = False Negative, F_aP = False Positive, T_rN =True Negative. It is based on the confusion matrix, which is represented in Figure 5,

Figure 5. Confusion Matrix for the Parameters

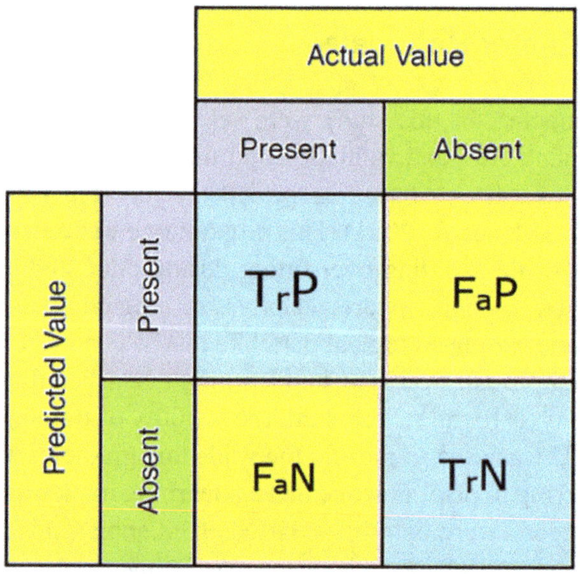

Accuracy is commonly used as a basic metric of assessment of sorting performance, but it may be misleading in emotion recognition when the dataset is imbalanced, if some emotions are significantly overrepresented compared to others. Relying solely on accuracy does not fully tell the model's real performance, especially in recognizing given subtle emotional states. It is represented in (22),

$$Accuracy = \frac{TrP + TrN}{TrP + TrN + FaP + FaN} \tag{22}$$

Precision compensates for many functions of emotional recognition; for example, false positives can be prevented because they lead to the misinterpretation of emotions, thereby injuring usability applications like human-computer interaction. It is represented in (23) as follows,

$$Precision = \frac{\text{TrP}}{\text{TrP} + \text{FaP}} \tag{24}$$

Recall, or otherwise known as sensitivity, pertains to how accurately the model can identify the presence of authentic emotional states. Recall is necessary to enhance the detection of emotional responses truly, and it is of great interest in the area of mental health monitoring. It is represented in (25),

$$Recall = \frac{\text{TrP}}{\text{TrP} + \text{FaN}} \tag{25}$$

The F1 score serves as an aggregate marker of precision and recall, to provide the model an objective assessment of performance, particularly when unbalanced data is concerned. It is represented in (26),

$$F1\ Score = \frac{2 \times Precision \times Recall}{Precision + Recall} \tag{26}$$

6.1 Result Table

EEGNet showed excellent classification capabilities for different emotional states for the DREAMER dataset. Classification statistics for emotional states for different emotional states are compiled in Table 1 below.

Table 1. Performance Evaluation Across Emotional States

Emotional State	Accuracy	Precision	Recall	F1-Score
Happiness	0.8712	0.8531	0.8857	0.8668
Sadness	0.8235	0.8059	0.8479	0.8280
Anger	0.8663	0.8323	0.8721	0.8502
Fear	0.8054	0.7850	0.8123	0.7942
Disgust	0.8472	0.8221	0.8567	0.8350

6.2 Precision-Recall Trade-off and F1-Score Analysis

The calculation of diverse metrics emphasizes the importance of using the F1 score to evaluate model performance across imbalanced classes. This is likely ideal for the emotion of happiness; the F1 score is 0.8668. This F1 measure gives a good balance between precision and recall, very much needed for the reliable detection of emotions. Although this trade-off will not result in missing essential emotional states, it also eliminates false positives (Zhang et al., 2023).

High F1 scores seen in each emotion demonstrate the generalization capability of the EEGNet model across emotional states. Nevertheless, in certain categories such as sadness and fear, the model could exhibit improved performance with further training and data augmentation. From the confusion matrix, one can calculate the following fine-granular values for happiness:

- True Positives (A): 160 (happiness is justly predicted as happiness)
- False Positives (C): 12 (sadness is wrongly predicted as happiness)
- False Negatives (D):15 (happiness is wrongly predicted as sadness)
- True Negatives (B): 130 (sadness is correctly predicted as sadness)

Figure 6. Confusion Matrix for Emotion Classification

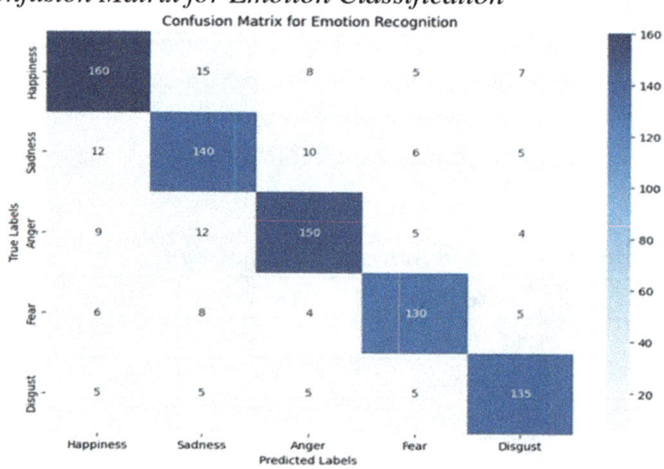

6.3 Details of Benchmark Models

To ensure transparency in the performance comparisons shown in Figure 7, the baseline models are explicitly defined as follows:

Traditional Machine Learning (ML):

Standard classifiers like Random Forest (100 trees), Support Vector Machine (SVM with RBF kernel), and Logistic Regression were used. Handcrafted features taken from EEG signals, such as wavelet coefficients, statistical moments, and power spectral density (PSD) features, were used to train these models. Their inclusion reflects popular classical methods for recognizing emotions based on EEG.

Convolutional Neural Network (CNN):

The two convolutional layers (32 and 64 filters, kernel size = 3) in a baseline CNN architecture were followed by max-pooling and ReLU activation. Prior to the SoftMax output layer, a fully connected layer with 128 units and a dropout of 0.3 was applied. According to Farokhah et al. (2023), this arrangement reflects a typical CNN structure that is frequently documented in EEG emotion recognition studies.

Long Short-Term Memory (LSTM):

Two stacked LSTM layers with 128 hidden units each made up the LSTM baseline model. A SoftMax classifier and a dense layer with dropout (0.3) came next. Temporal dependencies in sequential EEG signals were intended to be captured by the LSTM. Based on earlier research in EEG-based sequence modeling, hyperparameters were selected (Nakisa et al., 2018).

6.4 Comparison With Existing Methods

EEGNet was compared with a variety of conventional machine learning techniques and advanced deep learning approaches. The comparison results are presented below in Figure 7,

Figure 7. Performance Comparison of EEGNet with Other Models

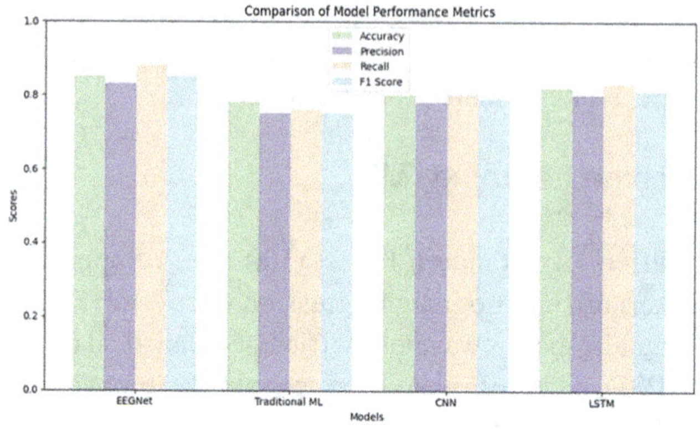

EEGNet outperforms conventional machine learning methods as well as other deep learning models in every criterion, especially in precision and F1 score (F. Wang et al., 2020). This shows that it can pick up better the complex and demanding patterns that characterize the emotionally loaded states.

6.5 Behavioural Analysis Of Model

The results suggest that the EEGNet model demonstrates excellent emotion recognition capacity of higher-order emotions such as happiness and anger, but it is less efficient when it comes to detecting distress and grief (Mai et al., 2021; Y. Wang et al., 2023). The model has high recall for these emotions as a result of its capability to capture temporal and spatial features from EEG and ECG signals. On the other hand, false positive issues arise due to the model's lower precision for sadness and fear. Their architecture adeptly merges CNN's ability for identifying spatial relationships with the power of temporal filtering; this allows the model to form sophisticated representations of emotional states imperative for mental health monitoring and HCI applications.

6.6 Real-World Implications

The EEGNet method presents many opportunities for practical emotional recognition applications. Once the emotional states are learned, then these models will enhance global user experience in several sectors such as virtual reality, game playing, and health care. With combined data from EEG and ECG, a more comprehensive view of an emotional reaction is given, which leads to better mental health diagnosis and individualized treatments. This approach expands access to mental health resources for a larger population, especially the underprivileged, due to the possibilities of telemedicine and mobile applications. This possibility is highly relevant for early emotional problem identification and for timely intervention.

Enhanced Human-Computer Interaction (HCI): One of the most immediate implications of EEGNet-based emotion recognition is in improving human-computer interaction. Emotion-aware systems can adapt their behavior based on a user's affective state. For example, virtual assistants, chatbots, and educational software can dynamically adjust their tone, pace, and content delivery depending on whether a user is stressed, confused, or relaxed.In gaming, this can lead to adaptive game design that responds to players' emotional engagement, enhancing immersion. In education, emotionally intelligent tutoring systems can offer better learning experiences by recognizing boredom or frustration and adapting content accordingly.

Healthcare and Mental Health Monitoring: Emotion recognition through EEG and ECG signals has transformative implications in mental health diagnosis and therapy. Patients suffering from conditions like depression, anxiety, PTSD, or bipolar disorder often exhibit specific physiological patterns.

Driver Monitoring and Road Safety: Fatigue and emotional instability are significant contributors to road accidents. EEGNet-based systems can be integrated into driver monitoring systems to assess stress, anger, or drowsiness by analyzing the driver's brain and heart signals in real-time.

Brain-Computer Interfaces (BCIs) and Assistive Technology: EEGNet-based multimodal emotion recognition also enhances brain-computer interface applications. BCIs aim to allow users—especially those with physical impairments—to interact with external devices using their brain activity.

Marketing and Consumer Behavior Analysis: In the world of neuromarketing, EEG and ECG signals offer a way to gauge emotional reactions to products, advertisements, or environments with high accuracy.

6.7 Limitations and Challenges

Although the EEGNet model yields promising results, it also has several limitations that need to be addressed for practical real-world implementations. A significant challenge is that the model may commit false positives and negatives while identifying between emotional states that are somewhat alike (Du et al., 2020). This can lead to incorrect interpretations of emotional expressions with resultant implications towards the treatments involving mental health and the experience by users. It is also apparent that because it uses the DREAMER dataset, indirect findings could be achieved, which might point out the contrary assertions against the generalized nature of it towards diverse populations and settings for the expression of emotion. The potential for a broader spectrum of emotional experiences may be limited by the current dataset's inability to capture rare or nuanced emotional states. Any of the identified gaps in the current understanding could be addressed by making targeted future research efforts. One that is of significant concern is the subject-independent classification, which is created to establish models that have learned to generalize across different individuals while dealing with emotional expression variations and physiological responses.

However, research should continue to incorporate multimodal data. A possible direction of improvement is by combining EEG biomarkers with other physiological cues, such as facial expression and GSR (Wani et al., 2021). provided more reliable results through thorough cross-validation of the methods. Accurate representation of time and changes in emotions are generating the requisite for longitudinal studies. Such studies would track the emotional states over time and hence present holistic dynamics fit for emotional states, thus making our models adaptive to changes over time. Finally, these techniques for explainable and interpretable models remain necessary as they make transparent the decision-making process in the model, thus generating trust in clinical applications. The use of deep learning models such as EEGNet requires significant computational resources, which can be challenging in resource-constrained environments during development and deployment.

7. CONCLUSION

It presents a deep learning system that uses multimodal physiological signals to enhance the accuracy and general robustness of emotion recognition. While using advanced fusion mechanisms to bring together raw EEG and ECG data, the modified DNN extracts aspects of emotion generation within both temporal and spatial contexts. The use of augmentation techniques, for example, adds impulsive noise to the signal or scales the signal to a large value so that the built-in convolutional filters generalize well under diverse conditions.

This methodology uses a loss function dedicated to multi-label classification, along with an intentionally adaptive learning rate scheduler and other regularization strategies to counter the chances of overfitting. All these enhancements together help maintain stable training across different sets. It is demonstrated by employing an array of metrics that cater to class distribution and accuracy that the new system proves very robust and competitive against any proposed benchmark.

It demonstrates considerable room where this framework could provide scaling and efficiency in healthcare, social robots, or interactive technologies, provided that the physiological feature points offer enough robustness. Emotion at real-time standalone behaviours would not fully cooperate unless multifaceted physiological signals are merged to limit any possibility of uncomplicated fraud in these real-world scenarios.

This system still has challenges in areas of data matching and scalability, particularly in cases where the impact of population-specific generalization has been questioned. Future works could analyse the blending of any type of data source, ranging from physical behaviour, facial expression, to text data for enhancing the fusion strategies even further. Furthermore, it could help in uplifting applicability for real-world scenarios by expanding the dataset with larger and more diverse samples.

REFERENCES

Can, Y. S., Mahesh, B., & André, E. (2023). Approaches, applications, and challenges in physiological emotion recognition—A tutorial overview. *Proceedings of the IEEE, 111*(10), 1287–1313. DOI: 10.1109/JPROC.2023.3286445

Deng, X., Zhang, B., Yu, N., Liu, K., & Sun, K. (2021). Advanced TSGL-EEGNet for motor imagery EEG-based brain-computer interfaces. *IEEE Access : Practical Innovations, Open Solutions, 9*, 25118–25130. DOI: 10.1109/ACCESS.2021.3056088

Ding, Y., Hu, X., Xia, Z., Liu, Y.-J., & Zhang, D. (2018). Inter-brain EEG feature extraction and analysis for continuous implicit emotion tagging during video watching. *IEEE Transactions on Affective Computing, 12*(1), 92–102. DOI: 10.1109/TAFFC.2018.2849758

Du, G., Zhou, W., Li, C., Li, D., & Liu, P. X. (2020). An emotion recognition method for game evaluation based on electroencephalogram. *IEEE Transactions on Affective Computing, 14*(1), 591–602. DOI: 10.1109/TAFFC.2020.3023966

Farokhah, L., Sarno, R., & Fatichah, C. (2023). Simplified 2D CNN architecture with channel selection for emotion recognition using EEG spectrogram. *IEEE Access : Practical Innovations, Open Solutions, 11*, 46330–46343. DOI: 10.1109/ACCESS.2023.3275565

García-Martínez, B., Martinez-Rodrigo, A., Alcaraz, R., & Fernández-Caballero, A. (2019). A review on nonlinear methods using electroencephalographic recordings for emotion recognition. *IEEE Transactions on Affective Computing, 12*(3), 801–820. DOI: 10.1109/TAFFC.2018.2890636

Gursesli, M. C., Lombardi, S., Duradoni, M., Bocchi, L., Guazzini, A., & Lanata, A. (2024). Facial emotion recognition (FER) through custom lightweight CNN model: Performance evaluation in public datasets. *IEEE Access : Practical Innovations, Open Solutions, 12*, 45543–45559. DOI: 10.1109/ACCESS.2024.3380847

Hasnul, M. A., Aziz, N. A. A., Alelyani, S., Mohana, M., & Aziz, A. A. (2021). Electrocardiogram-based emotion recognition systems and their applications in healthcare—A review. *Sensors (Basel), 21*(15), 5015. DOI: 10.3390/s21155015 PMID: 34372252

Huang, W., Xue, Y., Hu, L., & Liuli, H. (2020). S-EEGNet: Electroencephalogram signal classification based on a separable convolution neural network with bilinear interpolation. *IEEE Access : Practical Innovations, Open Solutions, 8*, 131636–131646. DOI: 10.1109/ACCESS.2020.3009665

Jafari, M., Shoeibi, A., Khodatars, M., Bagherzadeh, S., Shalbaf, A., García, D. L., Gorriz, J. M., & Acharya, U. R. (2023). Emotion recognition in EEG signals using deep learning methods: A review. *Computers in Biology and Medicine*, *165*, 107450. DOI: 10.1016/j.compbiomed.2023.107450 PMID: 37708717

Katsigiannis, S., & Ramzan, N. (2017). DREAMER: A database for emotion recognition through EEG and ECG signals from wireless low-cost off-the-shelf devices. *IEEE Journal of Biomedical and Health Informatics*, *22*(1), 98–107. DOI: 10.1109/JBHI.2017.2688239 PMID: 28368836

Kumawat, R., & Jain, M. (2021). EEG based emotion recognition and classification: A review. *International Research Journal on Advanced Science Hub*, *3*(Special Issue ICITCA-2021 5S, 5S), 1–10. DOI: 10.47392/irjash.2021.131

Liu, W., Qiu, J.-L., Zheng, W.-L., & Lu, B.-L. (2021). Comparing recognition performance and robustness of multimodal deep learning models for multimodal emotion recognition. *IEEE Transactions on Cognitive and Developmental Systems*, *14*(2), 715–729. DOI: 10.1109/TCDS.2021.3071170

Mai, N.-D., Long, N. M. H., & Chung, W.-Y. (2021). *1D-CNN-based BCI system for detecting Emotional states using a Wireless and Wearable 8-channel Custom-designed EEG Headset.* Paper presented at the 2021 IEEE International Conference on Flexible and Printable Sensors and Systems (FLEPS). DOI: 10.1109/FLEPS51544.2021.9469818

Mai, N.-D., Nguyen, H.-T., & Chung, W.-Y. (2023). Real-time on-chip machine-learning-based wearable behind-the-ear electroencephalogram device for emotion recognition. *IEEE Access : Practical Innovations, Open Solutions*, *11*, 47258–47271. DOI: 10.1109/ACCESS.2023.3276244

Nakisa, B., Rastgoo, M. N., Rakotonirainy, A., Maire, F., & Chandran, V. (2018). Long short term memory hyperparameter optimization for a neural network based emotion recognition framework. *IEEE Access : Practical Innovations, Open Solutions*, *6*, 49325–49338. DOI: 10.1109/ACCESS.2018.2868361

Pan, D., Zheng, H., Xu, F., Ouyang, Y., Jia, Z., Wang, C., & Zeng, H. (2023). MSFR-GCN: A multi-scale feature reconstruction graph convolutional network for EEG emotion and cognition recognition. *IEEE Transactions on Neural Systems and Rehabilitation Engineering*, *31*, 3245–3254. DOI: 10.1109/TNSRE.2023.3304660 PMID: 37578922

Raheel, A., Majid, M., Alnowami, M., & Anwar, S. M. (2020). Physiological sensors based emotion recognition while experiencing tactile enhanced multimedia. *Sensors (Basel)*, *20*(14), 4037. DOI: 10.3390/s20144037 PMID: 32708056

Rolls, E. T. (2023). Emotion, motivation, decision-making, the orbitofrontal cortex, anterior cingulate cortex, and the amygdala. *Brain Structure & Function, 228*(5), 1201–1257. DOI: 10.1007/s00429-023-02644-9 PMID: 37178232

Salama, E. S., El-Khoribi, R. A., Shoman, M. E., & Shalaby, M. A. W. (2021). A 3D-convolutional neural network framework with ensemble learning techniques for multi-modal emotion recognition. *Egyptian Informatics Journal, 22*(2), 167–176. DOI: 10.1016/j.eij.2020.07.005

Sepúlveda, A., Castillo, F., Palma, C., & Rodriguez-Fernandez, M. (2021). Emotion recognition from ECG signals using wavelet scattering and machine learning. *Applied Sciences (Basel, Switzerland), 11*(11), 4945. DOI: 10.3390/app11114945

Sharafi, M., Yazdchi, M., Rasti, R., & Nasimi, F. (2022). A novel spatio-temporal convolutional neural framework for multimodal emotion recognition. *Biomedical Signal Processing and Control, 78*, 103970. DOI: 10.1016/j.bspc.2022.103970

Subasi, A., Tuncer, T., Dogan, S., Tanko, D., & Sakoglu, U. (2021). EEG-based emotion recognition using tunable Q wavelet transform and rotation forest ensemble classifier. *Biomedical Signal Processing and Control, 68*, 102648. DOI: 10.1016/j.bspc.2021.102648

Toraman, S., & Dursun, Ö. O. (2021). GameEmo-CapsNet: Emotion Recognition from Single-Channel EEG Signals Using the 1D Capsule Networks. *TS. Traitement du Signal, 38*(6), 1689–1698. DOI: 10.18280/ts.380612

Wang, F., Wu, S., Zhang, W., Xu, Z., Zhang, Y., Wu, C., & Coleman, S. (2020). Emotion recognition with convolutional neural network and EEG-based EFDMs. *Neuropsychologia, 146*, 107506. DOI: 10.1016/j.neuropsychologia.2020.107506 PMID: 32497532

Wang, X., Ma, Y., Cammon, J., Fang, F., Gao, Y., & Zhang, Y. (2023). Self-supervised EEG emotion recognition models based on CNN. *IEEE Transactions on Neural Systems and Rehabilitation Engineering, 31*, 1952–1962. DOI: 10.1109/TNSRE.2023.3263570 PMID: 37015115

Wang, X., Zhang, J., He, C., Wu, H., & Cheng, L. (2023). A novel emotion recognition method based on the feature fusion of single-lead EEG and ECG signals. *IEEE Internet of Things Journal, 11*(5), 8746–8756. DOI: 10.1109/JIOT.2023.3320269

Wang, Y., Tian, W., Xu, J., Tian, Y., Xu, C., Ma, B., Hao, Q., Zhao, C., & Liu, H. (2023). Wearable wireless dual-channel EEG system for emotion recognition based on machine learning. *IEEE Sensors Journal, 23*(18), 21767–21775. DOI: 10.1109/JSEN.2023.3303441

Wani, T. M., Gunawan, T. S., Qadri, S. A. A., Kartiwi, M., & Ambikairajah, E. (2021). A comprehensive review of speech emotion recognition systems. *IEEE Access : Practical Innovations, Open Solutions*, *9*, 47795–47814. DOI: 10.1109/ACCESS.2021.3068045

Yu, C., & Wang, M. (2022). Survey of emotion recognition methods using EEG information. *Cognitive Robotics*, *2*, 132–146. DOI: 10.1016/j.cogr.2022.06.001

Zhang, L., Xia, B., Wang, Y., Zhang, W., & Han, Y. (2023). A fine-grained approach for EEG-based emotion recognition using clustering and hybrid deep neural networks. *Electronics (Basel)*, *12*(23), 4717. DOI: 10.3390/electronics12234717

KEY TERMS AND DEFINITIONS:

Convolutional Neural Network (CNN): It is a type of deep learning model designed to automatically and adaptively learn spatial hierarchies of features from input data, typically images, but also used for time-series and signal processing tasks like EEG and ECG analysis.

Electrocardiogram (ECG): It is a test that records the electrical activity of the heart over a period using electrodes placed on the skin. It reflects how the heart muscles depolarize and repolarize during each heartbeat.

Electroencephalogram (EEG): It is a non-invasive test that records the electrical activity of the brain using electrodes placed on the scalp. Neurons in the brain communicate via electrical impulses, and the EEG captures these signals as brain waves.

Recurrent Neural Network (RNN): It is a deep learning model designed for sequential data. Unlike CNNs, RNNs have loops that allow information to persist across time steps, making them ideal for time-dependent tasks like language modeling or emotion tracking over time.

Rolling Mean: It is also called Moving Average. It is a statistical technique used to smooth time-series data by calculating the average value of a variable over a specific number of consecutive data points (a window) that "rolls" forward one step at a time.

Spatial filtering: It is a signal processing technique used to enhance or suppress specific components of a signal based on their spatial properties, i.e., how the signal behaves across multiple sensors or channels (such as EEG electrodes placed on different scalp regions).

Temporal filtering: It is a signal processing technique used to modify or enhance signals based on their frequency content over time. It involves the use of mathematical filters to isolate, attenuate, or amplify specific frequency bands in a time-series signal, such as EEG, ECG, audio, or any time-dependent data.

Chapter 12

Decision-Making in Emerging Technologies Backed by Cognitive Analytics for Accurate ESG Reporting and Organizational Resilience

Jose De Jesus Reyes-Sánchez
https://orcid.org/0000-0002-3866-2363
TECNM Instituto Tecnologico Superior de Fresnillo, Mexico

ABSTRACT

The research explores how the synergy between cognitive analytics and emerging technologies reconfigures corporate decision-making processes to generate more accurate ESG reporting and strengthen organizational resilience. Through a qualitative analysis of emblematic cases in logistics, manufacturing and health, it is shown that technological modularity shortens learning cycles and, when integrated with transformational leadership and transparent data governance, increases supply chain traceability and reduces technical debt. The findings show that competitive value arises from converting the abundance of information into actionable knowledge, legitimizing iterative experimentation and aligning financial incentives with verifiable socio-environmental metrics.

DOI: 10.4018/979-8-3373-5702-7.ch012

INTRODUCTION

The convergence between cognitive analytics and emerging technologies has radically transformed the competitive logic, process architecture, and legitimacy narrative of organizations over the past decade. Data, previously a by-product of the transaction, is now the essential raw material for modeling volatile scenarios, detecting incipient patterns and articulating sustainable value propositions; however, the value of data does not lie in its mere accumulation, but in the company's ability to orchestrate explainable algorithms that reveal pertinent signals and, above all, to translate those signals into strategic decisions that strengthen the firm's financial, social, and environmental resilience. This epistemological shift occurs in parallel with an unprecedented expansion of regulatory frameworks on transparency and disclosure of environmental, social, and governance performance, so that competitive viability is no longer evaluated only in terms of profitability, but also of legitimacy to regulators, investors, and vulnerable communities (Chopra et al., 2024).

The rise of modular architectures has reduced experimentation costs and accelerated product iteration, allowing development teams to deploy features in shorter and shorter cycles without compromising operational stability. Such infrastructures, combined with blockchains hosted in the cloud, facilitate the traceability of assets and documents throughout the logistics chain, a phenomenon evident in the maritime industry by reducing customs frictions and strengthening inter-organizational trust (Wong et al., 2023). However, these technological benefits are fully realized only when data governance integrates principles of information sovereignty and cryptographic control mechanisms that harmonize the right to transparency with the protection of strategic information.

While the digital infrastructure mutates with exponential speed, the psychodynamics of organizations face the challenge of reconfiguring their work routines, their leadership frameworks and their trust pacts. Transformational leadership, based on narratives of purpose and openness to experimentation, is revealed as a catalyst for technological adoption in highly regulated sectors, where professional practice is subjected for the first time to the iterative logic of software development (Kludacz-Alessandri et al., 2025). Human capital is challenged by cross-border mobility, the automation of cognitive tasks and the need for continuous analytical literacy; Organizations that design flexible career paths and *data-driven upskilling* programs not only retain critical talent, but also amplify their ability to absorb external knowledge, creating epistemic communities that operate as psychosocial buffers against uncertainty (Evans et al., 2025).

In the field of ESG disclosure, regulatory pressure converges with societal demand for independent verification. Reports that were previously produced on an annual and static basis are evolving into live platforms that integrate data in near real-time, audited by smart contracts and subjected to satellite verification. Such metamorphosis requires interdisciplinary competencies that combine forensic accounting, data science, and strategic thinking to avoid information fragmentation and ensure narrative coherence between financial indicators and impact metrics (Cardoso et al., 2025). At the same time, the logic of double materiality forces us to reveal how socio-environmental impacts feed back into financial risks; consequently, the ESG report becomes a central device for responsible capital allocation and corporate reputation articulation (Vitale et al., 2019).

The circular economy paradigm also offers a fertile laboratory to study the integration of cognitive analytics and distributed governance. By tokenizing incentives for the recovery of materials and penalizing waste, value chains turn sustainability into a source of profitability and reconfigure information flows towards a collaborative model, where each link shares performance data in exchange for benefits proportional to its environmental contribution (Nandi et al., 2021). These practices show that distributed transparency, grounded in open standards and independent audits, can overcome reluctance to share sensitive information and create virtuous circles of open innovation, as long as the incentive architecture is designed with fairness and verifiability (Georgescu et al., 2024).

Organizational resilience, understood as the ability to detect disturbances, absorb their impact and reconfigure assets quickly, is nourished by the synergy between digital monitoring systems and cultures that normalize post-event reflection. IoT sensors and cognitive dashboards shorten the distance between the irruption of an anomaly and the corrective response, while learning rituals institutionalize the knowledge acquired and feed a virtuous cycle of continuous improvement (Fritzsch et al., 2022).

In this hybrid context, data-driven decision-making is emancipated from siloed intuition to embrace a continuous flow that combines descriptive, predictive, and prescriptive analytics. Algorithms, however, require explainability frameworks that allow decision-makers to question assumptions, measure the robustness of models, and ensure the fairness of results; otherwise, the risk of algorithmic opacity can undermine institutional trust and exacerbate historical biases (Davenport et al., 2020). The responsibility lies with senior management, called upon to orchestrate a balance between the speed of innovation and reflexive control, articulating codes of algorithmic ethics that guarantee the integrity of processes and the protection of fundamental rights (Butt et al., 2024).

Thus, the virtuous integration of cognitive analytics and emerging technologies should not be seen as an exclusively technical project, but as a sociotechnical process that involves redesigning data architecture, reconfiguring governance models, cultivating inclusive leadership, and forging learning-oriented cultures. Organizations that manage to align these pieces will be better positioned to meet the growing demand for ESG transparency, capture reputation premiums, and turn disruption into an opportunity for inclusive growth. On the other hand, those that limit themselves to incorporating digital tools without a holistic vision run the risk of multiplying technical debt, generating regulatory fatigue and deepening trust gaps with their stakeholders (Kaminski & Malgieri, 2021).

STATEMENT OF THE PROBLEM

The contemporary debate on the integration of emerging technologies with cognitive analytics to strengthen the accuracy of environmental, social, and governance reporting while strengthening organizational resilience is at a critical crossroads where regulatory interests, market pressures, and internal structural constraints converge; the literature agrees that strategic agility only materializes when the company transforms data into a learning lever and redesigns processes, an issue that is still distant for most firms operating in sectors characterized by disparate hierarchical structures and compliance standards (Cardoso et al., 2025).

Despite the enthusiasm generated by the adoption of microservices architectures and DevOps practices, evidence indicates that the technical debt accumulated in legacy systems slows down the digital transition, increases maintenance costs and hinders visibility on sustainability indicators, so that the organization is trapped in a cycle of superficial iterations that does not alter the core of its business model or the consistency of its ESG metrics. thus maintaining the gap between transformative aspiration and operational execution (Fritzsch et al., 2022).

The promise of cloud-hosted blockchain platforms to ensure traceability and data integrity also runs into a shared governance dilemma: while interoperability increases network value, the reluctance of peripheral actors to expose sensitive information reduces the critical mass needed to capitalize on distributed transparency and turns the initiative into a partial experiment that barely mitigates information asymmetries without reaching scale expected transformative (Wong et al., 2023).

In healthcare environments, where regulatory pressure and clinical risks raise the threshold of acceptance of change, the adoption of transformational leadership based on analytical evidence reveals the paradox of digitalization: technology intensifies the ability to predict and personalize services, but simultaneously requires redefining deep-rooted professional routines and managing the cultural resistance that emerges when medical practices are subjected to the logic of iterative experimentation typical of software development (Kludacz-Alessandri et al., 2025).

The growing regulatory density around ESG disclosure poses a scenario where reporting systems must evolve from static formats to dynamic platforms capable of integrating real-time data from multiple links in the value chain; however, the absence of interdisciplinary competencies that combine data engineering, forensic accounting, and strategic thinking means that reports are fragmented and often unable to meet stakeholders' expectations in terms of comparability and veracity (Chopra et al., 2024).

The global mobility of talent, in turn, is conditioned by migratory restrictions, digital skills gaps and new demands for meaning from the workforce; Organizations that fail to articulate purpose narratives and upskilling programs risk eroding their critical human capital, compromising absorptive capacity and slowing the speed with which they can reconfigure their competencies in the face of technological or health disruptions of planetary scope (Evans et al., 2025).

The literature on organizational resilience suggests that the intersection between sustainable HR practices and inclusive cultures constitutes a buffer against external shocks, but underlines that such practices only generate significant results when they are linked to performance metrics that transcend the financial short-term and recognize the creation of shared value as an indicator of strategic success; the difficulty lies in translating this vision into everyday management tools that guide decision-making at all hierarchical levels (Georgescu et al., 2024).

On the other hand, circular economy strategies driven by smart contracts demonstrate that it is possible to monetize return flows and penalize waste, but they also show that value capture depends on the ability of companies to synchronize internal economic incentives with external regulatory and social signals, a challenge that requires designing flexible governance mechanisms and, at the same time, sufficiently robust to ensure data integrity and equitable benefit-sharing (Nandi et al., 2021).

The problem hinges around a fundamental tension: while organizations recognize the need to integrate emerging technologies and cognitive analytics to meet ESG demands and strengthen their resilience, the combination of legacy technological barriers, human capability deficits, cultural resistances, and divergent regulatory mosaics limits the realization of such integration and maintains the distance between the transformative potential of digital innovation and the the operational reality of companies (Cardoso et al., 2025).

Three questions emerge from this diagnosis, the answers to which would contribute to closing the identified gap: how can modular digital architectures reconcile the need for interoperability with the protection of sensitive data without generating compliance overload? how can transformational leadership articulate narratives of purpose that mobilize the workforce toward ESG practices while also legitimizing the iterative experimentation inherent in technological innovation? And what collaborative governance mechanisms allow aligning economic and environmental incentives along global supply chains, overcoming the reluctance to share information and distributing benefits equitably?

BACKGROUND

Genesis of enterprise cognitive analytics

The genesis of enterprise cognitive analytics can be traced back to the convergence between deep learning models and the organizational need to translate increasingly complex operating environments into understandable decision schemes. In their early days, expert systems were limited to reproducing static rules; However, the maturity of neural networks made it possible to generate high-dimensional representations capable of anticipating fluctuations in demand, optimizing the instantaneous allocation of resources and uncovering correlations invisible to executive intuition. These pioneering applications, incubated in pioneering financial circles, cemented a new corporate grammar where algorithms went from being accessory tools to being central actors in everyday strategic dialogue (Butt et al., 2024; Buonocore et al., 2024).

The second trend of progress was underpinned by cross-sectoral initiatives that released large volumes of anonymized data and encouraged the training of transferable models. As compute costs fell thanks to elastic cloud infrastructure, analytics teams migrated from siloed labs to collaborative platforms that facilitate continuous experimentation and agile deployment. This ecosystem, bolstered by macroeconomic predictions about AI's potential contribution to global domestic product growth, persuaded boards to revamp legacy systems and drive microservices architectures that democratize model exploration across heterogeneous business units (Bernardo et al., 2024).

Academic dialogue accompanied this expansion with multidisciplinary frameworks that underscore the importance of data governance and algorithmic trust. Recent studies highlight how responsible adoption requires integrating ethical principles, explainability mechanisms, and safeguards against bias within the design, training, and validation cycle (Auld et al., 2022) . In this way, cognitive analytics moves from being a technical privilege to becoming an organizational catalyst that fosters the opening of silos and reinforces institutional memory, laying the foundations for dynamic capabilities that sustain competitive advantages in volatile contexts (Dwivedi et al., 2019).

Meanwhile, the proliferation of open-source frameworks and the emergence of automatic feature generation tools lowered the barrier to entry for non-specialized teams. This phenomenon transformed the decision-making culture: executive committees began to demand solid quantitative evidence before approving investments, and middle managers adopted interactive dashboards that turn algorithmic predictions into persuasive narratives. Analytical literacy was thus positioned as a decisive transversal competence, supported by internal training programs that combine visualization, statistical reasoning, and ethical awareness (Davenport et al., 2020).

Technical evolution continues to be driven by edge computing architectures and lightweight models that enable near-real-time inference close to the data generation source. Such advances, combined with neuro-inspired optimization algorithms, enable predictive maintenance systems and orchestration of energy resources that operate even with intermittent connectivity. By decentralizing processing, organizations increase both resiliency and privacy, while decreasing latency for critical decision-making in manufacturing and logistics environments (Sun et al., 2018).

Historical evolution of ESG standards

The evolution of ESG standards can be read as a process of progressive institutionalization that started from the spontaneity of disclosure towards an increasingly robust and convergent regulatory regime. During the nineties, the first voluntary environmental reports were born motivated by social pressure and the need to legitimize business activity in highly polluting industries; However, the absence of homogeneous criteria resulted in incomparable documents that were difficult to audit, a circumstance that stimulated academic research on the relationship between real environmental performance and disclosed content, showing that the quality of information varied according to reputational incentives and the ability of companies to manage their negative externalities (Clarkson et al., 2008).

The creation of the Global Reporting Initiative at the end of the last century constituted the first universal platform of indicators with the aim of standardizing non-financial information. Although its guidelines provided a common lexicon, coexistence with sectoral and regional frameworks preserved the initial heterogeneity. This scenario led various professional bodies to push for stricter materiality agendas, pointing out that the usefulness of the reports depended on the connection between the matters disclosed and the economic impacts that really concern investors, a perspective that cemented the transition from mere compliance to the strategic integration of sustainability (Eccles et al., 2012).

The regulatory conversation intensified as market-oriented standards and multilateral initiatives began to converge. On one side, the Sustainability Accounting Standards Board advanced industry-specific scorecards that helped investors separate financially material sustainability signals from generic rhetoric; on the other, the United Nations Global Compact scaled a principles-based commitment that thousands of companies adopted to frame responsible conduct. In parallel, the Global Reporting Initiative matured into a widely used disclosure backbone, and the dialogue between these streamsn crystallized a de facto doctrinal corpus. Even if nonbinding, it started to shape buy-side practice: large asset managers demanded disclosures that were decision-useful, consistent over time, and comparable across direct peers so that climate, reputational, and governance risks could be consistently priced into valuations (Xie et al., 2023; Orzes et al., 2020; Bais et al., 2024; Luo & Tang, 2023; Boulton et al., 2024).

Concretely, SASB's industry-based lens nudged issuers toward metrics that investors could compare within sectors and track longitudinally, reducing noise and improving peer benchmarking (Xie et al., 2023). UN Global Compact participation, while voluntary, signaled organizational alignment with baseline norms and has been associated with tangible performance effects in several contexts, reinforcing the idea that ESG commitments can be decision-relevant rather than purely symbolic (Orzes et al., 2020). Meanwhile, GRI's standardization push supplied a common language that, when actually followed, has been linked to more credible environmental outcomes(Bais et al., 2024; Luo & Tang, 2023). As regulators experiment with mandates, the thrust is the same: reduce dispersion in methods, increase temporal consistency, and help markets internalize sustainability risk with fewer measurement frictions (Boulton et al., 2024). The regulatory turning point was consolidated when the European Union enacted the Sustainable Finance Taxonomy, a classification system that defines with technical precision which economic activities contribute to climate mitigation and adaptation. By including quantitative thresholds and criteria of no significant harm, the taxonomy forced companies operating in different Member States to harmonize their data collection and verification processes with equivalent auditing standards, thereby reducing the trust gap between issuers and analysts and mobilizing capital towards projects with demonstrable impacts (Papa et al., 2022).

A decisive turn in global disclosure has been the embrace of "double materiality," under which reports must capture both how sustainability issues affect the firm and how the firm affects society and the environment. This shift has run alongside the creation of the International Sustainability Standards Board (ISSB) under the IFRS Foundation, whose standards aim to be investor-focused and interoperable across jurisdictions, even as Europe's ESRS keep impact materiality at the center. For multinational portfolios, this emerging two-track architecture nudges practice toward comparability while still accommodating broader stakeholder concerns. (Krivogorsky, 2024; de Villiers et al., 2024).

In parallel, the Task Force on Climate-related Financial Disclosures (TCFD) has normalized forward-looking climate communication by separating transition and physical risks and encouraging scenario analysis. Firms have started to stitch those narratives into valuation by linking scenario-conditioned pathways to cash-flow assumptions and capital costs—small steps that nonetheless raise the methodological floor of climate reporting (Andersson & Arvidsson, 2023). Where these practices travel into emerging markets, the long-standing North–South transparency gap narrows: recent evidence from a frontier market shows that more granular climate disclosure is associated with higher firm value, suggesting investors are rewarding credible climate information even outside developed markets (Pradhan et al., 2024).

Digitization has amplified these gains. Inline/XBRL taxonomies for sustainability data reduce manual handling, shrink reconciliation costs, and make peer benchmarking at scale feasible, especially when taxonomies are designed for machine readability and versioning (Suta et al., 2023). Inside firms, the same datafication is streamlining sustainability reporting workflows; integrated ERPs and rule-based engines are improving timeliness and internal controls, which, in turn, sharpens the signal available to markets (Pizzi et al., 2024). Assurance is the other leg of this stool: independent verification measurably dampens information asymmetry, strengthening the credibility of ESG signals once they enter capital-market pricing (Kim & Park, 2023). As methodologies mature, disclosure frontiers are widening beyond carbon to biodiversity, human rights, and algorithmic governance; here too, the research is converging on standardized metrics and materiality filters that can be embedded natively into accounting systems and dashboards rather than appended as voluntary extras (Elliot et al., 2024; Krueger et al., 2023).

Taken together, the trajectory is clear: standard-setting (ISSB/ESRS), climate risk practice (TCFD), digitized pipelines (XBRL), and credible assurance are coalescing into a disclosure infrastructure that is more comparable, more decision-useful, and increasingly global in scope. That architecture does not eliminate trade-offs. but it is steadily converting sustainability data into finance-grade information without losing sight of systemic effects (de Villiers et al., 2024; Krivogorsky, 2024).

Interrelationship between global crises and organizational resilience

The confluence of financial crises, health disruptions, and geopolitical tensions has revealed the inherent fragility of global supply chains and has forced us to rethink resilience as a systemic attribute and not just an operational one. Recent studies on organizational resilience agree that exogenous shocks act as stress tests that expose structural rigidities and collective learning deficits, so that the ability to anticipate, absorb, and recompose becomes a competitive advantage and, at the same time, social responsibility (Hillmann & Guenther, 2021).

Increasing macroeconomic volatility has driven the adoption of continuous surveillance platforms that integrate financial variables, epidemiological data, climate indicators, and policy information to generate early warnings. These systems, supported by deep learning algorithms and prescriptive visualizations, facilitate the dynamic reconfiguration of inventories and the reassignment of logistics routes before the latent risk materializes; empirical evidence demonstrates that organizations with this observation infrastructure reduce recovery times and disruption costs at the level of essential production (Ozdemir et al., 2022).

However, resilience goes beyond mere resistance to shocks and requires converting critical experience into new strategic resources. Research on dynamic capabilities underscores that the absorption of external knowledge, the process of controlled experimentation, and the institutionalization of post-event learning form a virtuous cycle that amplifies organizational agility and reinforces confidence in financial markets, even in capital-intensive sectors (Duchek, 2020).

Viable supply chain simulations show that the integration of real-time data and scenario analytics reduces the so-called aftershock effect, a phenomenon by which local disruptions spread throughout the network generating exponential losses. Computational models confirm that investment in flexible redundancy, rather than static redundancy, maintains the flow of materials without sacrificing efficiency and allows for a staggered restart of operations using alternative production configurations when critical nodes are disabled (Ivanov et al., 2019).

On the other hand, the pandemic stimulated reflections on the viability of the company as an adaptive system. When analyzing cases of accelerated recovery, it is observed that management teams that deployed decentralized decision-making mechanisms and promoted emotional transparency were able to maintain workforce engagement and preserve quality in remote work contexts, which supports the positive relationship between cultural resilience and post-crisis financial performance

Advances in Emerging Enabling Technologies

The expansion of the Industrial Internet of Things has enabled ubiquitous sensors to capture continuous data on energy consumption, mechanical vibrations, and atmospheric composition, nurturing digital twin models capable of forecasting microscopic deviations in operating parameters and suggesting microstructure adjustments that minimize raw material waste and extend the life of industrial assets (Fritz et al., 2018).

With the adoption of blockchain, organizations integrated a distributed and immutable ledger that records the origin of each batch of inputs and certifies in real time the fulfillment of socio-environmental commitments, so that sustainability reports are based on traceable empirical evidence and independent audits can verify the documentary trail without resorting to intermediaries. Edge computing brings processing power closer to the data generation source, decreases decision-making latency, and ensures operational continuity in enclaves with intermittent connectivity; this architecture is essential for facilities operating in rural areas, where predictive maintenance algorithms must execute inferences locally and coordinate with regional clouds only when bandwidth allows it.

The convergence between computer vision and low-orbit satellites makes it possible to externally validate land use metrics and gas emissions, raises the standard of verification and mitigates the risk of greenwashing; the combination of multispectral images with convolutional neural networks identifies spatial anomalies and generates early warnings that feed the dashboards of regulatory compliance areas (Kritzinger et al., 2018).

Quantum computing applied to logistics route optimization and energy planning looms on the horizon, while additive manufacturing and collaborative robotics introduce feedback loops in which the adjustment of production parameters is carried out in near real time, closing the loop between digital perception and physical action within the plant (Bughin et al., 2018).

Data-driven decision-making paradigm

The preeminence of managerial intuition gave way to organizational cultures that base their deliberations on broad, clean, and rigorously governed data sets; Executive boards now demand explanatory models that justify each strategic investment and demand that financial projections be integrated with environmental and social variables to reflect long-term costs and benefits (Ransbotham et al., 2017).

Analytical literacy was transformed into transversal competence: internal programs intersperse statistical theory, visualization techniques, and persuasive narrative workshops for staff to turn algorithmic findings into understandable strategic arguments; this democratization of quantitative knowledge accelerates the innovation cycle and reduces reliance on highly specialized data science units (Davenport et al., 2020).

Data governance frameworks have moved from descriptive catalogs to lineage controls that document the transformation of each attribute from its origin to its consumption in management dashboards, ensuring integrity, security, and regulatory compliance in distributed and multicloud processing environments. At the same time, decision intelligence combines machine learning, scenario simulations, and optimization techniques to recommend concrete actions and quantify their probabilistic impact; in this way, leaders weigh alternatives based on replicable evidence, reduce variability in the results of strategic projects, and strengthen investor confidence. This cultural shift is consolidated when algorithmic ethics committees monitor the fairness of models, review training biases, and validate the explainability of recommendations, so that data-driven decision-making aligns with corporate responsibility and preserves the social legitimacy of the organization (Dwivedi et al., 2019).

International regulatory trajectories

The European Union's Corporate Sustainability Reporting Directive and the sustainable finance taxonomy inaugurated a harmonized accounting language that obliges corporations to document climate risks and mitigation contributions using verifiable indicators, accelerating the channeling of capital into activities considered sustainable according to explicit technical criteria. In the United States, the Securities and Exchange Commission proposed rules that standardize the disclosure of scope one, two, and three issues, introduce climate scenario analysis, and require independent underwriting, with which issuers that lack reliable traceability face higher risk premiums in debt markets

The IFRS Foundation created the International Sustainability Standards Board to unify reporting frameworks and facilitate comparability across jurisdictions; its dual materiality approach recognizes that environmental and social impacts affect both the company and the ecosystem that surrounds it, reinforcing the connection between extra-financial performance and stock market valuation (Papa et al., 2022).

In Latin America, central banks and stock market superintendencies develop taxonomies that prioritize biodiversity and social justice, while designing differentiated credit incentives that reduce financing costs for firms with robust reporting systems; this regulatory convergence reduces information asymmetries and strengthens regional economic integration. In Asia Pacific, alliances between regulatory bodies and stock exchanges promote corporate governance codes that require diversity, workplace well-being, and climate neutrality metrics; such initiatives, added to the progressive adoption of the European taxonomy as a reference, make up an increasingly articulated regulatory mosaic that rewards transparency and penalises opacity (Andersson & Arvidsson, 2023).

Use of artificial intelligence in corporate reporting

The use of artificial intelligence in corporate reporting has evolved from basic text mining routines to real-time-generated narrative ecosystems. Natural language processing models monitor thousands of reports, transcripts of investor calls, and network conversations, turn those corpora into semantic vectors, and extract indicators that describe risks, opportunities, and compliance gaps. The combination of specialized embeddings with lightweight classifiers allows the generation of comparative tables that immediately reveal how each subsidiary performs against its sector peers and against the strategic commitments approved by the board. This architecture, when integrated with self-service dashboards, makes it easier for managers without technical training to measure the materiality of each indicator before validating public narratives, displacing the historical dependence on external analysts (Dwivedi et al., 2019; Barredo Arrieta et al., 2020).

The normalization of metrics advances when language engines are combined with accounting taxonomies that ensure grammatical, lexical and semantic traceability. By mapping sentences to templates based on the Global Reporting Initiative and the new European taxonomy, discrepancies are detected as context anomalies that the auditor can review prior to accounting closing. In this way, artificial intelligence reduces the costs of reconciling heterogeneous systems, strengthens cross-border comparability, and reduces the information asymmetry perceived by institutional investors (Clarkson et al., 2008; Ong et al., 2024).

External verification has been strengthened by merging computer vision models with very high-resolution satellite imagery. These convolutional networks identify thermal and chemical signatures associated with fugitive emissions and deforestation, contrast the findings with the reported figures, and generate alerts when the divergence exceeds regulatory thresholds. Unlike episodic audits, orbital observation provides ongoing empirical evidence that closes oversight gaps and mitigates the risk of greenwashing (Rolnick et al., 2023).

Likewise, the regulatory demand for transparency has placed algorithmic explainability at the center of system design. Methods such as SHAP and LIME are integrated into executive dashboards to expose which variables drive each prediction and with what relative weight, eliminating the opacity that traditionally surrounds black-box models. This visibility strengthens the confidence of auditors and supervisors, accelerates data-driven decision-making, and consolidates an organizational culture in which managerial intuition is complemented, never replaced, by advanced analytics (Ransbotham et al., 2017).

Evolution of dynamic capabilities in the strategic literature

The academic conversation about competitive advantages stopped focusing on the mere accumulation of resources when it was shown that the key lies in the organizational ability to renew them with discipline and opportunity. The first robust syntheses confirmed that identifying changes in the environment, recombining assets and shaping new business schemes are orchestrated processes that can be learned, which displaced the static vision of the company and opened the door to comparable agility indicators across sectors

With the empirical advance, the research revealed that the influence of these capabilities on performance depends on the adjustment with the internal structure and competitive pressure. Cross-sectional studies have shown that, in turbulent contexts, teams that master exploration routines, collective learning, and incentive realignment achieve higher operating margins than rivals focused only on efficiency (Wilden et al., 2013).

The microfoundational perspective specified the map of underlying actions: from technological surveillance mechanisms and rapid experimentation to the institutionalization of forums where senior management contrasts hypotheses with real-time data. Psychometric validation of ability scales to detect, leverage, and transform revealed that the intensity of each dimension varies according to the degree of digitalization and cultural openness to error as a source of learning.

More recently, it has been highlighted that strategic renewal does not depend solely on algorithms or cloud investments; It requires cultivating individual skills and networks of trust that accelerate the circulation of tacit knowledge. Companies that train their leaders to modulate the rhythms of exploitation and exploration are able to sustain continuous innovation cycles without compromising financial stability, thus integrating the logic of routines with that of internal entrepreneurial initiatives (Felin et al., 2012).

Emergence of impact metrics and shared value

Natural capital accounting has evolved from descriptive exercises to systems that assign shadow prices to ecosystem services, so that environmental degradation is reflected in projected cash flows. By incorporating these valuations, investment committees redirect resources towards projects with climate-adjusted returns, supported by analytical platforms that integrate field sensors and satellite data (Andersson & Arvidsson, 2023).

At the same time, the standardization of corporate reporting on water, soil and emissions has reduced the information asymmetry between issuers and analysts. The design of detailed guidelines for disclosure made it possible to compare sectors as diverse as food and technology, while empirical evidence confirmed the positive correlation between environmental transparency and access to capital on favorable terms (Kareiva et al., 2015).

In the social sphere, the measurement of community capital advanced from qualitative surveys to panels that quantify levels of trust, inclusion, and local resilience. Transdisciplinary studies have shown that firms that invest in social cohesion reduce absenteeism and increase productivity by integrating well-being indicators into their risk models and long-term goal setting (Liu et al., 2010).

Finally, the notion of shared value went from being a declaration of principles to becoming an operational framework with verifiable metrics. Recent research indicates that the simultaneous creation of economic and social value is enhanced when companies deliberate with stakeholders to define common goals and external verification mechanisms; this approach reinforces legitimacy and accelerates the attraction of impact-oriented capital (Kenter et al., 2016).

THEORETICAL FRAMEWORK

Theory of dynamic capabilities

The theory of dynamic capabilities has gone from being an intuitive metaphor to constituting an analytical framework with predictive power on corporate survival. At its core is a cycle that begins with the detection of changing signals from the environment, continues with the appropriation of emerging opportunities, and culminates with the deliberate transformation of the resource base. The firm that masters this cycle installs technological radars, uses real-time market analysis, and funds rapid validation experiments that reduce the risk of strategic bets, thereby reducing reliance on rigid trajectories and accelerating collective learning (Wilden et al., 2013).

Microfoundational research confirms that structural flexibility arises from the interaction between individual skills and coordination systems. Functional rotation programs, post-project review workshops and collaborative platforms encourage the transfer of tacit experiences, while variable incentive schemes linked to the generation of knowledge promote the responsible exploration of new ideas. This architecture maintains the efficiency of proven routines without stifling curiosity that opens paths to disruptive business models(Kenter et al., 2016).

The component of absorption of external knowledge is decisive when technological complexity exceeds internal research capacity. By partnering with universities, industry consortia, and open-source communities, organizations expand their scientific vocabulary and accelerate the recombination of solutions. Empirical findings show that firms that practice this systematic exchange translate patents into tradable products more quickly and sustain higher operating margins in turbulent contexts (Li et al., 2018).

A trend that is gaining strength in contemporary literature focuses on the synchrony between dynamic capacities and organizational resilience. When the environment becomes volatile, companies need to absorb impacts without disrupting the provision of value to customers. Evidence in volatile sectors indicates that companies that refine their sensing, appropriation, and transformation processes can reconfigure supply chains, rethink portfolios, and adapt cost structures with less productivity loss than their less agile competitors (Hillmann & Guenther, 2021; Chari et al., 2022; Garrido-Moreno et al., 2024).

The articulation of capacities is completed with the measurement of results. Indicators such as time to discover opportunities, speed of project deployment and financial value generated by innovations serve to feed back into decision systems. This learning loop makes it possible to detect bottlenecks, reallocate resources, and redefine success criteria before inefficiencies take hold, strengthening the culture of continuous improvement (Koh et al., 2024; Engelen et al., 2024; Björk et al., 2023).

Dynamic capacities have also been linked to sustainability and environmental, social and governance reporting. By internalizing the logic of analyzing risks and opportunities with real-time information, the company can integrate environmental and social impact metrics into strategic deliberations. Thus, the transformation is not limited to optimizing immediate financial returns, but incorporates variables of institutional legitimacy, reputation, and access to responsible capital, generating a virtuous circle between economic performance and public responsibility (Wilden et al., 2013).

The evolution of the theory also points to the coexistence of exploration and exploitation. Rather than privileging one over the other, high-performing organizations manage portfolios of initiatives that balance radical opportunity seeking with incremental refinement of core activities. This ambidexterity is coordinated through analytical platforms that allocate resources based on expected profitability and potential learning data, ensuring that exploratory successes translate into exploitable competencies at scale (Hillmann & Guenther, 2021; Chari et al., 2022; Garrido-Moreno et al., 2024). So, the theory of dynamic capabilities explains why some companies renew their advantages fluidly while others get caught up in their own historical success. The key lies in combining systematic surveillance, shared learning processes, and decision-making structures that reward disciplined experimentation. This synergy enables the organization to reposition itself with agility in the face of technological breakdowns, regulatory changes, and constantly changing social expectations, turning uncertainty into a source of sustainable differentiation (Li et al., 2018).

Knowledge-based perspective

The knowledge-based perspective describes the company as a living network of meanings that flows between people, processes, and technologies. The heart of this network beats with the dialogue between tacit knowledge and active knowledge codified as patents, manuals or algorithms. When both poles are connected through communities of practice, interactive repositories, and cross-mentoring rounds, intellectual capital expands organically: individual inventiveness finds collective discipline and germinal ideas become prototypes that feed diversified innovation portfolios (Martín de Castro et al., 2013).

The socio-technical architecture that sustains this flow requires more than digital platforms; It requires norms of conversation, shared taxonomies, and rituals of reflection. Knowledge leaders map internal networks, identify nodes of expertise, and orchestrate encounters where stories are contrasted with data. At the same time, dashboards capture exchange frequency, argumentative quality, and the incidence of learning in output cycle metrics, so that creativity is anchored to verifiable outcomes (Kianto et al., 2014).

The absorption of external knowledge becomes an advantage when the firm can recognize value in scattered signals, assimilate those findings, and reconfigure routines without friction. Partnerships with university labs, cross-licensing, and open-source exploration expand the available technology vocabulary. Recent studies show that the combination of potential and realized absorption capacity multiplies innovative performance in green processes and products, a sign that learning from abroad is no longer an option but a condition for strategic survival (Albort Morant et al., 2018).

The effective circulation of knowledge also depends on relational density. Teams that share purpose, trust, and a sense of reciprocity trigger micro-feedback loops that speed up the transfer of lessons and reduce the reinvention of solutions. In volatile environments, this social stapler explains why certain organizations reconfigure their portfolios with agility while others are trapped in costly historical inertia (Tortoriello et al., 2012; Gittell, 2002).

Rigorously measuring the contribution of these flows to corporate value requires systems that combine hard metrics with soft signals such as interdisciplinary conversation density or reputation in scientific-technical networks. The integrated dashboard provides continuous feedback to senior management and allows for rapid reallocation of talent and budget when exploration paths lose traction (Wuchty et al., 2007; Borgatti et al., 2009).

In short, digitalization redefines the very perimeter of the community. Cloud collaboration tools, machine translation and immersive reality make it easy for specialists dispersed in different time zones to meet. Small and medium-sized enterprises that adopt these infrastructures from a capacity logic are able to integrate into global value chains and accelerate their learning curve without sacrificing regional identity (Leonardi, 2014).

Dependence on resources and information power

The theory of resource dependence is based on the premise that every organization is inserted in networks where physical, financial, and symbolic inputs circulate, whose access determines its bargaining power and strategic resilience. When management identifies potential bottlenecks in critical raw materials or low-cost liquidity, it designs supplier portfolios that combine global partners and local suppliers, develops tactical reserves, and signs flexible purchasing agreements. In this way, it reduces the risk of disruption and strengthens its position in the value chain, while increasing its ability to capture extraordinary rents during periods of sudden shortages, a trait that explains the advantage of firms with dense and diversified procurement structures (Martín de Castro et al., 2013).

The rise of real-time data analytics has expanded the tactical possibilities of the theory by turning information from futures markets, emerging regulations, and political climates into early warning signals. Deep learning algorithms project demand scenarios, calculate substitution elasticities, and recommend portfolio adjustments even before disruption is visible to competitors. Organizations that integrate these flows with executive panels can renegotiate contracts with price variation clauses, secure critical inventories, or redirect investment to less exposed regions, increasing their room for maneuver and reducing cash flow volatility (Hillmann & Guenther, 2021; Chari et al., 2022; Garrido-Moreno et al., 2024).

Reputational capital has become as critical a resource as physical infrastructure. Actively participating in sectoral forums, publishing environmental performance data and collaborating with specialized NGOs gives legitimacy that translates into preferential access to green credits and greater regulatory tolerance. This symbolic power does not arise from isolated advertising campaigns, but from the congruence between narrative and practice, an unavoidable requirement in an environment where institutional investors demand auditable evidence of social and environmental commitment before allocating long-term capital (Kenter et al., 2016).

Relational intelligence is expressed when the company shares strategic information with its stakeholders to reduce asymmetries and build mutual trust. Local supplier development programs transfer knowledge of processes, quality standards and sustainability metrics, generating social barriers that reduce the probability of contractual breakdown in times of crisis. In parallel, community dialogues about expansion plans and open financial reporting consolidate support networks that cushion pressure from external actors and facilitate the obtaining of licenses and permits in shorter periods of time (Liu et al., 2010).

Technology transparency adds an extra layer of security by enabling near-instantaneous traceability of resource flows. Blockchain-based platforms record every transaction in the supply chain and allow buyers, regulators, and the public to verify the legitimacy of provenance. This distributed ledger increases the cost of opacity, deters opportunistic practices, and reinforces the reputation of participants who meet rigorous ethical and environmental standards, multiplying opportunities for collaboration in increasingly interdependent industrial ecosystems (Andersson & Arvidsson, 2023).

Rigorously measuring the effect of these strategies on performance requires indicators that integrate traditional variables such as operating profitability with risk exposure metrics. Network analysis tools identify critical nodes and calculate the centrality of the firm within the constellation of alliances, while simulation models quantify the financial impact of a potential disruption at each supply point.

With empirical evidence in hand, senior management can redistribute investments, renegotiate contract terms, or accelerate technology substitution initiatives before the market penalizes their inaction (Felin et al., 2012).

Organizational resilience

Corporate resilience is today conceived as a multi-layered capability that integrates forecasting, absorbing and transforming disturbances, overcoming the mere passive resistance associated with physical redundancy. Qualitative research in commercial aviation shows that maintaining liquidity buffers, creating high-quality employment links, and deploying two-way communication systems dampen collective anxiety, reduce layoffs, and facilitate critical human capital retention during severe operational crises, confirming that internal social capital acts as a psychological and operational shield against extreme events (Gittell et al., 2006).

Beyond the financial cushion, resilience is materialized through exploration routines that register weak signals, warn of impending misalignments, and mobilize resources before disruption crystallizes. Continuous monitoring models that integrate meteorological data, epidemiological markers, and input prices allow for simulating impact trajectories and suggest asset redeployment strategies, making analytical adaptability a competitive differential in sectors where operational continuity is critical to investor reputation and confidence (Van der Vegt et al., 2015).

Post-critical incident management becomes a catalyst for structural improvements when the organization institutionalizes learning mechanisms that transcend local fault correction. Narrative reconstruction workshops, multidepartmental drills, and prospective scenario laboratories transform traumatic experiences into inputs for redesigning processes, redefining investment priorities, and adjusting talent policies, establishing a virtuous cycle where each crisis feeds the capacity for future anticipation

The relational dimension of resilience is especially relevant in global supply networks, where logistical interdependence generates cascading vulnerabilities. Recent studies show that companies that cultivate collaborative contracts, share idle capacity information, and establish mutual aid protocols reduce recovery times and loss of stock market value after major disruptions, highlighting the need to invest in interorganizational governance that formalizes commitments to productive solidarity. Digitalization enables resilience metrics that transcend conventional financial indicators. Dashboards that record the average time to return to operation, the recovery of the customer satisfaction index and the degree of redundancy utilization provide continuous feedback to senior management, enabling adjustments in procurement policies, training plans and communication strategies before entropy compromises business continuity.

ESG-oriented corporate governance

Corporate governance evolves from a shareholder-centric fiduciary vision to a holistic model that explicitly incorporates environmental and social interests. Evidence from the U.S. banking sector indicates that more functionally and gender-diverse boards take a closer look at climate and social risks, moderate short-term decisions, and promote durable value-oriented strategies, resulting in lower stock volatility and reduced risk premiums

Linking executive compensation to indicators of greenhouse gas emissions, job security, and customer satisfaction shifts the focus from quarterly profitability to sustainable value creation. Such incentives are supported by institutional investors who manage long investment horizons and seek to minimize future costs associated with litigation and regulatory fines, configuring a financial ecosystem that rewards transparency and penalizes opportunism (Clarkson et al., 2008).

Emerging international regulations, such as the International Sustainability Standards Board guidelines, require the publication of climate scenarios, human rights policies, and pay equity metrics. Organizations that integrate these requirements into digital reporting systems generate understandable narratives for analysts and regulators, accelerate the approval of responsible investment projects, and improve access to loans linked to ESG performance, strengthening a virtuous circle between transparency, trust, and cost of capital (Eccles et al., 2012).

The convergence between inclusive governance and independent verification strengthens the credibility of the reports. External audits that validate environmental and social metrics, specialized committees that monitor data quality, and disclosure platforms in interoperable formats facilitate the traceability of information and encourage evidence-based decision-making, consolidating corporate legitimacy in markets that are increasingly sensitive to information integrity.

Therefore, the voluntary adoption of global codes of conduct, participation in multi-stakeholder initiatives and collaboration with civil society organizations raise reputation and create symbolic barriers that hinder the entry of competitors less committed to sustainability. This reputational advantage translates into customer loyalty, greater capacity to attract and retain talent, and resilience in the face of activism campaigns, showing that ESG-oriented corporate governance represents both a regulatory obligation and a strategic focus of differentiation

Analytic Decision Architecture

The contemporary decision-making fabric combines descriptive systems that chronicle the historical trajectory of performance, predictive algorithms that outline future scenarios, and prescriptive engines that recommend courses of action, thus generating a continuous flow of evidence that runs from plant sensors to the board table and replaces isolated intuition with informed deliberation (Ransbotham et al., 2017).

The promise of this analytical scaffolding only materializes when the data is coherent, complete and rigorously governed; Standardized semantic catalogs, real-time lineage controls, and anonymization protocols ensure that inputs maintain their fidelity throughout the lifecycle and prevent leaks that would compromise the legitimacy of the analysis (Davenport et al., 2010).

In addition to infrastructure, analytical literacy stands as a strategic variable: training programs that combine visualization, statistical logic, and critical thinking enable teams to formulate hypotheses, interpret patterns, and question biases, making analytics a collective capability and not the exclusive domain of specialists (Bogers, et al., 2017).

Algorithmic risk and ethics of artificial intelligence

The deployment of complex models exposes companies to legal and reputational contingencies when bias or lack of transparency are detected, which is why algorithmic ethics committees oversee data selection, training hyperparameters, and results intelligibility, anchoring innovation to principles of justice and accountability (Rudin et al., 2021).

Interpretability tools such as Shapley values or linear local explanations make it possible to relate each prediction to its input variables and facilitate auditing by regulators, clients and partners, increasing the social acceptance of artificial intelligence and mitigating the risk of massive litigation arising from opaque decisions (Doshi-Velez et al., 2017).

Along with transparency, technical robustness requires stress testing against adversarial tampering and cross-validation in diverse environments; standardized documentation makes it possible for third parties to reproduce experiments and verify consistency, integrating algorithmic ethics into comprehensive risk management and turning digital reliability into a source of competitive differentiation (Floridi et al., 2018).

Open innovation and digital ecosystems

The accelerated pace of technology hinders innovation confined to internal laboratories; Co-creation platforms, corporate incubators, and community meetings broaden the radar of opportunities and accelerate the transition from concept to prototype, as corroborated by bibliometric analyses that show higher patent density and shorter commercialization cycles in firms immersed in open innovation networks (Bogers et al., 2020).

The value of digital ecosystems lies in the orchestration of complementary roles: anchor companies, young start-ups, universities and public bodies share exploration, testing and scaling activities, while shared data platforms ensure the traceability of contributions and foster collective learning, attracting philanthropic and venture capital investment towards sustainability solutions. To safeguard intellectual property without stifling collaboration, participants turn to flexible licensing, graduated trust clauses, and contracts that link knowledge input to benefit-sharing, striking a delicate balance between openness and protection that sustains the ecosystem's vitality in the long term (Hillmann & Guenther, 2021; Chari et al., 2022; Garrido-Moreno et al., 2024).

Institutional Theory and Legitimacy

Organizations operate within normative fields that delineate what is socially appropriate; the adoption of international reporting standards, global codes of conduct, and third-party certifications reduces investor uncertainty and confers legitimacy with governments and communities. Isomorphism mechanisms explain the spread of responsible practices: coercive pressure from regulators, mimetic competition among peers, and normative persuasion from professional associations encourage homogeneity, while strategic firms calibrate their response to demonstrate conformity without sacrificing differentiating features that underpin competitive advantage. Corporate narratives consistent with dominant social values attract regulatory support and tolerance for experimentation, enabling pilot projects that would otherwise meet with political resistance; in this way, legitimacy becomes an intangible asset that facilitates the exploration of emerging opportunities (Kenter et al., 2016).

Quantifying sustainable value

Triple-bottom line accounting integrates financial flows, social impacts, and environmental footprint; By monetizing ecosystem services through shadow pricing, firms internalize previously outsourced costs and reveal the true risk-adjusted return of their investment projects. Natural capital valuation instruments, such as water and emissions balances, are combined with well-being surveys and citizen participation panels to measure social capital; cognitive analytics uncovers correlations between labor inclusion and productivity, confirming that social investment generates indirect returns in reputation, customer loyalty, and litigation reduction. The integration of these metrics into executive dashboards allows investment alternatives to be weighted under the criterion of shared value, where private utility is aligned with collective benefit, broadening the planning horizon and discouraging decisions that transfer risk to society (Hillmann & Guenther, 2021; Chari et al., 2022; Garrido-Moreno et al., 2024).

Technology-based organizational change

Digital transformation is conceived as a systemic process that encompasses physical infrastructure, human competencies and corporate culture; Maturity diagnoses, job retraining plans, and well-defined data governance frameworks increase the speed of implementation and shorten the learning curve, optimizing the return on technological investment (Caputo et al., 2021).

The decentralization of decisions supported by live dashboards and recommendation algorithms redistributes internal power: senior management adopts a strategic curation role that aligns local experiments with the global vision, while operational teams design, test and scale solutions with supervised autonomy, which translates into greater customer satisfaction and shorter cycle times. Organizations that consolidate this integrative digital layer report superior resilience to external disruptions, as data transparency allows bottlenecks to be detected early, resources to be reallocated, and operational continuity to be maintained even when regulatory, health, or geopolitical environments become unpredictable (Ransbotham et al., 2017).

CHALLENGES FACED BY ORGANIZATIONS OR INSTITUTIONS

Systemic volatility due to the convergence of health crises, geopolitical fractures, and technological disruptions forces organizations to operate in environments characterized by ever-shorter cycles of change and the simultaneous coexistence of threats and opportunities. This complexity blurs industry boundaries, erodes traditional positional advantages, and requires firms to articulate digital architectures capable of orchestrating real-time information flows and reconfiguring assets with strategic agility. The literature shows that resilience depends on the incorporation of innovation capabilities and the adoption of data-driven business models that allow for the synchronization of investment and long-term organizational learning, thus creating buffers against uncertainty (Cardoso et al., 2025; Georgescu et al., 2024).

The speed of digital transformation poses a double challenge: on the one hand, it demands robust technological infrastructures; on the other, it imposes a leadership capable of catalyzing processes of profound cultural change. Empirical evidence in health services reveals that digital-oriented transformational leadership increases digital intensity when linked to organizational agility and institutional contexts that legitimize innovation, underscoring that technology alone is not enough to capture value if it is not integrated with new work and collaboration logics (Kludacz-Alessandri et al., 2025).

At the same time, the regulatory and reputational pressures associated with the disclosure of environmental, social, and governance performance pose considerable challenges. The mosaic of voluntary standards, the demands for traceability in the supply chain, and the resource asymmetries between large corporations and medium-sized suppliers trigger tensions that, if not managed, lead to legitimacy risks and administrative overloads. Recent studies suggest that the absence of interdisciplinary capacities limits the quality of reporting and hinders the integration of the Sustainable Development Goals into corporate strategy (Chopra et al., 2024).

In the talent space, cross-border mobility has been affected by migration restrictions, skills gaps, and changes in workforce expectations. Global talent management faces the paradox of combining cultural consistency with local adaptation, while balancing social purposes and profitability objectives. Organizations need collaborative networks that allow for the design of more flexible career paths that favor the rotation of knowledge without losing identity cohesion (Evans et al., 2025)

Likewise, resilience is not built with technology alone, but through human resources practices that reinforce continuous learning, participation and trust. Evidence indicates that sustainable human resource strategies and inclusive cultures enhance absorptive capacity and favor adaptation during crises, encouraging the co-creation of solutions and expanding the robustness of organizational systems (Georgescu et al., 2024).

However, there is a lack of alignment between the strategic vision and the operational reality. Many companies implement digital initiatives in a fragmented way without articulating an architecture that connects data, processes, and competencies. Research indicates that when digital transformation is conceived as a purely technological project, resistance is generated, technical debt is multiplied, and human capital is underutilized, which undermines the strategic agility needed to respond to external shocks (Cardoso et al., 2025).

In addition, the governance of information in sustainability requires integrated measurement and verification systems, capable of translating complex metrics into narratives understandable by the different stakeholders. The lack of interoperability between platforms, the lack of analytical talent, and regulatory dispersion make cross-sectional comparison and decision-making based on robust indicators difficult, compromising the credibility of reports (Chopra et al., 2024).

In short, today's challenges are intertwined in a framework where the recursive dynamics between digitalization, sustainability, talent and resilience define future competitive advantage. Organizations that articulate transformational leadership, learning-centered HR practices, and interoperable data platforms will be better positioned to turn disruption into opportunity, capitalize on collective intelligence, and align purpose and economic performance (Kludacz-Alessandri et al., 2025; Evans et al., 2025).

EXAMPLES OF IMPLEMENTATION IN COMPANIES OR ORGANIZATIONS

The intensification of global competition and the pressure to integrate environmental, social and governance criteria have led many companies to move from rhetorical declaration to the practical execution of digital and sustainable strategies; this leap implies reconfiguring technological architectures, redesigning operational processes, and cultivating a culture of experimentation that legitimizes informed risk-taking, since only in this way is it possible to capitalize on the synergies that emerge from data and translate them into differential value propositions (Cardoso et al., 2025).

The experience of A. P. Møller-Maersk, a pioneer in the adoption of *cloud-hosted blockchain* platforms interconnected with machine learning algorithms, demonstrates how document traceability can become a strategic asset: the reduction of administrative friction and the optimization of sea routes not only decreased logistics costs, but also strengthened supply chain transparency; however, the network value was directly proportional to the number of ports and shippers willing to share data, so that interoperability and distributed governance emerged as essential enabling conditions (Wong et al.,, 2023).

In advanced manufacturing, Siemens illustrates the transition from monolithic architectures to microservices ecosystems managed under DevOps practices: by breaking functionality down into independent components, the firm was able to accelerate release cycles and develop real-time A/B tests on production lines, generating continuous learning and reducing technical debt; this success, however, was conditional on the creation of performance metrics that rewarded cross-functional collaboration and the establishment of psychological pacts that normalized controlled failure as a prior step to iterative improvement (Fritzsch et al., 2022).

In the European health sector, several primary care entities adopted transformational leadership based on analytical evidence to increase digital intensity; the deployment of interoperable medical records and telemonitoring systems led to more preventive and personalized care models, while the legitimacy of these changes was sustained by directive narratives that articulated the technological promise with the humanistic vocation of the medical profession (Kludacz-Alessandri et al., 2025).

ESG disclosure obligations have led multinational conglomerates to implement integrated measurement systems that capture Scope Three emissions, workforce diversity, and supplier safety; evidence indicates that the effort becomes profitable when the indicators are aligned with internal economic incentives and external regulatory expectations, as only in this way is information fragmentation avoided and stakeholder confidence strengthened (Chopra et al., 2024).

Attracting and retaining talent has also become a laboratory for organisational innovation: cross-border mobility policies based on digital platforms and upskilling programmes based on learning analytics are redefining more flexible career paths, which reinforces the capacity to absorb knowledge and favours the creation of global communities of practice; such an approach, when accompanied by an inclusive purpose and distributed leadership frameworks, increases resilience to critical skills supply crises (Evans et al., 2025).

From the perspective of the circular economy, consumer goods consortia have redesigned their value chains through smart contracts that monetize the recovery of materials and penalize waste; experimentation with incentive tokenization has shown that it is possible to link environmental metrics with financial rewards, so that sustainability is no longer a cost center and becomes a source of sustainable competitive advantage (Nandi et al., 2021).

Although each case is framed in specific sectoral and cultural realities, they all show that value creation is enhanced when three vectors converge: inter-organizational co-creation, technological modularity and enabling leadership; the harmonious articulation of these factors makes it possible to translate data into strategic decisions, reduce information asymmetries, and channel innovation towards tangible economic and social results (Georgescu et al., 2024).

RESULTS

The cross-sectional analysis of the cases reveals that the incorporation of cognitive analytics and emerging technologies translates into qualitative improvements in organizational resilience and the accuracy of environmental, social, and governance reports, provided that firms manage to articulate a strategic vision based on data and a modular architecture that facilitates safe experimentation (Cardoso et al., 2025).

In the maritime industry, the cloud blockchain platform implemented by A. P. Møller-Maersk showed that decentralized transparency reduces traceability blind spots while also generating reputational capital that strengthens partnerships with port players; this qualitative improvement was visible in the reduction of document disputes and mutual trust between shippers, elements that reinforce the resilience of the logistics network even without standardized quantitative metrics (Wong et al., 2023).

The migration from monolithic architectures to microservices in the German manufacturing sector showed that continuous release cycles increase organizational learning capacity because each iteration works as a microexperiment where feedbacks are captured in real time; qualitatively, teams reported greater autonomy, less cross-functional friction, and a more favorable willingness to share tacit knowledge, cultural indicators that precede sustainable performance improvements (Fritzsch et al., 2022).

In the European healthcare field, it was observed that transformational leadership, when supported by predictive analytics, generates an atmosphere of trust that drives the adoption of interoperable clinical records; qualitatively, professionals highlight greater diagnostic safety and participation in digital communities of practice, symptoms of a profound cultural change that transcends the mere digitalization of processes (Kludacz-Alessandri et al., 2025).

ESG reporting experiences indicate that organizations that develop interdisciplinary capabilities produce more coherent and credible narratives; interviews suggest that such consistency reduces domestic regulatory fatigue and boosts investor confidence, even before direct financial benefits materialize (Chopra et al., 2024).

In terms of talent, cross-border mobility programmes based on digital platforms generated epistemic communities that accelerate the circulation of good practices and mitigate the dependence on localised experts; participants reported a greater sense of purpose and a perception of global equity that strengthens retention, qualitative factors associated with the organization's psychosocial resilience (Evans et al., 2025).

The examination of inclusive cultures and sustainable human resources practices showed that interpersonal trust and openness to error are consolidated as buffers against external shocks; managers and collaborators agreed that continuous feedback and participation in decision-making boost morale and speed up adaptation to regulatory or logistical shocks (Georgescu et al., 2024).

From the perspective of the circular economy, initiatives that tokenize incentives for the recovery of materials not only align financial and environmental interests, but also generate a narrative of co-responsibility that strengthens the link with consumers and suppliers; qualitatively, managers highlight the ease of tracking commitments and the reduction of disputes over the quality of returned materials (Nandi et al., 2021).

The cognitive dashboards introduced in companies with a high data footprint shortened the time between anomaly detection and corrective action, an effect that the interviewees relate to the possibility of carrying out simulations in natural language and visualizing impact scenarios without requiring the intermediation of specialists, which democratizes analytics and reinforces collective accountability (Kenter et al., 2016).

Finally, the valuation of natural and social capital through impact metrics allowed organizations to build shared value narratives that resonated with external stakeholders; the investment committees highlighted that visibility on ecosystem services and community cohesion facilitates the integration of extra-financial factors in strategic deliberations, thus consolidating a common language between sustainability and profitability (Hillmann & Guenther, 2021; Chari et al., 2022; Garrido-Moreno et al., 2024).

Taken together, these qualitative findings suggest that the main lever for success lies not only in technological sophistication, but in the ability to intertwine data governance, inclusive leadership and organizational learning. Organizations that conceive of cognitive analytics as a sociotechnical process are able to translate information into actionable insights, strengthen the legitimacy of their ESG reporting, and cultivate resilience in the face of volatile environments.

CONCLUSIONS

The trajectories analyzed point to an axial conclusion: sustainable value creation in the digital economy emerges when organizations intertwine, in a single socio-technical fabric, cognitive analytics, modular data architectures, transformational leadership, and collaborative governance along global supply chains, as only in this way can they translate information abundance into financial resilience. social legitimacy and lasting competitive advantages (Cardoso et al., 2025). Qualitative evidence confirms that algorithms do not generate impact by themselves; what is decisive lies in the human capacity to orchestrate its deployment with criteria of equity, transparency and strategic sense, integrating operational efficiency and commitment to the sustainable development goals into the same equation (Ransbotham, et al., 2017).

The convergence between microservices and DevOps practices demonstrated that technological modularity shortens learning cycles, reduces technical debt, and amplifies cultural adaptability, which translates into a tangible reduction in reaction times to external disruptions and an increase in the accuracy of ESG reporting, as the lineage of data becomes traceable in real time (Fritzsch et al., 2022). However, Maersk's experience reveals that any promise of interoperability collides with the reluctance of peripheral actors to expose strategic information; hence, cryptographic protection, granular permission control, and standardization of semantic taxonomies are essential to avoid compliance overload without sacrificing transparency (Wong et al., 2023).

Findings in European primary care show that transformational leadership, when leveraged on shared-purpose narratives, manages to mobilize highly regulated professionals towards the adoption of intelligent clinical records and predictive dashboards, dismantling cultural resistance by subjecting everyday practice to algorithmic scrutiny and legitimizing iteration as a natural mechanism of improvement (Kludacz-Alessandri et al., 2025). This same pattern appears in advanced manufacturing and consumer chains, where leaders who link the ESG agenda to the promise of inclusive growth and continuous learning reap higher talent retention rates and more active workforce participation in generating analytical insights (Evans et al., 2025).

In ESG reporting, the deployment of natural language processing engines and self-service cognitive dashboards reduced misclassification, improved narrative consistency, and increased investor and regulatory confidence, provided that the teams responsible for combining forensic accounting, data science, and strategic thinking skills to avoid information fragmentation. Organizations that integrate satellite verification and smart contracts into their assurance cycle extend traceability to the most remote links, reinforcing legitimacy and mitigating the risk of greenwashing(Chopra et al., 2024).

The tokenized circular economy illustrates how collaborative governance mechanisms align economic and environmental incentives: by monetizing material recovery through smart contracts, consortia turn sustainability into a source of profitability, while distributing benefits equitably between suppliers and recyclers, reducing reluctance to share strategic data and generating communities of practice oriented towards systemic efficiency. Similarly, cross-sector alliances observed in health and logistics projects show that distributed trust, based on open standards and independent audits, accelerates the adoption of shared infrastructures without compromising the protection of sensitive data (Georgescu et al., 2024).

Reflections on human capital indicate that analytical literacy and upskilling programs are sine qua non conditions for democratizing cognitive analytics and turning collaborators into co-producers of knowledge; companies that integrate data challenges into their career trajectories report a greater sense of belonging and a superior ability to absorb external knowledge, a fact that reinforces psychosocial resilience and strategic agility (Lee et al., 2024). When these practices are combined with inclusive cultures that reward transparency and horizontal feedback, the organization will have buffers that reduce anxiety about disruption and transform every crisis into a learning opportunity (Kareiva et al., 2015).

The results also underscore the importance of institutional legitimacy: regulatory convergence around double materiality and the proliferation of independent assurance standards generate an ecosystem of transparency that favors access to responsible capital and reduces the opportunity cost of detailed reporting; firms that adopt these frameworks proactively position themselves as sectoral benchmarks and capture reputation premiums that are difficult to imitate. mThis legitimacy is reinforced when senior management links executive compensation to environmental and social indicators, sending a clear signal of strategic alignment between purpose and financial performance (Lewis, 2003).

In light of the evidence collected, it can be stated that modular digital architectures allow reconciling interoperability and data protection through the segregation of functionalities in microservices, end-to-end encryption, and the use of semantic standards that automate compliance, thereby reducing the administrative burden without sacrificing transparency. Transformational leadership succeeds in mobilizing the workforce when it articulates narratives of purpose that connect ESG ambition with professional identity and offers safe spaces for iterative experimentation, legitimizing error as a necessary step for continuous improvement (Martín de Castro et al., 2013).

Finally, the most effective collaborative governance mechanisms are those that combine distributed ledgers, benefit-sharing clauses, and independent verification, as they foster transparency, overcome reluctance to share data, and align economic and environmental incentives along the global chain, generating virtuous circles of open innovation and shared value (Jean & Kim, 2020).

RECOMMENDATIONS FOR FUTURE RESEARCH

The revised empirical and conceptual body confirms that the conjunction of cognitive analytics and emerging technologies opens up a fertile, although still insufficiently explored, field to deepen the links between organizational resilience, ESG transparency and the creation of shared value. A first line of research should investigate how the maturity of microservices architectures influences the longitudinal consistency of ESG indicators, considering that modularity reduces technical debt but also introduces complexities in data governance when services are orchestrated across multi-stakeholder ecosystems. The challenge lies in establishing methodological frameworks that measure, in a comparative way, the transactional carbon footprint and the volatility of the supply chain under differentiated technical configurations, so that the findings can be generalized beyond emblematic cases (Fritzsch et al., 2022).

A second area calls for longitudinal studies that connect the evolution of transformational leadership with the progressive appropriation of predictive analytics by highly regulated professionals, such as clinical teams. There is evidence that narratives of purpose strengthen technology adoption, but trajectory models that describe how epistemological trust in algorithms is consolidated or eroded over time and under what circumstances patterns of cultural resistance re-emerge, especially when the automation of diagnoses and treatments is increased (Kludacz-Alessandri et al., 2025).

The literature on tokenized circular economy suggests that smart contract-based incentives can align environmental and financial goals, but most studies focus on isolated value chains. Cross-sectional research comparing platforms from different industries and regions is therefore recommended to assess the portability of governance mechanisms and the robustness of benefit-sharing models. This approach would make it possible to determine the extent to which distributed transparency is sufficient to overcome the reluctance to share strategic data and how the results vary when the power asymmetries between suppliers and corporate anchors are modified (Nandi et al., 2021).

A fourth front concerns the intersection between global talent mobility and psychosocial resilience. While positive effects of digital learning platforms on knowledge retention and absorption have been identified, studies that analyze the costs of cross-cultural coordination and the risks of cognitive fatigue derived from constant exposure to real-time performance metrics are lacking. It is suggested to address these issues through mixed methods that combine social network analysis, digital ethnographies, and organizational well-being metrics, in order to build predictive models on the relationship between collaborative density and sustainability of work engagement (Evans et al., 2025).

It is also necessary to delve into the ethical limits of cognitive analytics applied to ESG disclosure. Natural language processing engines have proven efficient in reducing classification errors, but there is a gap in the biases that can be introduced when trained on historical data dominated by corporate narratives. It is proposed to deploy comparative algorithmic audits that contrast outputs generated by different models and assess their sensitivity to variations in the corpora, with a view to designing safeguards that allow detecting and correcting distortions before they reach official reports (Chopra et al., 2024).

Another line suggests investigating the synergies between forensic accounting and satellite remote sensing in the verification of environmental commitments. Although high-resolution imagery and hyperspectral sensors multiply monitoring capacity, the best way to integrate this evidence with real-time financial reporting systems is still unknown. Studies combining data fusion techniques and dual materiality analysis could clarify which metrics are most sensitive for predicting climate risk events and how to communicate them in a way that is understandable to investors and regulators (Borgatti et al., 2009; Leonardi, 2014).

On a methodological level, there is an urgent need to develop scales of dynamic capabilities that capture the speed with which organizations learn to feel, do and transform opportunities derived from generative artificial intelligence. The existing scales measure isolated components and do not yet incorporate the ethical dimension or co-creation with external communities of practice. Designing validated instruments in multiple sectors will make it possible to compare rates of evolution and explore how institutional maturity conditions strategic reconfiguration.

Finally, it is recommended that future studies analyze the relationship between institutional legitimacy and the cost of responsible capital, studying companies that have linked executive remuneration to environmental and social objectives. Investigating whether this practice generates sustainable reputation premiums or if, on the contrary, it leads to corporate governance imbalances when objectives are too ambitious, will allow us to refine signaling theories and create more realistic implementation routes for medium-scale firms (Kenter et al., 2016).

REFERENCES

Albort Morant, G., Henseler, J., Cepeda Carrión, G., & Leal Rodríguez, A. L. (2018). Potential and realized absorptive capacity as complementary drivers of green product and process innovation performance. *Sustainability (Basel), 10*(2), 381. DOI: 10.3390/su10020381

Andersson, F. N. G., & Arvidsson, S. (2023). Understanding, mapping and reporting of climate-related risks among listed firms in Sweden. *Climate Policy, 23*(8), 945–958. DOI: 10.1080/14693062.2022.2116383

Auld, G., Casovan, A., Clarke, A., & Faveri, B. (2022). Governing AI through ethical standards: Learning from the experiences of other private governance initiatives. *Journal of European Public Policy, 29*(11), 1822–1844. DOI: 10.1080/13501763.2022.2099449

Bais, B., Nassimbeni, G., & Orzes, G. (2024). Global Reporting Initiative: Literature review and research directions. *Journal of Cleaner Production, 471*, 143428. DOI: 10.1016/j.jclepro.2024.143428

Barredo Arrieta, A., Díaz Rodríguez, N., Del Ser, J., Bennetot, A., Tabik, S., Barbado, A., Garcia, S., Gil-Lopez, S., Molina, D., Benjamins, R., Chatila, R., & Herrera, F. (2020). Explainable Artificial Intelligence (XAI): Concepts, taxonomies, opportunities and challenges toward responsible AI. *Information Fusion, 58*, 82–115. DOI: 10.1016/j.inffus.2019.12.012

Bernardo, B. M. V., Mamede, H. S., Barroso, J. M. P., & Santos, V. M. P. D. (2024). Data governance & quality management—Innovation and breakthroughs across different fields. *Journal of Innovation & Knowledge, 9*(4), 100598. DOI: 10.1016/j.jik.2024.100598

Björk, J., Magnusson, M., & Karlsson, C. (2023). Measuring innovation effectively—Nine critical lessons. *Research Technology Management, 66*(2), 24–33. DOI: 10.1080/08956308.2022.2151232

Bogers, M., Zobel, A.-K., Afuah, A., Almirall, E., Brunswicker, S., Dahlander, L., Frederiksen, L., Gawer, A., Gruber, M., Haefliger, S., Hagedoorn, J., Hilgers, D., Laursen, K., Magnusson, M. G., Majchrzak, A., McCarthy, I. P., Moeslein, K. M., Nambisan, S., Piller, F. T., & Ter Wal, A. L. J. (2017). The open innovation research landscape: Established perspectives and emerging themes across different levels of analysis. *Industry and Innovation, 24*(1), 8–40. DOI: 10.1080/13662716.2016.1240068

Boulton, T. J., Iqbal, U., Khaksari, S., & Qiu, S. (2024). Mandatory ESG disclosure, information asymmetry, and IPO underpricing. *European Financial Management*. Advance online publication. DOI: 10.1111/eufm.12494

Buonocore, F., Annosi, M. C., De Gennaro, D., & Riemma, F. (2024). Digital transformation and social change: Leadership strategies for responsible innovation. *Journal of Engineering and Technology Management*, *74*, 101843. DOI: 10.1016/j.jengtecman.2024.101843

Butt, A., Imran, F., Helo, P., & Kantola, J. (2024). Strategic design of culture for digital transformation. *Long Range Planning*, *57*(4), 102415. DOI: 10.1016/j.lrp.2024.102415

Caputo, A., Pizzi, S., Pellegrini, M. M., & Dabić, M. (2021). Digitalization and business models: Where are we going. *Journal of Business Research*, *123*, 489–501. DOI: 10.1016/j.jbusres.2020.09.053

Cardoso, A., Figueiredo, J., Oliveira, I., & Pocinho, M. (2025). From crisis to opportunity: Digital transformation, digital business models, and organizational resilience in the post-pandemic era. *Administrative Sciences*, *15*(6), 193. Advance online publication. DOI: 10.3390/admsci15060193

Chari, A., Niedenzu, D., Despeisse, M., Machado, C. G., Azevedo, J. D., Boavida-Dias, R., & Johansson, B. (2022). Dynamic capabilities for circular manufacturing supply chains—Exploring the role of Industry 4.0 and resilience. *Business Strategy and the Environment*, *31*(5), 2500–2517. DOI: 10.1002/bse.3040

Chopra, S. S., Senadheera, S. S., Dissanayake, P. D., & Withana, P. A. (2024). Navigating the challenges of environmental, social and governance reporting: The path to broader sustainable development. *Sustainability, 16*(2). *Section, 606*. Advance online publication. DOI: 10.3390/su16020606

Clarkson, P. M., Li, Y., Richardson, G. D., & Vasvari, F. P. (2008). Revisiting the relation between environmental performance and environmental disclosure: An empirical analysis. *Accounting, Organizations and Society*, *33*(4), 303–327. DOI: 10.1016/j.aos.2007.05.003

Davenport, T., Guha, A., Grewal, D., & Bressgott, T. (2020). How artificial intelligence will change the future of marketing. *Journal of the Academy of Marketing Science*, *48*(1), 24–42. DOI: 10.1007/s11747-019-00696-0

De Villiers, C., Hsiao, P., Maroun, W., Ng, J., Roussey, S., & Vasquez, D. (2024). The International Sustainability Standards Board's (ISSB) past, present and future: Critical reflections and a research agenda. *Pacific Accounting Review*, *36*(3), 437–457. DOI: 10.1108/PAR-02-2024-0038

Doshi-Velez, F., Kim, B., Buffoni, L., & Goodwin, T. (2017). Towards a rigorous science of interpretable machine learning. *arXiv Preprint*, 1702.08608. https://doi.org/arXiv.1702.08608DOI: 10.48550

Duchek, S. (2020). Organizational resilience: A capability-based conceptualization. *Business Research*, *13*(1), 215–246. DOI: 10.1007/s40685-019-0085-7

Dwivedi, Y. K., Hughes, L., Ismagilova, E., Aarts, G., & Williams, M. D. (2019). Artificial intelligence: Multidisciplinary perspectives on emerging challenges, opportunities, and agenda for research, practice and policy. *International Journal of Information Management*, *57*, 101994. DOI: 10.1016/j.ijinfomgt.2019.08.002

Eccles, R. G., Krzus, M. P., Rogers, J., & Serafeim, G. (2012). The need for sector specific materiality and sustainability reporting standards. *The Bank of America Journal of Applied Corporate Finance*, *24*(2), 65–71. DOI: 10.1111/j.1745-6622.2012.00380.x

Elliot, V., Jonäll, K., Paananen, M., Bebbington, J., & Michelon, G. (2024). Biodiversity reporting: Standardization, materiality, and metrics. *Current Opinion in Environmental Sustainability*, *64*, 101435. DOI: 10.1016/j.cosust.2024.101435

Engelen, A., Klyver, K., & de Mol, E. (2024). Building a resilient organization through a pre-shock strategic emphasis on innovation. *Journal of Product Innovation Management*, *41*(1), 36–61. Advance online publication. DOI: 10.1111/jpim.12697

Evans, P. A. L., Pucik, V., Stahl, G. K., & Björkman, I. (2025). Global talent management in an age of disruption. *Organizational Dynamics*, *54*(3), 101149. Advance online publication. DOI: 10.1016/j.orgdyn.2025.101149

Felin, T., Foss, N. J., Heimeriks, K. H., & Madsen, T. L. (2012). Microfoundations of routines and capabilities: Individuals, processes, and structure. *Journal of Management Studies*, *49*(8), 1351–1374. DOI: 10.1111/j.1467-6486.2012.01052.x

Floridi, L., Cowls, J., Beltrametti, M., Chatila, R., Chazerand, P., Dignum, V., Luetge, C., Madelin, R., Pagallo, U., Rossi, F., Schafer, B., Valcke, P., & Vayena, E. (2018). AI4People—An ethical framework for a good AI society. *Minds and Machines*, *28*(4), 689–707. DOI: 10.1007/s11023-018-9482-5 PMID: 30930541

Fritzsch, J., Bogner, J., Haug, M., & Franco da Silva, A. C. (2022). Adopting microservices and DevOps in the cyber-physical systems domain: A rapid review and case study. *Software, Practice & Experience*. Advance online publication. DOI: 10.1002/spe.3169

Garrido-Moreno, A., Martín-Rojas, R., & García-Morales, V. J. (2024). The key role of innovation and organizational resilience in improving business performance: A mixed-methods approach. *International Journal of Information Management*, *77*, 102777. DOI: 10.1016/j.ijinfomgt.2024.102777

Georgescu, I., Bocean, C. G., Vărzaru, A. A., Rotea, C. C., Mangra, M. G., & Mangra, G. I. (2024). Enhancing organizational resilience: The transformative influence of strategic human resource management practices and organizational culture. *Sustainability, 16*(10). *Sustainability (Basel)*, *4315*(10), 4315. Advance online publication. DOI: 10.3390/su16104315

Gittell, J. H., Cameron, K., Lim, S., & Rivas, V. (2006). Relationships, layoffs, and organizational resilience: Airline industry responses to September eleven. *The Journal of Applied Behavioral Science*, *42*(3), 300–329. DOI: 10.1177/0021886306286466

Hillmann, J., & Guenther, E. (2021). Organizational resilience: A valuable construct for management research. *International Journal of Management Reviews*, *23*(1), 7–44. DOI: 10.1111/ijmr.12239

Ivanov, D., Dolgui, A., & Sokolov, B. (2019). The impact of digital technology and Industry 4.0 on the ripple effect and supply chain risk analytics. *International Journal of Production Research*, *57*(3), 829–846. DOI: 10.1080/00207543.2018.1488086

Jean, R.-J., & Kim, D. (2020). Internet and SMEs' internationalization: The role of platform and website capabilities. *Journal of International Management*, *26*(1), 100690. DOI: 10.1016/j.intman.2019.100690

Kaminski, M. E., & Malgieri, G. (2021). Algorithmic impact assessments under the GDPR: Producing multi-layered explanations. *International Data Privacy Law*, *11*(2), 125–144. DOI: 10.1093/idpl/ipaa020

Kareiva, P. M., McNally, B. W., McCormick, S., Miller, T., & Ruckelshaus, M. (2015). Improving global environmental management with standard corporate reporting. *Proceedings of the National Academy of Sciences of the United States of America*, *112*(24), 7375–7382. DOI: 10.1073/pnas.1408120111 PMID: 26082543

Kenter, J. O., Bryce, R., Christie, M., Cooper, N., Hockley, N., Irvine, K. N., Fazey, I., O'Brien, L., Orchard-Webb, J., Ravenscroft, N., Raymond, C. M., Reed, M. S., Tett, P., & Watson, V. (2016). Shared values and deliberative valuation: Future directions. *Ecosystem Services*, *21*, 358–371. DOI: 10.1016/j.ecoser.2016.10.006

Kianto, A., Ritala, P., Spender, J. C., & Vanhala, M. (2014). The interaction of intellectual capital assets and knowledge management practices in organizational value creation. *Journal of Intellectual Capital*, *15*(3), 362–375. DOI: 10.1108/JIC-05-2014-0059

Kim, J. W., & Park, C. K. (2023). Can ESG performance mitigate information asymmetry? Moderating effect of assurance services. *Applied Economics*, *55*(26), 2993–3007. DOI: 10.1080/00036846.2022.2107991

Kludacz-Alessandri, M., Hawrysz, L., Żak, K., & Zhang, W. (2025). The impact of digital transformational leadership on digital intensity among primary healthcare entities: A moderated mediation model. *BMC Health Services Research*, *25*(1), 117. DOI: 10.1186/s12913-025-12283-x PMID: 39838353

Kludacz-Alessandri, M., Hawrysz, L., Żak, K., & Zhang, W. (2025). The impact of digital transformational leadership on digital intensity among primary healthcare entities: A moderated mediation model. *BMC Health Services Research*, *25*(1), 117. DOI: 10.1186/s12913-025-12283-x PMID: 39838353

Koh, S. C. L., Leung, J. Y.-T., Gunasekaran, A., & Simpson, R. J. (2024). Quantifying organisational resilience: An integrated indicator and validation in manufacturing. *International Journal of Production Research*. Advance online publication. DOI: 10.1080/00207543.2023.2296018

Kritzinger, W., Karner, M., Traar, G., Henjes, J., & Sihn, W. (2018). Digital twin in manufacturing: A categorical literature review and classification. *IFAC-PapersOnLine*, *51*(11), 1016–1022. DOI: 10.1016/j.ifacol.2018.08.474

Krivogorsky, V. (2024). Sustainability reporting with two different voices: The European Union and the International Sustainability Standards Board. *Journal of International Accounting, Auditing & Taxation*, *56*, 100635. DOI: 10.1016/j.intaccaudtax.2024.100635

Krueger, P., Matos, P., & Starks, L. T. (2023). The effects of mandatory ESG disclosure around the world. *Journal of Accounting Research*, *61*(6), 2117–2175. DOI: 10.1111/1475-679X.12548

Leonardi, P. M. (2014). Social media, knowledge sharing, and innovation: Toward a theory of communication visibility. *Information Systems Research, 25*(4), 796–816. DOI: 10.1287/isre.2014.0536

Lewis, K. (2003). Measuring transactive memory systems in the field: Scale development and validation. *The Journal of Applied Psychology, 88*(4), 587–604. DOI: 10.1037/0021-9010.88.4.587 PMID: 12940401

Liu, S., Costanza, R., Farber, S., & Troy, A. (2010). Valuing ecosystem services: Theory, practice, and the need for a transdisciplinary synthesis. *Annals of the New York Academy of Sciences, 1185*(1), 54–78. DOI: 10.1111/j.1749-6632.2009.05167.x PMID: 20146762

Luo, L., & Tang, Q. (2023). The real effects of ESG reporting and GRI standards on carbon mitigation: International evidence. *Business Strategy and the Environment, 32*(6), 2985–3000. DOI: 10.1002/bse.3281

Martín de Castro, G., Delgado Verde, M., Navas López, J. E., & Cruz González, J. (2013). The moderating role of innovation culture in the relationship between knowledge assets and product innovation. *Technological Forecasting and Social Change, 80*(2), 351–363. DOI: 10.1016/j.techfore.2012.08.012

Nandi, S., Sarkis, J., Hervani, A. A., & Helms, M. M. (2021). Redesigning supply chains using blockchain-enabled circular economy and COVID-19 experiences. *Sustainable Production and Consumption, 27*, 10–22. DOI: 10.1016/j.spc.2020.10.019 PMID: 33102671

Ong, K., Mao, R., Satapathy, R., & Shirota Filho, R. (2024). Explainable natural language processing for corporate sustainability analysis. *Information Fusion, 115*, 102726. DOI: 10.1016/j.inffus.2024.102726

Orzes, G., Moretto, A. M., Moro, M., Rossi, M., Sartor, M., Caniato, F., & Nassimbeni, G. (2020). The impact of the United Nations Global Compact on firm performance: A longitudinal analysis. *International Journal of Production Economics, 227*, 107664. DOI: 10.1016/j.ijpe.2020.107664

Ozdemir, D., Sharma, M., Dhir, A., & Daim, T. (2022). Supply chain resilience during the COVID-19 pandemic. *Technology in Society, 68*, 101847. DOI: 10.1016/j.techsoc.2021.101847 PMID: 35075312

Papa, M., Carrassi, M., Muserra, A. L., & Wieczorek Kosmala, M. (2022). The impact of the European Union non financial information directive on environmental disclosure: Evidence from Italian environmentally sensitive industries. *Meditari Accountancy Research, 30*(7), 87–120. DOI: 10.1108/MEDAR-03-2021-1247

Pizzi, S., Mastroleo, G., Venturelli, A., & Caputo, F. (2024). The digitalization of sustainability reporting processes: A conceptual framework. *Business Strategy and the Environment, 33*(2), 1040–1050. DOI: 10.1002/bse.3544

Pradhan, R. P., Samarakoon, S. M. R. K., Maradana, R. P., & Sahoo, P. (2024). Climate change disclosure and firm value in a frontier market: Exploring the determinants. *Natural Resources Forum*. https://doi.org/DOI: 10.1111/1477-8947.12462

Rolnick, D., Donti, P. L., Kaack, L. H., Kochanski, K., Lacoste, A., Sankaran, K., Ross, A. S., Milojevic-Dupont, N., Jaques, N., Waldman-Brown, A., Luccioni, A. S., Maharaj, T., Sherwin, E. D., Mukkavilli, S. K., Kording, K. P., Gomes, C. P., Ng, A. Y., Hassabis, D., Platt, J. C., & Bengio, Y. (2023). Tackling climate change with machine learning. *ACM Computing Surveys, 55*(2), 1–96. DOI: 10.1145/3485128

Rudin, C., Chen, C., Chen, Z., Huang, H., Semenova, L., & Zhong, C. (2021). Interpretable machine learning: Fundamental principles and ten grand challenges. *Statistics Surveys, 16*(none), 1–85. DOI: 10.1214/21-SS133

Suta, A., Pintes, O., Molnár, P., Lukács, B., Kedves, L., & Tóth, Á. (2023). Sustainability and financial reporting in the XBRL era. *Chemical Engineering Transactions, 107*, 577–582. DOI: 10.3303/CET23107097

Tortoriello, M., Reagans, R., & McEvily, B. (2012). Bridging the knowledge gap: The influence of strong ties, network cohesion, and network range on the transfer of knowledge between organizational units. *Organization Science, 23*(4), 1024–1039. DOI: 10.1287/orsc.1110.0688

Van der Vegt, G. S., Essens, P., Wahlström, M., & George, G. (2015). Managing risk and resilience. *Academy of Management Journal, 58*(4), 971–980. DOI: 10.5465/amj.2015.4004

Wilden, R., Gudergan, S. P., Nielsen, B. B., & Lings, I. (2013). Dynamic capabilities and performance: Strategy, structure and environment. *Long Range Planning, 46*(1-2), 72–96. DOI: 10.1016/j.lrp.2012.12.001

Wong, S., Yeung, J. K.-W., Lau, Y.-Y., & Kawasaki, T. (2023). A case study of how Maersk adopts cloud-based blockchain integrated with machine learning for sustainable practices. *Sustainability, 15*(9). *Sustainability (Basel), 7305*(9), 7305. Advance online publication. DOI: 10.3390/su15097305

Wuchty, S., Jones, B. F., & Uzzi, B. (2007). The increasing dominance of teams in production of knowledge. *Science, 316*(5827), 1036–1039. DOI: 10.1126/science.1136099 PMID: 17431139

Xie, J., Nozawa, W., & Managi, S. (2023). Do investors incorporate financial materiality? Remapping environmental information for carbon-sensitive investment. *Corporate Social Responsibility and Environmental Management, 30*(6), 2881–2899. DOI: 10.1002/csr.2524

KEY TERMS AND DEFINITIONS

Accurate ESG reporting. Integration of environmental: social, and governance metrics into verifiable accounting flows thanks to platforms that merge blockchain and machine learning, ensuring integrity and comparability

Cognitive analytics. Organizational capacity to transform large volumes of heterogeneous data into actionable knowledge for strategic decisions: using interpretable deep learning models.

Data-driven decision-making. Replacing siloed intuition with a continuous cycle of descriptive: predictive, and prescriptive analytics that democratizes insights and accelerates innovation

Emerging technologies. A set of scalable resources that make it easier to experiment and reconfigure value chains without compromising operational stability:

Organizational resilience. Ability to detect disturbances: absorb their impact, and turn experience into learning, combining digital monitoring and a culture of post-event reflection.

Compilation of References

Abu Issa Gazi, M., Mohd Faizal Yusof, M., & Aminul Islam, M. (2024). Analyzing the impact of employee job satisfaction on their job behavior in the industrial setting. *Journal of Open Innovation, 10*(4), 100–427. DOI: 10.1016/j.joitmc.2024.100427

ACS Code of Professional Ethics. (2023). Source: https://www.acs.org.au/memberships/professional-ethics-conduct-and-complaints.html

Agrawal, A., Gans, J., & Goldfarb, A. (2018). Prediction, judgment, and complexity: a theory of decision-making and artificial intelligence. In *The economics of artificial intelligence: An agenda* (pp. 89–110). University of Chicago Press.

Ahmad, M., & Wu, Q. (2024). Heuristic-driven biases as mental shortcuts in investment management activities: A qualitative study. *Qualitative Research in Financial Markets, 16*(2), 291–309. DOI: 10.1108/QRFM-10-2022-0167

Ahmed, W. (2025). The Role of Machine Learning and Deep Learning in Revolutionizing Healthcare: A Review. *Premier Journal of Artificial Intelligence, 2025*(3), 100010. DOI: 10.70389/pjai.100010

Ajzen, I. (1991). The theory of planned behavior. *Organizational Behavior and Human Decision Processes, 50*(2), 179–211. DOI: 10.1016/0749-5978(91)90020-T

Al-Adwan, A. S., Al-Debei, M. M., & Dwivedi, Y. K. (2022). E-commerce in high uncertainty avoidance cultures: The driving forces of repurchase and word-of-mouth intentions. *Technology in Society, 71*, 102083. Advance online publication. DOI: 10.1016/j.techsoc.2022.102083

Albort Morant, G., Henseler, J., Cepeda Carrión, G., & Leal Rodríguez, A. L. (2018). Potential and realized absorptive capacity as complementary drivers of green product and process innovation performance. *Sustainability (Basel), 10*(2), 381. DOI: 10.3390/su10020381

Aleessawi, N., & Djaghrouri, L. (2025). Artificial Intelligence in Decision-Making: Literature Review. *Journal of Association of Arab Universities for Research of Higher Education*, *45*(1), 263–278. https://doi.org/10.36024/1248-045-001-013

Alhalalmeh, M., Alkhawaldah, R. A., Mohammad, A., Al-Quran, A., Hijjawi, G., & Al-Hawary, S. (2022). The Effect of Selected Marketing Activities and Promotions on the Consumers Buying Behavior. *Business: Theory and Practice*, *23*(1), 79–87. DOI: 10.3846/btp.2022.13929

Ali, M., Khan, T., Khattak, M., & Şener, İ. (2024). Synergizing AI and business: Maximizing innovation, creativity, decision precision, and operational efficiency in high-tech enterprises. *Journal of Open Innovation*, *10*(3), 100–352. DOI: 10.1016/j.joitmc.2024.100352

Alimadadi, A., Aryal, S., Manandhar, I., Munroe, P. B., Joe, B., & Cheng, X. (2020). Artificial Intelligence and Machine Learning to Fight COVID-19. *Physiological Genomics*, *52*(4), 200–202. DOI: 10.1152/physiolgenomics.00029.2020 PMID: 32216577

Alkatheiri, M. S. (2022). Artificial Intelligence Assisted Improved Human-Computer Interactions for Computer Systems. *Computers & Electrical Engineering*, *101*, 108024. Advance online publication. DOI: 10.1016/j.compeleceng.2022.107950

Alqurashi, D. R., Alkhaffaf, M., Daoud, M. K., & Ahmad, J., Al-Gasawneh, & Alghizzawi, M. (2023). Exploring the Impact of Artificial Intelligence in Personalized Content Marketing: A Contemporary Digital Marketing. *Migration Letters : An International Journal of Migration Studies*, *20*(S8), 548–560.

Amin, M., & Lee, C. (2022). Hydrogen production through renewable and non-renewable energy processes and their impact on climate change. *International Journal of Hydrogen Energy*, *47*(48), 27174–27189. DOI: 10.1016/j.ijhydene.2022.07.172

Amodio, D. M., & Frith, C. D. (2006). Meeting of minds: The medial frontal cortex and social cognition. *Nature Reviews. Neuroscience*, *7*(4), 268–277. DOI: 10.1038/nrn1884

Anderson, J. R., Bothell, D., Byrne, M. D., Douglass, S., Lebiere, C., & Qin, Y. (2004). An Integrated Theory of the Mind. *Psychological Review*, *111*(4), 1036–1060. DOI: 10.1037/0033-295X.111.4.1036 PMID: 15482072

Andersson, F. N. G., & Arvidsson, S. (2023). Understanding, mapping and reporting of climate-related risks among listed firms in Sweden. *Climate Policy*, *23*(8), 945–958. DOI: 10.1080/14693062.2022.2116383

Antomarioni, S., Lucantoni, L., Ciarapica, F. E., & Bevilacqua, M. (2021). Data-Driven Decision Support System for Managing Item Allocation in an ASRS: A Framework Development and a Case Study. *Expert Systems with Applications, 185*, 115622. Advance online publication. DOI: 10.1016/j.eswa.2021.115622

Aquino, G. J., Chamberlin, J., Yacoub, B., Kocher, M. R., Kabakus, I., Akkaya, S., Mercer, M., Waltz, J., Fiegel, M., Leaphart, N., Jacob, A., Gulsun, M. A., Gilkes, J., Stephenson, J., Sharma, P., Sahbaee, P., Schoepf, J., Zimmerman, S., Field, M. E., & Burt, J. R. (2022). Diagnostic accuracy and performance of artificial intelligence in measuring left atrial volumes and function on multiphasic CT in patients with atrial fibrillation. *European Radiology, 32*(8), 5256–5264. DOI: 10.1007/s00330-022-08657-y

Ara, J., Ghodke, S., Akter, J., & Roy, A. (2025). Optimizing E-Commerce Platforms with AI-Enabled Visual Search: Assessing User Behavior, Interaction Metrics, and System Accuracy. *Journal of Economics. Finance and Accounting Studies, 7*(3), 9–17. DOI: 10.32996/jefas

Arefeen, M. A., Debnath, B., & Chakradhar, S. (2024). Lean Context: Cost-efficient domain-specific question answering using LLMs. *Proceedings of the 2024 Conference on Empirical Methods in Natural Language Processing*, 2301-2312.

Arif, S. (2025). Cross-Cultural perspectives on AI in education: Case studies from global classrooms. *AI EDIFY Journal, 2*(1), 12–20.

Ashraf, A. R., Thongpapanl, N. T., & Auh, S. (2014). The Application of the Technology Acceptance Model under Different Cultural Contexts: The Case of Online Shopping Adoption. *Journal of International Marketing, 22*(3), 68–93. DOI: 10.1509/jim.14.0065

Augstein, M., Herder, E., & Wörndl, W. (2023). *Personalized Human-Computer Interaction*. De Gruyter., DOI: 10.1515/9783110988567

Auld, G., Casovan, A., Clarke, A., & Faveri, B. (2022). Governing AI through ethical standards: Learning from the experiences of other private governance initiatives. *Journal of European Public Policy, 29*(11), 1822–1844. DOI: 10.1080/13501763.2022.2099449

Awan, U., Shamim, S., Khan, Z., Zia, N. U., Shariq, S. M., & Khan, M. N. (2021). Big data analytics capability and decision-making: The role of data-driven insight on circular economy performance. *Technological Forecasting and Social Change, 168*, 120766. Advance online publication. DOI: 10.1016/j.techfore.2021.120766

Awwad, B. (2024). *The AI Revolution: Driving Business Innovation and Research.* Springer.

Babatunde, O., Akintayo, B., Ighravwe, D., & Olanrewaju, O. (2025). Transdisciplinary approach to accelerate the adoption of hybrid renewable energy systems through sustainable design. *Frontiers in Built Environment, 1*, 1520883. Advance online publication. DOI: 10.3389/fbuil.2025.1520883

Bailey, M. (2024). Inside Google Marketing: How we're using AI. https://business.google.com/aunz/think/ai-excellence/how-google-uses-ai-marketing-tools- and-experiements/

Bairaktarova, D., & Woodcock, A. (2017). Engineering Student's Ethical Awareness and Behavior: A New Motivational Model. *Science and Engineering Ethics, 23*(4), 1129–1157. DOI: 10.1007/s11948-016-9814-x PMID: 27752963

Bais, B., Nassimbeni, G., & Orzes, G. (2024). Global Reporting Initiative: Literature review and research directions. *Journal of Cleaner Production, 471*, 143428. DOI: 10.1016/j.jclepro.2024.143428

Bala, M. Y., & Damla, K. (2021). "A Review of Human-Computer Interaction Design Approaches Towards Information Systems Development." BRAIN. *Broad Research in Artificial Intelligence and Neuroscience, 12*(1), 229–250. DOI: 10.70594/brain/12.1/180

Barbey, A. K., Koenigs, M., & Grafman, J. (2012). Dorsolateral prefrontal contributions to human working memory. *Cortex, 48*(1), 62–70. DOI: 10.1016/j.cortex.2011.06.012

Barnes, A. J., Zhang, Y., & Valenzuela, A. (2024). AI and culture: Culturally dependent responses to AI systems. *Current Opinion in Psychology, 58*, 101838. DOI: 10.1016/j.copsyc.2024.101838

Barnes, R., & Zvarikova, K. (2021). Artificial Intelligence-enabled Wearable Medical Devices, Clinical and Diagnostic Decision Support Systems, and Internet of Things-based Healthcare Applications in COVID-19 Prevention, Screening, and Treatment. *American Journal of Medical Research (New York, N.Y.), 8*(2), 9–22. DOI: 10.22381/ajmr8220211

Barney, J. B. (1991). Firm resources and sustained competitive advantage. *Journal of Management, 17*(1), 99–120. DOI: 10.1177/014920639101700108

Barocas, S., Hardt, M., & Narayanan, A. (2019). *Fairness and Machine Learning: Limitations and Opportunities*. MIT Press. https://fairmlbook.org

Barocas, S., & Selbst, A. (2016). Big data's disparate impact. *SSRN*, *104*(3), 671–732. DOI: 10.2139/ssrn.2477899

Barredo Arrieta, A., Díaz Rodríguez, N., Del Ser, J., Bennetot, A., Tabik, S., Barbado, A., Garcia, S., Gil-Lopez, S., Molina, D., Benjamins, R., Chatila, R., & Herrera, F. (2020). Explainable Artificial Intelligence (XAI): Concepts, taxonomies, opportunities and challenges toward responsible AI. *Information Fusion*, *58*, 82–115. DOI: 10.1016/j.inffus.2019.12.012

Bastian, B., & Zucchella, A. (2022). Entrepreneurial metacognition: A study on nascent entrepreneurs. *The International Entrepreneurship and Management Journal*, *18*(4), 1775–1805. DOI: 10.1007/s11365-022-00799-1

Batko, K., & Ślęzak, A. (2022). The use of Big Data Analytics in healthcare. *Journal of Big Data*, *9*(1), 3–3. DOI: 10.1186/s40537-021-00553-4 PMID: 35013701

Bawack, R. E., Wamba, S. F., Carillo, K. D. A., & Akter, S. (2022). Artificial intelligence in E-Commerce: A bibliometric study and literature review. *Electronic Markets*, *32*(1), 297–338. DOI: 10.1007/s12525-022-00537-z

Baxter, P., & Jack, S. (2010). Qualitative Case Study Methodology: Study Design and Implementation for Novice Researchers. *The Qualitative Report.*

Beaton, D. E., Bombardier, C., Guillemin, F., & Ferraz, M. B. (2000). Guidelines for the process of cross-cultural adaptation of self-report measures. *Spine*, *25*(24), 3186–3191. DOI: 10.1097/00007632-200012150-00014

Bechara, A., Damasio, H., Tranel, D., & Damasio, A. R. (2005). The Iowa Gambling Task and the somatic marker hypothesis: Some questions and answers. *Trends in Cognitive Sciences*, *9*(4), 159–162. DOI: 10.1016/j.tics.2005.02.002

Beichert, M., Bayerl, A., Goldenberg, J., & Lanz, A. (2024). Revenue Generation Through Influencer Marketing. *Journal of Marketing*, *88*(4), 40–63. DOI: 10.1177/00222429231217471

Bernardo, B. M. V., Mamede, H. S., Barroso, J. M. P., & Santos, V. M. P. D. (2024). Data governance & quality management—Innovation and breakthroughs across different fields. *Journal of Innovation & Knowledge*, *9*(4), 100598. DOI: 10.1016/j.jik.2024.100598

Bernovskis, A., Sceulovs, D., & Stibe, A. (2024). Society 5.0: Shaping the future of e-commerce. *Journal of Open Innovation*, *10*(4), 100391. Advance online publication. DOI: 10.1016/j.joitmc.2024.100391

Bin Ahsan, W. (2024). *The EDIT UX Framework: A User-Centered Approach to Effective Product Redesign*. Userhub., DOI: 10.58947/PBSY-TGJR

Birge, J. R., & Louveaux, F. (2011). *Introduction to Stochastic Programming* (2nd ed.). Springer. DOI: 10.1007/978-1-4614-0237-4

Björk, J., Magnusson, M., & Karlsson, C. (2023). Measuring innovation effectively—Nine critical lessons. *Research Technology Management*, *66*(2), 24–33. DOI: 10.1080/08956308.2022.2151232

Blackley, C., Redmond, P., & Peel, K. (2021). Teacher Decision-Making in the Classroom: The Influence of Cognitive Load and Teacher Affect. *Journal of Education for Teaching*, *47*(4), 548–561. DOI: 10.1080/02607476.2021.1902748

Blanco, F. (2017). Cognitive bias. In Vonk, J., & Shackelford, T. (Eds.), *Encyclopedia of animal cognition and behavior*. Springer., DOI: 10.1007/978-3-319-47829-6_1244-1

Bodenhausen, G. V., Gabriel, S., & Lineberger, M. (2000). Sadness and Susceptibility to Judgmental Bias: The Case of Anchoring. *Psychological Science*, *11*(4), 320–323. DOI: 10.1111/1467-9280.00263 PMID: 11273392

Bogers, M., Zobel, A.-K., Afuah, A., Almirall, E., Brunswicker, S., Dahlander, L., Frederiksen, L., Gawer, A., Gruber, M., Haefliger, S., Hagedoorn, J., Hilgers, D., Laursen, K., Magnusson, M. G., Majchrzak, A., McCarthy, I. P., Moeslein, K. M., Nambisan, S., Piller, F. T., & Ter Wal, A. L. J. (2017). The open innovation research landscape: Established perspectives and emerging themes across different levels of analysis. *Industry and Innovation*, *24*(1), 8–40. DOI: 10.1080/13662716.2016.1240068

Bóna, P., & Lippert, R. (2020). The effect of hard and soft factors on the success of a company. *Problems of Management in the 21st century*, 10(1), 6-17. https://www.proquest.com/publiccontent/docview/2343771276

Botvinick, M. M., Cohen, J. D., & Carter, C. S. (2004). Conflict monitoring and anterior cingulate cortex: An update. *Trends in Cognitive Sciences*, *8*(12), 539–546. DOI: 10.1016/j.tics.2004.10.003

Boukrouh, I., & Azmani, A. (2024). Artificial Intelligence Applications in E-Commerce: A Bibliometric Study from 1995 to 2023 Using Merged Data Sources. *International Journal of Professional Business Review*, *9*(4), e4537. Advance online publication. DOI: 10.26668/businessreview/2024.v9i4.4537

Boulton, T. J., Iqbal, U., Khaksari, S., & Qiu, S. (2024). Mandatory ESG disclosure, information asymmetry, and IPO underpricing. *European Financial Management*. Advance online publication. DOI: 10.1111/eufm.12494

Brette, R., & Gerstner, W. (2005). Adaptive exponential integrate-and-fire model as an effective description of neuronal activity. *Journal of Neurophysiology, 94*(5), 3637–3642. DOI: 10.1152/jn.00686.2005

Brown, T., Mann, B., Ryder, N., Subbiah, M., Kaplan, J. D., Dhariwal, P., & Amodei, D. (2020). Language models are few-shot learners. *Advances in Neural Information Processing Systems, 33*, 1877–1901. DOI: 10.48550/arXiv.2005.14165

Brynjolfsson, E., & McAfee, A. (2017). The business of artificial intelligence. *Harvard Business Review, 95*(7), 3–11.

Bullough, A., Cotton, R., Dastmalchian, A., Dorfman, P., & Egri, C. (2020). GLOBE 2020: The Latest Findings on Cultural Practices, Culture Change, and Leadership Ideals. *Proceedings - Academy of Management, 17374*(1), 17374. Advance online publication. DOI: 10.5465/AMBPP.2020.17374symposium

Buonocore, F., Annosi, M. C., De Gennaro, D., & Riemma, F. (2024). Digital transformation and social change: Leadership strategies for responsible innovation. *Journal of Engineering and Technology Management, 74*, 101843. DOI: 10.1016/j.jengtecman.2024.101843

Busemeyer, J. R., & Townsend, J. T. (1993). Decision field theory: A dynamic-cognitive approach to decision making in an uncertain environment. *Psychological Review, 100*(3), 432–459. DOI: 10.1037/0033-295X.100.3.432 PMID: 8356185

Busse, K., Tahaei, M., Krombholz, K., Zezschwitz, E. v., Smith, M., & Tian, J. (2020). Cash, Cards or Cryptocurrencies? A Study of Payment Culture in Four Countries. *2020 IEEE European Symposium on Security and Privacy Workshops*, 200-209. https://doi.org/DOI: 10.1109/EuroSPW51379.2020.00035

Butt, A., Imran, F., Helo, P., & Kantola, J. (2024). Strategic design of culture for digital transformation. *Long Range Planning, 57*(4), 102415. DOI: 10.1016/j.lrp.2024.102415

Calvo, R. A., Milne, D. N., Hussain, M. S., & Christensen, H. (2020). Natural language processing in mental health applications using non-clinical texts. *Natural Language Engineering, 23*(5), 649–685. DOI: 10.1017/S1351324916000383

Can, Y. S., Mahesh, B., & André, E. (2023). Approaches, applications, and challenges in physiological emotion recognition—A tutorial overview. *Proceedings of the IEEE, 111*(10), 1287–1313. DOI: 10.1109/JPROC.2023.3286445

Capgemini. (2020). Johnson & Johnson transforms its demand planning and external manufacturing processes. https://www.capgemini.com/news/client-stories/johnson-and-johnson-transforms-its-demand-planning-and-external-manufacturing-processes/

Caputo, A., Pizzi, S., Pellegrini, M. M., & Dabić, M. (2021). Digitalization and business models: Where are we going. *Journal of Business Research, 123*, 489–501. DOI: 10.1016/j.jbusres.2020.09.053

Cardoso, A., Figueiredo, J., Oliveira, I., & Pocinho, M. (2025). From crisis to opportunity: Digital transformation, digital business models, and organizational resilience in the post-pandemic era. *Administrative Sciences, 15*(6), 193. Advance online publication. DOI: 10.3390/admsci15060193

Chancellor, S., & De Choudhury, M. (2020). Methods in predictive diagnostics for mental health on social media: A survey. In *Proceedings of the 26th ACM SIGKDD International Conference on Knowledge Discovery & Data Mining* (pp. 3473-3474). ACM.

Chandra Reddy, B. R. V. N., Konakalla, L., Kumar, G. P., Ganapathi, K. S., Santhosh, M., & Rao, M. N. (2025). AI-Powered E-commerce Platform for Small Businesses. *International Journal of Innovative Science and Research Technology, 10*(4), 495–501. DOI: 10.38124/ijisrt/25apr420

Chari, A., Niedenzu, D., Despeisse, M., Machado, C. G., Azevedo, J. D., Boavida-Dias, R., & Johansson, B. (2022). Dynamic capabilities for circular manufacturing supply chains—Exploring the role of Industry 4.0 and resilience. *Business Strategy and the Environment, 31*(5), 2500–2517. DOI: 10.1002/bse.3040

Chen, D., Esperanca, J. P., & Wang, S. (2022). The Impact of Artificial Intelligence on Firm Performance: An Application of the Resource-Based View to e-Commerce Firms. *Frontiers in Psychology, 13*, 884830. DOI: 10.3389/fpsyg.2022.884830

Chen, L., Chen, Z., Zhang, Y., Liu, Y., Osman, A. I., Farghali, M., Hua, J., Al-Fatesh, A., Ihara, I., Rooney, D. W., & Yap, P. S. (2023). Artificial intelligence-based solutions for climate change: A review. *Environmental Chemistry Letters, 21*(5), 2525–2557. DOI: 10.1007/s10311-023-01617-y

Chen, R. J., Wang, J. J., Williamson, D. F. K., Chen, T. Y., Lipkova, J., Lu, M. Y., Sahai, S., & Mahmood, F. (2023). Algorithmic fairness in artificial intelligence for medicine and healthcare. *Nature Biomedical Engineering, 7*(6), 719–742. DOI: 10.1038/s41551-023-01056-8 PMID: 37380750

Chkirbene, Z., Hamila, R., Gouissem, A., & Devrim, U. (2024). *Large Language Models (LLM) in Industry: A Survey of Applications, Challenges, and Trends. In 2024, IEEE 21st International Conference on Smart Communities: Improving Quality of Life using AI, Robotics and IoT (HONET)*. IEEE., https://ieeexplore.ieee.org/abstract/document/10822885

Chodak, G. (2024). *Future of E-commerce*. Springer. DOI: 10.1007/978-3-031-55225-0

Chopra, S. S., Senadheera, S. S., Dissanayake, P. D., & Withana, P. A. (2024). Navigating the challenges of environmental, social and governance reporting: The path to broader sustainable development. *Sustainability, 16*(2). *Section, 606*. Advance online publication. DOI: 10.3390/su16020606

Chowdhary, S., Selvakumar, P., Sangeetha, B., Sarvaani, V. V., Anute, N., & Sharma, M. (2025). The Role of Social Capital in Entrepreneurial Development. In Martínez-Falcó, J., Sánchez-García, E., & Marco-Lajara, B. (Eds.), *Support Networks for Entrepreneurs: Social Capital and Strategic Collaborations* (pp. 311–338). IGI Global Scientific Publishing., DOI: 10.4018/979-8-3693-9836-4.ch013

Clarkson, P. M., Li, Y., Richardson, G. D., & Vasvari, F. P. (2008). Revisiting the relation between environmental performance and environmental disclosure: An empirical analysis. *Accounting, Organizations and Society*, *33*(4), 303–327. DOI: 10.1016/j.aos.2007.05.003

Clemen, R. T., & Reilly, T. (2013). *Making Hard Decisions with DecisionTools*. Cengage Learning.

Coates, J. M., Santhanalakshmi, K., Dinesh, N., Gandhimathi, S., & Chandel, P. S. Mohit & Selvakumar P, (2025). "Assessment and Evaluation in Transnational Staff Development." In G Morris & J Ye (Eds.), Training and Development in Transnational Higher Education (pp. 387–406). IGI Global Scientific Publishing. https://doi.org/DOI: 10.4018/979-8-3373-1897-4.ch019

Cohen, S., Kamarck, T., & Mermelstein, R. (1983). A global measure of perceived stress. *Journal of Health and Social Behavior*, *24*(4), 385–396. DOI: 10.2307/2136404

Csaszar, F., Ketkar, H., & Kim, H. (2024). Artificial Intelligence and Strategic Decision-Making: Evidence from Entrepreneurs and Investors. *Strategy Science*, *9*(4), 322–345. Advance online publication. DOI: 10.1287/stsc.2024.0190

Damasio, A. R. (1994). *Descartes' error: Emotion, reason, and the human brain*. Putnam.

Damasio, A. R., Everitt, B. J., & Bishop, D. (1997). The somatic marker hypothesis and the possible functions of the prefrontal cortex. *Philosophical Transactions of the Royal Society of London. Series B, Biological Sciences, 351*(1346), 1413–1420. DOI: 10.1098/rstb.1996.0125 PMID: 8941953

Danziger, S., Levav, J., & Avnaim-Pesso, L. (2011). Extraneous factors in judicial decisions. *Proceedings of the National Academy of Sciences of the United States of America, 108*(17), 6889–6892. DOI: 10.1073/pnas.1018033108 PMID: 21482790

Darney, P. E.P. Ebby Darney. (2022). Power flow optimization of a hybrid energy system with salp swarm algorithm. *Journal of Electronics and Informatics, 4*(4), 266–274. DOI: 10.36548/jei.2022.4.006

Davenport, T. H., Brynjolfsson, E., McAfee, A., & Wilson, H. J. (2019). *Artificial intelligence: The insights you need from Harvard Business Review*. Harvard Business Press.

Davenport, T., Guha, A., Grewal, D., & Bressgott, T. (2020). How artificial intelligence will change the future of marketing. *Journal of the Academy of Marketing Science, 48*(1), 24–42. DOI: 10.1007/s11747-019-00696-0

Davies, M., Wild, A., Orchard, G., Sandamirskaya, Y., Guerra, G. A. F., Joshi, P., Ploeg, F., Risbud, S. R., & Wang, R. M. (2021). Advancing neuromorphic computing with Loihi: A survey of results and outlook. *Proceedings of the IEEE, 109*(5), 911–934. DOI: 10.1109/JPROC.2021.3067593

Davis, F. D. (1989). Perceived usefulness, perceived ease of use, and user acceptance of information technology. *Management Information Systems Quarterly, 13*(3), 319–340. DOI: 10.2307/249008

Davis, F. D. (1989). Technology acceptance model: TAM. Al-Suqri, MN, Al-Aufi, AS. *Information Seeking Behavior and Technology Adoption, 205*(219), 5.

Davis, M. H. (1983). Measuring individual differences in empathy: Evidence for a multidimensional approach. *Journal of Personality and Social Psychology, 44*(1), 113–126. DOI: 10.1037/0022-3514.44.1.113

De Villiers, C., Hsiao, P., Maroun, W., Ng, J., Roussey, S., & Vasquez, D. (2024). The International Sustainability Standards Board's (ISSB) past, present and future: Critical reflections and a research agenda. *Pacific Accounting Review, 36*(3), 437–457. DOI: 10.1108/PAR-02-2024-0038

Delétang, G., Ruoss, A., Duquenne, P. A., Catt, E., Genewein, T., Mattern, C., & Veness, J. (2023). Language modeling is compression. arXiv preprint arXiv:2309.10668.

Deloitte. (2024). Generative AI and the future of work. https://www2.deloitte.com/content/dam/Deloitte/us/Documents/consulting/us-ai-institute- generative-ai-and-the-future-of-work.pdf

Demajorovic, J., Lopes, J. C., & Santiago, A. L. F. (2019). The Samarco dam disaster: A grave challenge to social license to operate discourse. *Resources Policy, 61*, 273–282.

Deng, X., Zhang, B., Yu, N., Liu, K., & Sun, K. (2021). Advanced TSGL-EEGNet for motor imagery EEG-based brain-computer interfaces. *IEEE Access : Practical Innovations, Open Solutions, 9*, 25118–25130. DOI: 10.1109/ACCESS.2021.3056088

Deufel, P., Kemper, J., & Brettel, M. (2019). Pay now or pay later: A cross-cultural perspective on online payments. *Journal of Electronic Commerce Research, 20*(3), 141–154.

Diehl, P. U., Neil, D., Binas, J., Cook, M., Liu, S. C., & Pfeiffer, M. (2015). Fast-classifying, high-accuracy spiking deep networks through weight and threshold balancing. *2015 International Joint Conference on Neural Networks (IJCNN)*, 1–8. DOI: 10.1109/IJCNN.2015.7280696

Ding, Y., Hu, X., Xia, Z., Liu, Y.-J., & Zhang, D. (2018). Inter-brain EEG feature extraction and analysis for continuous implicit emotion tagging during video watching. *IEEE Transactions on Affective Computing, 12*(1), 92–102. DOI: 10.1109/TAFFC.2018.2849758

Ding, Z., Ji, Y., Gan, Y., Wang, Y., & Xia, Y. (2024). Current Status and Trends of Technology, Methods, and Applications of Human–Computer Intelligent Interaction (HCII): A Bibliometric Research. *Multimedia Tools and Applications, 83*(27), 69111–69144. DOI: 10.1007/s11042-023-18096-6

Dlugos, K. V., Mazwi, M., Lao, R., & Honjo, O. (2025). Noise is an underrecognized problem in medical decision making and is known by other names: A scoping review. *BMC Medical Informatics and Decision Making, 25*(1), 86. DOI: 10.1186/s12911-025-02905-z PMID: 39962496

Dragan, A. D., Lee, K. C. T., & Srinivasa, S. S. (2013). Legibility and predictability of robot motion. *2013 8th ACM/IEEE International Conference on Human-Robot Interaction (HRI)*, 301–308. DOI: 10.1109/HRI.2013.6483603

Dror, I. E. (2020). Cognitive and human factors in expert decision making: Six fallacies and the eight sources of bias. *Analytical Chemistry, 92*(12), 7998–8004. DOI: 10.1021/acs.analchem.0c00704 PMID: 32508089

Duan, Y., Edwards, J. S., & Dwivedi, Y. K. (2019). Artificial intelligence for decision making in the era of Big Data – evolution, challenges and research agenda. *International Journal of Information Management, 48*, 63–71. DOI: 10.1016/j.ijinfomgt.2019.01.021

Duchek, S. (2020). Organizational resilience: A capability-based conceptualization. *Business Research, 13*(1), 215–246. DOI: 10.1007/s40685-019-0085-7

Du, G., Zhou, W., Li, C., Li, D., & Liu, P. X. (2020). An emotion recognition method for game evaluation based on electroencephalogram. *IEEE Transactions on Affective Computing, 14*(1), 591–602. DOI: 10.1109/TAFFC.2020.3023966

Dungan, R. (2024, August 30). Google pioneers AI leadership meetings, but staff aren't happy. HR Grapevine. https://www.hrgrapevine.com/us/content/article/2024-08-30-google-pioneers- ai-leadership-meetings

Dwivedi, Y. K., Hughes, L., Ismagilova, E., Aarts, G., Coombs, C., Crick, T., Duan, Y., Dwivedi, R., Edwards, J., Eirug, A., Galanos, V., Ilavarasan, P. V., Janssen, M., Jones, P., Kar, A. K., Kizgin, H., Kronemann, B., Lal, B., Lucini, B., & Williams, M. D. (2021). Artificial Intelligence (AI): Multidisciplinary perspectives on emerging challenges, opportunities, and agenda for research, practice and policy. *International Journal of Information Management, 57*, 101994. Advance online publication. DOI: 10.1016/j.ijinfomgt.2019.08.002

Dwivedi, Y. K., Kshetri, N., Hughes, L., Slade, E. L., Jeyaraj, A., Kar, A. K., Baabdullah, A. M., Koohang, A., Raghavan, V., Ahuja, M., Albanna, H., Albashrawi, M. A., Al-Busaidi, A. S., Balakrishnan, J., Barlette, Y., Basu, S., Bose, I., Brooks, L., Buhalis, D., & Wright, R. (2023). "So what if ChatGPT wrote it?" Multidisciplinary perspectives on opportunities, challenges and implications of generative conversational AI for research, practice and policy. *International Journal of Information Management, 71*(102642), 102642. Advance online publication. DOI: 10.1016/j.ijinfomgt.2023.102642

Dwivedi, Y. K., Rana, N. P., Jeyaraj, A., Clement, M., & Williams, M. D. (2019). Re-examining the Unified Theory of Acceptance and Use of Technology (UTAUT): Towards a Revised Theoretical Model. *Information Systems Frontiers, 21*(3), 719–734. DOI: 10.1007/s10796-017-9774-y

Eccles, R. G., Krzus, M. P., Rogers, J., & Serafeim, G. (2012). The need for sector specific materiality and sustainability reporting standards. *The Bank of America Journal of Applied Corporate Finance, 24*(2), 65–71. DOI: 10.1111/j.1745-6622.2012.00380.x

Eisenhardt, K., & Graebner, M. (2007). Theory building from cases: Opportunities and challenges. *Academy of Management Journal*, *50*(1), 25–32. DOI: 10.5465/amj.2007.24160888

Elias, F., Ekpo, S., Alabi, S., Uko, M., Enahoro, S., Ijaz, M., Ji, H., Unnikrishnan, R., & Olasunkanmi, N. (2025). Design of multi-sourced MIMO multiband hybrid wireless RF-perovskite photovoltaic energy harvesting subsystems for IoTs applications in smart cities. *Technologies*, *13*(3), 92. Advance online publication. DOI: 10.3390/technologies13030092

Elliot, V., Jonäll, K., Paananen, M., Bebbington, J., & Michelon, G. (2024). Biodiversity reporting: Standardization, materiality, and metrics. *Current Opinion in Environmental Sustainability*, *64*, 101435. DOI: 10.1016/j.cosust.2024.101435

Enescu, D. (2019). Thermoelectric energy harvesting: Basic principles and applications. In *Green Energy Advances*. IntechOpen., DOI: 10.5772/intechopen.83495

Engelen, A., Klyver, K., & de Mol, E. (2024). Building a resilient organization through a pre-shock strategic emphasis on innovation. *Journal of Product Innovation Management*, *41*(1), 36–61. Advance online publication. DOI: 10.1111/jpim.12697

Englich, B., & Soder, K. (2009). Moody experts—How mood and expertise influence judgmental anchoring. *Judgment and Decision Making*, *4*(1), 41–50. DOI: 10.1017/S1930297500000693

Ertel, W. (2024). *Introduction to Artificial Intelligence*. Springer Nature., DOI: 10.1007/978-3-658-43102-0

European Commission. (2021). *Artificial intelligence*. European Commission – Shaping Europe's Digital Future. https://digital-strategy.ec.europa.eu/en/policies/artificial-intelligence

European Commission. (2024). AI Act: Regulatory framework for artificial intelligence. Retrieved from https://commission.europa.eu/index_en

Evans, P. A. L., Pucik, V., Stahl, G. K., & Björkman, I. (2025). Global talent management in an age of disruption. *Organizational Dynamics*, *54*(3), 101149. Advance online publication. DOI: 10.1016/j.orgdyn.2025.101149

Faith, D. O., & Agwu, E. (2018). A Review of the Effect of Pricing Strategies on the Purchase of Consumer Goods. *International Journal of Research in Management. Science & Technology*, *2*(2), 88–102.

Fang, W., Yu, Z., Chen, Y., Masquelier, T., Huang, T., & Tian, Y. (2021). SpikingJelly: A deep-learning framework for spiking neural networks. *Frontiers in Neuroscience*, *15*, 774067. Advance online publication. DOI: 10.3389/fnins.2021.774067

Fan, X., Zhou, Q., Liu, Z., & Xie, F. (2015). Electroencephalogram assessment of mental fatigue in visual search. *Bio-Medical Materials and Engineering*, *26*(1, suppl), S1455–S1463. DOI: 10.3233/BME-151444 PMID: 26405908

Farokhah, L., Sarno, R., & Fatichah, C. (2023). Simplified 2D CNN architecture with channel selection for emotion recognition using EEG spectrogram. *IEEE Access : Practical Innovations, Open Solutions*, *11*, 46330–46343. DOI: 10.1109/ACCESS.2023.3275565

Felin, T., Foss, N. J., Heimeriks, K. H., & Madsen, T. L. (2012). Microfoundations of routines and capabilities: Individuals, processes, and structure. *Journal of Management Studies*, *49*(8), 1351–1374. DOI: 10.1111/j.1467-6486.2012.01052.x

Fernandes, S. F., & Panda, R. (2018). Social reference group influence on women buying behaviour: A review. *Journal of Commerce and Management Thought*, *9*(2), 273–291. DOI: 10.5958/0976-478X.2018.00019.8

Filippucci, F., Gal, P., Jona-Lasinio, C., Leandro, A., & Nicoletti, G. (2024). The impact of artificial intelligence on productivity, distribution and growth: Key mechanisms, initial evidence and policy challenges. *OECD Artificial Intelligence Papers*. Financial Stability Board (FSB). (2017). Artificial intelligence and machine learning in financial services. Retrieved from https://www.fsb.org/DOI: 10.1787/dee339a8-en

Finucane, M. L., Alhakami, A., Slovic, P., & Johnson, S. M. (2000). The affect heuristic in judgments of risks and benefits. *Journal of Behavioral Decision Making*, *13*(1), 1–17. DOI: 10.1002/(SICI)1099-0771(200001/03)13:1<1::AID-BDM333>3.0.CO;2-S

Fishbein, M., & Ajzen, I. (1975). *Belief, Attitude, Intention, and Behavior: An Introduction to Theory and Research*. Addison-Wesley Publishing Company., https://books.google.com.au/books?id=8o0QAQAAIAAJ

Fiske, S. T., & Taylor, S. E. (1991). *Social Cognition*. McGraw-Hill Humanities/Social Sciences/Languages.

Floridi, L., Cowls, J., Beltrametti, M., Chatila, R., Chazerand, P., Dignum, V., Luetge, C., Madelin, R., Pagallo, U., Rossi, F., Schafer, B., Valcke, P., & Vayena, E. (2018). AI4People—An ethical framework for a good AI society: Opportunities, risks, principles, and recommendations. *Minds and Machines*, *28*(4), 689–707. DOI: 10.1007/s11023-018-9482-5 PMID: 30930541

Forcina, A., Silvestri, L., De Felice, F., & Falcone, D. (2024). Exploring Industry 4.0 Technologies to Improve Manufacturing Enterprise Safety Management: A TOPSIS-Based Decision Support System and Real Case Study. *Safety Science, 169*, 106351. Advance online publication. DOI: 10.1016/j.ssci.2023.106351

Fritzsch, J., Bogner, J., Haug, M., & Franco da Silva, A. C. (2022). Adopting microservices and DevOps in the cyber-physical systems domain: A rapid review and case study. *Software, Practice & Experience*. Advance online publication. DOI: 10.1002/spe.3169

Fu, Z., Barbara, A., & Scupelli, P. (2025). *Digital Futures in Human-Computer Interaction: Design Thinking for Digital Transformation*. CRC Press., DOI: 10.1201/9781032693606

García-Martínez, B., Martinez-Rodrigo, A., Alcaraz, R., & Fernández-Caballero, A. (2019). A review on nonlinear methods using electroencephalographic recordings for emotion recognition. *IEEE Transactions on Affective Computing*, *12*(3), 801–820. DOI: 10.1109/TAFFC.2018.2890636

Garrido-Moreno, A., Martín-Rojas, R., & García-Morales, V. J. (2024). The key role of innovation and organizational resilience in improving business performance: A mixed-methods approach. *International Journal of Information Management, 77*, 102777. DOI: 10.1016/j.ijinfomgt.2024.102777

Georgescu, I., Bocean, C. G., Vărzaru, A. A., Rotea, C. C., Mangra, M. G., & Mangra, G. I. (2024). Enhancing organizational resilience: The transformative influence of strategic human resource management practices and organizational culture. *Sustainability, 16*(10). *Sustainability (Basel), 4315*(10), 4315. Advance online publication. DOI: 10.3390/su16104315

Ghasemaghaei, M., & Kordzadeh, N. (2024). Understanding how algorithmic injustice leads to making discriminatory decisions: An obedience to authority perspective. *Information & Management*, *61*(2), 103921. Advance online publication. DOI: 10.1016/j.im.2024.103921

Gigerenzer, G., & Gaissmaier, W. (2011). Heuristic Decision Making. *Annual Review of Psychology*, *62*(1), 451–482. DOI: 10.1146/annurev-psych-120709-145346 PMID: 21126183

Gilanie, G., Hassan, M. U., Asghar, M., Qamar, A. M., Ullah, H., Khan, R. U., Aslam, N., & Khan, I. U. (2022). An automated and real-time approach of depression detection from facial micro-expressions. *Computers, Materials & Continua*, *73*(2), 2514–2530. DOI: 10.32604/cmc.2022.028229

Gilovich, T., Griffin, D., & Kahneman, D. (Eds.). (2013). *Heuristics and Biases: The Psychology of Intuitive Judgment*. Cambridge University Press.

Gioia, G. A., Isquith, P. K., Guy, S. C., & Kenworthy, L. (2000). *Behavior Rating Inventory of Executive Function* [Assessment instrument]. Psychological Assessment Resources.

Gittell, J. H., Cameron, K., Lim, S., & Rivas, V. (2006). Relationships, layoffs, and organizational resilience: Airline industry responses to September eleven. *The Journal of Applied Behavioral Science*, *42*(3), 300–329. DOI: 10.1177/0021886306286466

Golubev, V. A., Verbnikova, V. A., Lopyrev, I. A., Voznesenskaya, D. D., Alimov, R. N., Novikova, O. V., & Konnikov, E. A. (2021). Energy evolution: Forecasting the development of non-conventional renewable energy sources and their impact on the conventional electricity system. *Sustainability (Basel)*, *13*(22), 12919. DOI: 10.3390/su132212919

González Cusa, Y., Moya Rodríguez, J. L., Vieira de Melo, S. A. B., & Andrade Torres, E. (2025). Optimization of hybrid renewable energy systems for power generation: A bibliometric review. *International Journal for Innovation Education and Research*, *9*(3), 369–380. DOI: 10.31686/ijier.vol9.iss3.3015

Google. (2024). YouTube transparency report. https://transparencyreport.google .com/youtube- policy/removals?hl=en

Google. (2025). How Google anonymizes data. https://policies.google.com/ technologies/anonymization?hl=en-US

Gopal, S., & Clarke, T. (2015). *System mapping: A guide to developing actor maps*. FSG.

Goswami, S. S., & Behera, D. K. (2021). Evaluation of the best smartphone model in the market by integrating fuzzy-AHP and PROMETHEE decision-making approach. *Decision (Washington, D.C.)*, *48*(1), 71–96. DOI: 10.1007/s40622-020-00260-8

Goswami, S. S., Behera, D. K., Afzal, A., Kaladgi, A. R., Khan, S. A., Rajendran, P., Subbiah, R., & Asif, M. (2021). Analysis of a robot selection problem using two newly developed hybrid MCDM models of TOPSIS-ARAS and COPRAS-ARAS. *Symmetry*, *13*(8), 1331. DOI: 10.3390/sym13081331

Griffiths, T. L., Lieder, F., & Goodman, N. D. (2015). Rational Use of Cognitive Resources: Levels of Analysis Between the Computational and the Algorithmic. *Topics in Cognitive Science*, *7*(2), 217–229. DOI: 10.1111/tops.12142 PMID: 25898807

Gross, J. J., & John, O. P. (2003). Individual differences in two emotion-regulation processes: Implications for affect, relationships, and well-being. *Journal of Personality and Social Psychology*, *85*(2), 348–362. DOI: 10.1037/0022-3514.85.2.348

Gudmundsson, E., & Lönner, W. J. (2009). Cross-cultural test adaptation. In Weiner, B., & Craighead, E. J. (Eds.), *The Corsini encyclopedia of psychology* (4th ed., Vol. 1, pp. 448–450). John Wiley & Sons.

Guerreiro, N. M., Voita, E., & Martins, A. F. (2022). A Comprehensive Study of Hallucinations in Neural Machine Translation. In *Proceedings of the 17th Conference of the European Chapter of the Association for Computational Linguistics* (pp. 1059–1075). Association for Computational Linguistics., https://aclanthology.org/2023.eacl-main.75

Guha, R., & Manyika, J. (2023, September 13). Data Commons: AI-powered data access at scale. Google. https://blog.google/technology/ai/google-data-commons-ai

Guida, M., Caniato, F., Moretto, A., & Ronchi, S. (2023). Artificial intelligence for supplier scouting: An information processing theory approach. *International Journal of Physical Distribution & Logistics Management*, *53*(4), 387–423. DOI: 10.1108/IJPDLM-12-2021-0536

Gursesli, M. C., Lombardi, S., Duradoni, M., Bocchi, L., Guazzini, A., & Lanata, A. (2024). Facial emotion recognition (FER) through custom lightweight CNN model: Performance evaluation in public datasets. *IEEE Access : Practical Innovations, Open Solutions*, *12*, 45543–45559. DOI: 10.1109/ACCESS.2024.3380847

Haenlein, M., & Kaplan, A. (2019). A brief history of artificial intelligence: On the past, present, and future of artificial intelligence. *California Management Review*, *61*(4), 5–14. DOI: 10.1177/0008125619864925

Haerpfer, C., Inglehart, R., Moreno, A., Welzel, C., Kizilova, K., Diez-Medrano, J., Lagos, M., Norris, P., Ponarin, E., & Puranen, B. (2022). *World Values Survey Trend File (1981-2022) Cross-National Data-Set*. JD Systems Institute & WVSA Secretariat. https://doi.org/DOI: 10.4232/1.14021

Hall, E. T. (1976). *Beyond culture*. Anchor.

Hancock, T. O., Hess, S., Choudhury, C. F., & Tsoleridis, P. (2024). Decision field theory: An extension for real-world settings. *Journal of Choice Modelling*, *52*, 100495. DOI: 10.1016/j.jocm.2024.100495

Hanford, S. D., Janrathitikarn, O., & Long, L. N. (2009). Control of Mobile Robots Using the Soar Cognitive Architecture. *Journal of Aerospace Computing, Information, and Communication*, *6*(2), 69–91. DOI: 10.2514/1.37056

Han, Y., Diao, Y., Yin, Z., Jin, R., Kangwa, J., & Ebohon, O. J. (2021). Immersive Technology-Driven Investigations on Influence Factors of Cognitive Load Incurred in Construction Site Hazard Recognition, Analysis and Decision Making. *Advanced Engineering Informatics*, *48*, 101298. Advance online publication. DOI: 10.1016/j.aei.2021.101298

Hartmann, K. V., Primc, N., & Rubeis, G. (2022). Lost in translation? Conceptions of privacy and independence in the technical development of AI-based AAL. *Medicine, Health Care, and Philosophy*, *26*(1). Advance online publication. DOI: 10.1007/s11019-022-10126-8

Hasnul, M. A., Aziz, N. A. A., Alelyani, S., Mohana, M., & Aziz, A. A. (2021). Electrocardiogram-based emotion recognition systems and their applications in healthcare—A review. *Sensors (Basel)*, *21*(15), 5015. DOI: 10.3390/s21155015 PMID: 34372252

Helfat, C. E., & Peteraf, M. A. (2003). The dynamic resource-based view: Capability lifecycles. *Strategic Management Journal*, *24*(10), 997–1010. DOI: 10.1002/smj.332

Hicham, N., Nassera, H., & Karim, S. (2023). Strategic Framework for Leveraging Artificial Intelligence in Future Marketing Decision-Making. *Journal of Intelligent Management Decision*, *2*(3), 139–150. DOI: 10.56578/jimd020304

Hillmann, J., & Guenther, E. (2021). Organizational resilience: A valuable construct for management research. *International Journal of Management Reviews*, *23*(1), 7–44. DOI: 10.1111/ijmr.12239

Hindarto, D. (2023). The Role of E-Commerce in Increasing Sales Using Unified Modeling Language. [IJSECS]. *International Journal Software Engineering and Computer Science*, *3*(2), 120–129. DOI: 10.35870/ijsecs.v3i2.1503

Hofstede, G. (2020). The 6-D model of national culture. *Geert Hofstede. http://geerthofstede. com/culture-geert-hofstede-gert-jan-hofstede/6d-model-of-national-culture/(accessed: 19.02. 2020.).*

Hofstede, G. (2011). Dimensionalizing Cultures: The Hofstede Model in Context. *Online Readings in Psychology and Culture*, *2*(1). Advance online publication. DOI: 10.9707/2307-0919.1014

Holzinger, A., Goebel, R., Fong, R., Moon, T., Müller, K. R., & Samek, W. (2020). xxAI - Beyond Explainable Artificial Intelligence. In: Holzinger, A., Goebel, R., Fong, R., Moon, T., Müller, KR., Samek, W. (eds) xxAI - Beyond Explainable AI. xxAI 2020. Lecture Notes in Computer Science, vol 13200. Springer, Cham. DOI: 10.1007/978-3-031-04083-2_1

Hooda, M., Rana, C., Dahiya, O., Rizwan, A., & Hossain, M. S. (2022). Artificial Intelligence for Assessment and Feedback to Enhance Student Success in Higher Education. *Mathematical Problems in Engineering*, *2022*, 5215722. Advance online publication. DOI: 10.1155/2022/5215722

Hossain, J., Sadad Ovi, N., & Monirujjaman Khan, M. (2021). Design and investigation of energy harvesting system from noise. *Energy and Power Engineering*, *13*(8), 307–321. DOI: 10.4236/epe.2021.138021

Howard, R. A., & Matheson, J. E. (2005). Influence diagrams. In Decision Analysis (Jensen & Nielsen, 2007) (pp. 719–762). Springer. DOI: 10.1287/deca.1050.0020

Howie, E. E., Dharanikota, H., Gunn, E., Ambler, O., Dias, R., Wigmore, S. J., Skipworth, R. J. E., & Yule, S. (2023). Cognitive Load Management: An Invaluable Tool for Safe and Effective Surgical Training. *Journal of Surgical Education*, *80*(3), 311–322. DOI: 10.1016/j.jsurg.2022.12.010 PMID: 36669990

Huang, L., Yu, W., Ma, W., Zhong, W., Feng, Z., Wang, H., Chen, Q., Peng, W., Feng, X., Qin, B., & Liu, T. (2024). A survey on hallucination in large language models: Principles, taxonomy, challenges, and open questions. *ACM Transactions on Information Systems*.

Huang, W., Xue, Y., Hu, L., & Liuli, H. (2020). S-EEGNet: Electroencephalogram signal classification based on a separable convolution neural network with bilinear interpolation. *IEEE Access : Practical Innovations, Open Solutions*, *8*, 131636–131646. DOI: 10.1109/ACCESS.2020.3009665

Hubballi, R. B., Selvakumar, P., Seenivasan, R., Basava Aradhya, S. G., Dinesh, N., & Seelam, P. K. (2025). Overview of Current AI Technologies in Education. In Kyei-Blankson, L., & Ntuli, E. (Eds.), *Transformative AI Practices for Personalized Learning Strategies* (pp. 1–26). IGI Global Scientific Publishing., DOI: 10.4018/979-8-3693-8744-3.ch001

Hunkenschroer, A. L., & Kriebitz, A. (2022). Is AI recruiting (un)ethical? A human rights perspective on the use of AI for hiring. *AI and Ethics*, *3*(1), 199–213. DOI: 10.1007/s43681-022-00166-4

Hwangbo, H., Kim, Y. S., & Cha, K. J. (2018). Recommendation system development for fashion retail e-commerce. *Electronic Commerce Research and Applications*, *28*, 94–101. DOI: 10.1016/j.elerap.2018.01.012

Insights, R. E. S. O. (2024). Revolutionizing pharma: AI integration across the value chain. https://resoinsights.com/insight/revolutionizing-pharma-ai-integration -across-the-value- chain/

International Energy Agency. (2024).. . *World Energy Outlook, 2024*. Advance online publication. DOI: 10.1787/9789266829873-en

International Organization for Standardization (ISO). (2023). Information technology — Artificial intelligence — Management system (ISO/IEC 27001:2023). Retrieved from https://www.iso.org/standard/42001

IQVIA. (2024). Improving Predictive Precision of Oncology Patient Modeling with Enriched Data and Integrated Field Alerts. https://www.iqvia.com/library/case-studies/client-case-study- improving-predictive-precision-of-oncology-patient-modeling-with-enriched-data

Islam, M. M., Hassan, S., Akter, S., Jibon, F. A., & Sahidullah, M. (2024). A comprehensive review of predictive analytics models for mental illness using machine learning algorithms. *Healthcare Analytics*, *6*, 100350. DOI: 10.1016/j.health.2024.100350

Ivanov, D., Dolgui, A., & Sokolov, B. (2019). The impact of digital technology and Industry 4.0 on the ripple effect and supply chain risk analytics. *International Journal of Production Research*, *57*(3), 829–846. DOI: 10.1080/00207543.2018.1488086

Jafari, M., Shoeibi, A., Khodatars, M., Bagherzadeh, S., Shalbaf, A., García, D. L., Gorriz, J. M., & Acharya, U. R. (2023). Emotion recognition in EEG signals using deep learning methods: A review. *Computers in Biology and Medicine*, *165*, 107450. DOI: 10.1016/j.compbiomed.2023.107450 PMID: 37708717

Jagannathan, J., & Udaykumar, U. (2020). *Predictive modeling for improving healthcare using IoT: Role of predictive models in healthcare using IoT. Incorporating the internet of things in healthcare applications and wearable devices*. IGI Global., DOI: 10.4018/978-1-7998-1090-2.ch015

Jain, J., Walia, N., Singla, H., Singh, S., Sood, K., & Grima, S. (2023). Heuristic biases as mental shortcuts to investment decision-making: A mediation analysis of risk perception. *Risks*, *11*(4), 72. Advance online publication. DOI: 10.3390/risks11040072

Jakkal, V. (2024). How AI is Transforming Cybersecurity: Tackling the surge in Cyber Threats. Microsoft. https://news.microsoft.com/source/canada/features/ai/how-ai-is-transforming- cybersecurity-tackling-the-surge-in-cyber-threats/

Jakubik, J., Vössing, M., Kühl, N., Walk, J., & Satzger, G. (2024). Data-Centric Artificial Intelligence. *Business & Information Systems Engineering*, *66*(4), 507–515. DOI: 10.1007/s12599-024-00857-8

Jamieson, G. A., & Skraaning, G.Jr. (2017). Levels of Automation in Human Factors Models for Automation Design: Why We Might Consider Throwing the Baby Out with the Bathwater. *Journal of Cognitive Engineering and Decision Making*, 1555343417732856.

Jarrahi, M. H. (2018). Artificial intelligence and the future of work: Human-AI symbiosis in organizational decision making. *Business Horizons*, *61*(4), 577–586. DOI: 10.1016/j.bushor.2018.03.007

Jean, R.-J., & Kim, D. (2020). Internet and SMEs' internationalization: The role of platform and website capabilities. *Journal of International Management*, *26*(1), 100690. DOI: 10.1016/j.intman.2019.100690

Jenke, L., Bansak, K., Hainmueller, J., & Hangartner, D. (2021). Using eye-tracking to understand decision-making in conjoint experiments. *Political Analysis*, *29*(1), 75–101. DOI: 10.1017/pan.2020.11

Jensen, F. V., & Nielsen, T. D. (2007). *Bayesian (Spiegelhalter et al., 1993) Networks and Decision Graphs* (2nd ed.). Springer. DOI: 10.1007/978-0-387-68282-2

Jin, B., Yong Park, J., & Kim, J. (2008). Cross-cultural examination of the relationships among firm reputation, e-satisfaction, e-trust, and e-loyalty. *International Marketing Review*, *25*(3), 324–337. DOI: 10.1108/02651330810877243

Ji, Z., Lee, N., Frieske, R., Yu, T., Su, D., Xu, Y., Ishii, E., Bang, Y. J., Madotto, A., & Fung, P. (2023). Survey of hallucination in natural language generation. *ACM Computing Surveys*, *55*(12), 1–38.

Johnson & Johnson. (2024, June 20). MedTech announces the Polyphonic™ digital ecosystem for a more connected surgical experience. https://www.jnj.com/media-center/press-releases/johnson-johnson-medtech-announces-the-polyphonic-digital-ecosystem-for-a-more- connected-surgical-experience

Kahneman, D., Sibony, O., & Sunstein, C. R. (2022). *Noise: A Flaw in Human Judgment*. Little, Brown Spark.

Kahneman, D. (1982). *Judgment under uncertainty: Heuristics and biases*. Cambridge University. DOI: 10.1017/CBO9780511809477

Kalogiannidis, S., Kalfas, D., Papaevangelou, O., Giannarakis, G., & Chatzitheodoridis, F. (2024). The Role of Artificial Intelligence Technology in Predictive Risk Assessment for Business Continuity: A Case Study of Greece. *Risks*, *12*(2), 19. DOI: 10.3390/risks12020019

Kalra, N., Verma, P., & Verma, S. (2024). Advancements in AI based healthcare techniques with FOCUS ON diagnostic techniques. *Computers in Biology and Medicine*, *179*, 108917–108917. DOI: 10.1016/j.compbiomed.2024.108917 PMID: 39059212

Kaminski, M. E., & Malgieri, G. (2021). Algorithmic impact assessments under the GDPR: Producing multi-layered explanations. *International Data Privacy Law*, *11*(2), 125–144. DOI: 10.1093/idpl/ipaa020

Kang, K., & Alnefaie, A. E. A. (2024). Cultural Transformations of E-commerce Consumer Behavior and Intention Toward Using Artificial Intelligence (AI) Assistants. *The 57th Hawaii International Conference on System Sciences.*

Kang, K., & Li, L. (2021). *Exploring the Relationships between Cultural Content and Viewers' Watching Interest: A Study of Tiktok Videos Produced by Chinese Ethnic Minority Groups* Proceedings of the 18th International Conference on e-Business, Kang, K., Li, L., & Namisango, F. (2023). Dissimilar Social Settings Impact on User Motivations and Activities on Live-Streaming Digital Platforms. In *E-service Digital Innovation*. IntechOpen. https://doi.org/DOI: 10.5772/intechopen.112787

Kang, K., & Sohaib, O. (2016). Individualists vs. Collectivists in B2C E-Business Purchase Intention. *Journal of Internet and e-business Studies*, 1-11. https://doi.org/DOI: 10.5171/2016.948644

Kareiva, P. M., McNally, B. W., McCormick, S., Miller, T., & Ruckelshaus, M. (2015). Improving global environmental management with standard corporate reporting. *Proceedings of the National Academy of Sciences of the United States of America*, *112*(24), 7375–7382. DOI: 10.1073/pnas.1408120111 PMID: 26082543

Katsigiannis, S., & Ramzan, N. (2017). DREAMER: A database for emotion recognition through EEG and ECG signals from wireless low-cost off-the-shelf devices. *IEEE Journal of Biomedical and Health Informatics*, *22*(1), 98–107. DOI: 10.1109/JBHI.2017.2688239 PMID: 28368836

Kellington, A. (2025). Elevating internal communications at Microsoft with AI. Microsoft. https://www.microsoft.com/insidetrack/blog/elevating-internal-communications-at-microsoft- with-ai/

Kennedy, B., Mohammad, A., & Wyandt, M. (2025). Beyond Intelligence: The Synthetic Cognitive Augmentation Network Using Experts. In Zangana, H., & Omar, M. (Eds.), *Ensuring Secure and Ethical STM Research in the AI Era* (pp. 257–286). IGI Global Scientific Publishing., DOI: 10.4018/979-8-3373-4252-8.ch007

Kenter, J. O., Bryce, R., Christie, M., Cooper, N., Hockley, N., Irvine, K. N., Fazey, I., O'Brien, L., Orchard-Webb, J., Ravenscroft, N., Raymond, C. M., Reed, M. S., Tett, P., & Watson, V. (2016). Shared values and deliberative valuation: Future directions. *Ecosystem Services, 21*, 358–371. DOI: 10.1016/j.ecoser.2016.10.006

Khan, N. A., Kumar, A., & Rao, N. (2024). A hybrid robot selection model for efficient decisive support system using fuzzy logic and genetic algorithm. *International Journal of System Assurance Engineering and Management, 15*(6), 2120–2129. DOI: 10.1007/s13198-023-02224-4

Khrais, L. T. (2020). Role of Artificial Intelligence in Shaping Consumer Demand in E-Commerce. *Future Internet, 12*(12), 226. Advance online publication. DOI: 10.3390/fi12120226

Kianto, A., Ritala, P., Spender, J. C., & Vanhala, M. (2014). The interaction of intellectual capital assets and knowledge management practices in organizational value creation. *Journal of Intellectual Capital, 15*(3), 362–375. DOI: 10.1108/JIC-05-2014-0059

Kiciman, E., Dillon, E., Edge, D., Foster, A., Hilmkil, A., Jennings, J., Ma, C., Ness, R. O., Pawlowski, N., Sharma, A., & Zhang, C. (2022, December). A causal AI suite for decision-making. NeurIPS 2022 Workshop on Causality for Real-world Impact. https://www.microsoft.com/en-us/research/publication/a-causal-ai-suite-for-decision- making/

Kim, C., Johnson, N. F., & Gold, B. T. (2011). Cingulo-frontal network activates differentially to conscious and non-conscious conflict. *Proceedings of the National Academy of Sciences of the United States of America, 108*(42), 17547–17552. DOI: 10.1073/pnas.1103627108

Kim, J. W., & Park, C. K. (2023). Can ESG performance mitigate information asymmetry? Moderating effect of assurance services. *Applied Economics, 55*(26), 2993–3007. DOI: 10.1080/00036846.2022.2107991

Kimmel, A. J., & Kitchen, P. J. (2014). Word of mouth and social media. *Journal of Marketing Communications, 20*(1-2), 2–4. DOI: 10.1080/13527266.2013.865868

Kim, N., & Nam, C. S. (2020). Adaptive Control of Thought-Rational (ACT-R): Applying a Cognitive Architecture to Neuroergonomics. In Nam, C. S. (Ed.), *Neuroergonomics: Principles and Practice* (pp. 105–114). Springer International Publishing., DOI: 10.1007/978-3-030-34784-0_6

Kim, Y., & Yoo, H. (2021). Cross-cultural comparison of preferences for the external appearance of artificial intelligence agents. *Social Behavior and Personality, 49*(11), 1–23. DOI: 10.2224/sbp.10824

Kludacz-Alessandri, M., Hawrysz, L., Żak, K., & Zhang, W. (2025). The impact of digital transformational leadership on digital intensity among primary healthcare entities: A moderated mediation model. *BMC Health Services Research, 25*(1), 117. DOI: 10.1186/s12913-025-12283-x PMID: 39838353

Koh, S. C. L., Leung, J. Y.-T., Gunasekaran, A., & Simpson, R. J. (2024). Quantifying organisational resilience: An integrated indicator and validation in manufacturing. *International Journal of Production Research*. Advance online publication. DOI: 10.1080/00207543.2023.2296018

Kramer, T., Spolter-Weisfeld, S., & Thakkar, M. (2007). The Effect of Cultural Orientation on Consumer Responses to Personalization. *Marketing Science, 26*(2), 246–258. DOI: 10.1287/mksc.1060.0223

Kritzinger, W., Karner, M., Traar, G., Henjes, J., & Sihn, W. (2018). Digital twin in manufacturing: A categorical literature review and classification. *IFAC-PapersOnLine, 51*(11), 1016–1022. DOI: 10.1016/j.ifacol.2018.08.474

Krivogorsky, V. (2024). Sustainability reporting with two different voices: The European Union and the International Sustainability Standards Board. *Journal of International Accounting, Auditing & Taxation, 56*, 100635. DOI: 10.1016/j.intaccaudtax.2024.100635

Krueger, P., Matos, P., & Starks, L. T. (2023). The effects of mandatory ESG disclosure around the world. *Journal of Accounting Research, 61*(6), 2117–2175. DOI: 10.1111/1475-679X.12548

Kumar, A., Hussain, S. A. I., & Rai, R. N. (2019). Optimization by AHP-ARAS of EDM process parameters on machining AA7050-10% B 4 C composite. In Advances in Industrial and Production Engineering: Select Proceedings of FLAME 2018 (pp. 285-296). Springer Singapore.

Kumaran, N. (2023). New Gmail protections for a safer, less spammy inbox. Google. https://blog.google/products/gmail/gmail-security-authentication-spam-protection/

Kumawat, R., & Jain, M. (2021). EEG based emotion recognition and classification: A review. *International Research Journal on Advanced Science Hub, 3*(Special Issue ICITCA-2021 5S, 5S), 1–10. DOI: 10.47392/irjash.2021.131

Lamm, A. J., Roberts, T. G., Irani, T. A., Snyder, L. J. U., & Brendemuhl, J. (2012). The Influence of Cognitive Diversity on Group Problem Solving Strategy. *Journal of Agricultural Education*, *53*(1), 18–30. DOI: 10.5032/jae.2012.01018

Langley, H. (2024). Layoffs, pivots, and the ongoing race for AI supremacy inside Google's tumultuous 2023. Business Insider. https://www.businessinsider.com/layoffs-pivots-race-for- ai-supremacy-inside-google-tumultuous-2023-2023-12

Langley, H. (2024, February 15). Google's Goose AI model shows how internal tools can change the AI race. Business Insider. https://www.businessinsider.com/google-goose-ai-model- language-ai-coding-2024-2

Langley, P., Laird, J. E., & Rogers, S. (2009). Cognitive architectures: Research issues and challenges. *Cognitive Systems Research*, *10*(2), 141–160. DOI: 10.1016/j.cogsys.2006.07.004

Laradi, S., Alrawad, M., Lutfi, A., & Agag, G. (2024). Understanding factors affecting social commerce purchase behavior: A longitudinal perspective. *Journal of Retailing and Consumer Services*, *78*, 103751. Advance online publication. DOI: 10.1016/j.jretconser.2024.103751

Law, A. M., & Kelton, W. D. (2015). *Simulation (Birge & Louveaux, 2011) Modeling and Analysis* (5th ed.). McGraw-Hill Education.

Lazić, A., Milić, S., & Vukmirović, D. (2024). The Future of Electronic Commerce in the IoT Environment. *Journal of Theoretical and Applied Electronic Commerce Research*, *19*(1), 172–187. DOI: 10.3390/jtaer19010010

LeCun, Y., Bengio, Y., & Hinton, G. (2015). Deep learning. *Nature*, *521*(7553), 436–444. DOI: 10.1038/nature14539 PMID: 26017442

Lee, H., & Lalwani, A. K. (2024). Power Distance Belief and Consumer Purchase Avoidance: Exploring the Role of Cultural Factors in Retail Dynamics. *JMR, Journal of Marketing Research*, *61*(2), 349–367. DOI: 10.1177/00222437231182600

Lee, M. H., Clark, A., Rupp, J., Wietelman, D. C., & Graham, J. D. (2019). Public opinion toward hydraulic fracturing: The effect of beyond compliance and voluntary third-party certification. *Energy Policy*, *128*, 306–315.

Lee, S., Oh, J., & Moon, W.-K. (2022). Adopting Voice Assistants in Online Shopping: Examining the Role of Social Presence, Performance Risk, and Machine Heuristic. *International Journal of Human-Computer Interaction*, *39*(14), 2978–2992. DOI: 10.1080/10447318.2022.2089813

Leidner, D. E., & Kayworth, T. (2006). Review: A Review of Culture in Information Systems Research: Toward a Theory of Information Technology Culture Conflict. *Management Information Systems Quarterly, 30*(2), 357–399. DOI: 10.2307/25148735

Leimstoll, U., & Wölfle, R. (2021). Direct to Consumer (D2C) E-Commerce: Goals and Strategies of Brand Manufacturers. In *New Trends in Business Information Systems and Technology* (pp. 237-250). https://doi.org/DOI: 10.1007/978-3-030-48332-6_16

Leonardi, P. M. (2014). Social media, knowledge sharing, and innovation: Toward a theory of communication visibility. *Information Systems Research, 25*(4), 796–816. DOI: 10.1287/isre.2014.0536

Lewis, K. (2003). Measuring transactive memory systems in the field: Scale development and validation. *The Journal of Applied Psychology, 88*(4), 587–604. DOI: 10.1037/0021-9010.88.4.587 PMID: 12940401

Li, J., Chen, J., Ren, R., Cheng, X., Zhao, W. X., Nie, J. Y., & Wen, J. R. (2024). The dawn after the dark: An empirical study on factuality hallucination in large language models. arXiv preprint arXiv:2401.03205.

Li, L., & Kang, K. (2020). Analyzing shopping behavior of the middle-aged users in tiktok live streaming platform. *AMCIS 2020 Proceedings.*

Li, L., & Kang, K. (2022b). The Role of Cultural Attractors in Live Streaming Content: Regional Cultural Perspective Using Multi-Group Analysis. *PACIS 2022 Proceedings, 49.*

Li, L., Kang, K., & Sohaib, O. (2023). An Analysis of Cultural Content on Short Video Platforms: Cultural Protection Perspective. In *E-Business and Telecommunications* (pp. 29-50). https://doi.org/DOI: 10.1007/978-3-031-36840-0_2

Li, J., Chen, L., Yi, J., Mao, J., & Liao, J. (2019). Ecosystem-specific advantages in international digital commerce. *Journal of International Business Studies, 50*(9), 1448–1463. DOI: 10.1057/s41267-019-00263-3

Li, J., Liu, P., & Li, Z. (2020). Optimal design and techno-economic analysis of a solar-wind-biomass off-grid hybrid power system for remote rural electrification: A case study of west China. *Energy, 208*, 118387. Advance online publication. DOI: 10.1016/j.energy.2020.118387

Li, L., Chen, X., & Zhu, P. (2024). How do e-commerce anchors' characteristics influence consumers' impulse buying? An emotional contagion perspective. *Journal of Retailing and Consumer Services, 76*, 103587. Advance online publication. DOI: 10.1016/j.jretconser.2023.103587

Li, L., & Kang, K. (2022a). Impact of opportunity and capability on e-entrepreneurial motivation: A comparison of urban and rural perspectives. *Journal of Entrepreneurship in Emerging Economies, 16*(4), 932–953. DOI: 10.1108/JEEE-06-2022-0178

Li, L., & Kang, K. (2022c). Understanding the real-time interaction between middle-aged consumers and online experts based on the COM-B model. *Journal of Marketing Analytics, 12*(3), 570–582. DOI: 10.1057/s41270-022-00196-1

Li, L., & Kang, K. (2024a). Discovering online Chinese consumers' impulse buying in live streaming by the theory of planned behavior. *Journal of Economic Analysis, 3*(2), 60. Advance online publication. DOI: 10.58567/jea03020008

Li, L., & Kang, K. (2024b). Discussing Social Media Using Affordance of Young Chinese Ethnic Minority Group Users. *Cultural Arts Research and Development, 4*(1), 12–23. DOI: 10.55121/card.v4i1.64

Li, L., Kang, K., Feng, Y., & Zhao, A. (2022). Factors affecting online consumers' cultural presence and cultural immersion experiences in live streaming shopping. *Journal of Marketing Analytics, 12*(2), 250–263. DOI: 10.1057/s41270-022-00192-5

Li, L., Kang, K., & Namisango, F. (2024). Online younger viewers' motivation to purchase virtual gifts: Multi-group analysis of urban-rural backgrounds. *Aslib Journal of Information Management*. Advance online publication. DOI: 10.1108/AJIM-03-2024-0201

Li, L., & Pugalia, S. (2024). How Wanghong Live Streamers Affect Online Users' Purchasing Behavior on Live Streaming Platforms: A PLS-SEM Approach. *Journal of Global Information Technology Management, 27*(4), 277–292. DOI: 10.1080/1097198X.2024.2410675

Ling, Y., Zhang, J., & Wang, K. (2022). Computer-adaptive forms for multidimensional questionnaires: Methodological advances and simulation results. *Applied Psychological Measurement, 46*(5), 401–417. DOI: 10.1177/01466216211073012

Lin, H., & Chen, Q. (2024). Artificial intelligence (AI)-integrated educational applications and college students' creativity and academic emotions: Students and teachers' perceptions and attitudes. *BMC Psychology, 12*(1), 487.

Li, S., Wang, X., Sun, Y., Wu, P., & Chen, A. (2021). The ventromedial prefrontal cortex in emotion regulation: Combined TMS-fMRI evidence. *Cerebral Cortex (New York, N.Y.), 31*(12), 5482–5496. DOI: 10.1093/cercor/bhab139

Li, S., Zhu, B., & Yu, Z. (2023). The Impact of Cue-Interaction Stimulation on Impulse Buying Intention on Virtual Reality Tourism E-commerce Platforms. *Journal of Travel Research, 63*(5), 1256–1279. DOI: 10.1177/00472875231183163

Little, A. S., & Wu, S. J. (2021). Cognitive bias and neurosurgical decision making. *Journal of Neurosurgery, 137*(1), 307–312. DOI: 10.3171/2021.9.JNS212058 PMID: 34972092

Liu, Y., Yao, Y., Ton, J. F., Zhang, X., Cheng, R. G. H., Klochkov, Y., Taufiq, M. F., & Li, H. (2023). Trustworthy LLMs: A survey and guideline for evaluating large language models' alignment. arXiv preprint arXiv:2308.05374.

Liu, H., Zhang, P., Cheng, H., Hasan, N., & Chiong, R. (2025). Impact of AI-generated virtual streamer interaction on consumer purchase intention: A focus on social presence and perceived value. *Journal of Retailing and Consumer Services, 85*, 104290. Advance online publication. DOI: 10.1016/j.jretconser.2025.104290

Liu, S., Costanza, R., Farber, S., & Troy, A. (2010). Valuing ecosystem services: Theory, practice, and the need for a transdisciplinary synthesis. *Annals of the New York Academy of Sciences, 1185*(1), 54–78. DOI: 10.1111/j.1749-6632.2009.05167.x PMID: 20146762

Liu, W., Qiu, J.-L., Zheng, W.-L., & Lu, B.-L. (2021). Comparing recognition performance and robustness of multimodal deep learning models for multimodal emotion recognition. *IEEE Transactions on Cognitive and Developmental Systems, 14*(2), 715–729. DOI: 10.1109/TCDS.2021.3071170

Liu, Z. (2024). New Dynamics in Cross-Cultural Exchange: A Study on the Role and Impact of Artificial Intelligence Technology in the Global Cultural Market. *Academic Journal of Business & Management, 6*(12), 137–143.

Liu, Z., & Park, S. (2015). What makes a useful online review? Implication for travel product websites. *Tourism Management, 47*, 140–151. DOI: 10.1016/j.tourman.2014.09.020

Lu, Z., Li, X., Cai, D., Yi, R., Liu, F., Zhang, X., Lane, N. D., & Xu, M. (2024). Small language models: Survey, measurements, and insights. ArXiv (Cornell University). https://doi.org//arxiv.2409.15790.DOI: 10.48550

Luo, L., & Tang, Q. (2023). The real effects of ESG reporting and GRI standards on carbon mitigation: International evidence. *Business Strategy and the Environment, 32*(6), 2985–3000. DOI: 10.1002/bse.3281

Luo, X., Cheah, J.-H., Hollebeek, L. D., & Lim, X.-J. (2024). Boosting customers' impulsive buying tendency in live-streaming commerce: The role of customer engagement and deal proneness. *Journal of Retailing and Consumer Services, 77*, 103644. Advance online publication. DOI: 10.1016/j.jretconser.2023.103644

Madaan, A., Tandon, N., & Gupta, P.. (2023). Self-refine: Iterative Refinement with Self-Feedback. *Advances in Neural Information Processing Systems*, *36*, 88–102. DOI: 10.48550/arXiv.2303.17651

Magazine, C. D. O. (2025, January 18). How AI is shaping career pathways for J&J's 150,000-strong workforce. https://www.cdomagazine.tech/talent-development/deep -dive-how-ai-is-shaping- career-pathways-for-jjs-150000-strong-workforce

Mahapatra, D. M., Patra, S. K., & Baral, S. K. (2022). Unleashing the Potential of Artificial Intelligence (AI) in Customer Journey of Cognitive Marketing and Consciousness Intention in E-Commerce Websites. *Proceedings of the 2nd International Conference on Sustainability and Equity (ICSE-2021)*, 44-51. https://doi.org/DOI: 10.2991/ahsseh.k.220105.006

Maheswaran, K., Selvakumar, P., Shaikh, M., Shashank, G., Gupta, S., & Manjunath, T. C. (2025). The Evolution of Smart Urban Mobility With Emerging Technologies. In Ponnusamy, S., Assaf, M., Youssef, B., Jeon, G., & Kalyanaraman, S. (Eds.), *Leveraging VANETs and Blockchain Technology for Urban Mobility* (pp. 1–28). IGI Global Scientific Publishing., DOI: 10.4018/979-8-3373-0265-2.ch001

Mai, N.-D., Long, N. M. H., & Chung, W.-Y. (2021). *1D-CNN-based BCI system for detecting Emotional states using a Wireless and Wearable 8-channel Custom-designed EEG Headset.* Paper presented at the 2021 IEEE International Conference on Flexible and Printable Sensors and Systems (FLEPS). DOI: 10.1109/FLEPS51544.2021.9469818

Maier-Hein, L., Eisenmann, M., Sarikaya, D., März, K., Collins, T., Malpani, A., Fallert, J., Feussner, H., Giannarou, S., Mascagni, P., Nakawala, H., Park, A., Pugh, C., Stoyanov, D., Vedula, S. S., Cleary, K., Fichtinger, G., Forestier, G., Gibaud, B., & Speidel, S. (2022). Surgical data science - from concepts toward clinical translation. *Medical Image Analysis*, *76*, 102306. DOI: 10.1016/j.media.2021.102306 PMID: 34879287

Mai, N.-D., Nguyen, H.-T., & Chung, W.-Y. (2023). Real-time on-chip machine-learning-based wearable behind-the-ear electroencephalogram device for emotion recognition. *IEEE Access : Practical Innovations, Open Solutions*, *11*, 47258–47271. DOI: 10.1109/ACCESS.2023.3276244

Malamuthu, B. K., Balakrishnan, T. P., Kumar, R. P. N., Venkataramanaiah, B., & Malathy, V. (2025a). Explainable AI for Decision-Making: A Hybrid Approach to Trustworthy Computing. *International Journal of Computational and Experimental Science and Engineering*, *11*(2), 2. Advance online publication. DOI: 10.22399/ijcesen.1684

Malan, R. (2010). *The role of shared mental models of strategic thinking in the development of organisational strategy* [PhD Thesis]. University of Southern Queensland.

Mangus, C. W., & Mahajan, P. (2022). Decision making: Healthy heuristics and betraying biases. *Critical Care Clinics*, *38*(1), 37–49. DOI: 10.1016/j.ccc.2021.07.002 PMID: 34794630

Mantello, P., Ho, M.-T., Nguyen, M.-H., & Vuong, Q.-H. (2023). Bosses without a heart: Socio-demographic and cross-cultural determinants of attitude toward Emotional AI in the workplace. *AI & Society*, *38*(1), 97–119. DOI: 10.1007/s00146-021-01290-1

Marciano, J. (2024). How Microsoft is using AI to win at forecasts and help their FP&A team. Datarails. https://www.datarails.com/daniel-sousa-lennox-how-microsoft-is-using-ai-to-win- at-forecasts-and-help-their-fpa-team/

Maroua, B., Laid, Z., Benbouhenni, H., Elbarbary, Z. M. S., Colak, I., & Alammar, M. M. (2025). Z. M. S. Elbarbary, Ilhami Colak, and Mohammed M. Alammer. "Genetic algorithm type 2 fuzzy logic controller of microgrid system with a fractional-order technique.". *Scientific Reports*, *15*(1), 6318. DOI: 10.1038/s41598-025-90239-1

Martín de Castro, G., Delgado Verde, M., Navas López, J. E., & Cruz González, J. (2013). The moderating role of innovation culture in the relationship between knowledge assets and product innovation. *Technological Forecasting and Social Change*, *80*(2), 351–363. DOI: 10.1016/j.techfore.2012.08.012

McKinsey & Company. (2021). The state of AI in 2021. https://www.mckinsey.com/~/media/McKinsey/Business%20Functions/McKinsey%20Analytic s/Our%20Insights/Global%20survey%20The%20state%20of%20AI%20in%202021/Global-survey-The-state-of-AI-in-2021.pdf

Mehrabian, A., & Russell, J. A. (1974). *An approach to environmental psychology*. The MIT Press.

Meilatinova, N. (2021). Social commerce: Factors affecting customer repurchase and word-of-mouth intentions. *International Journal of Information Management*, *57*, 102300. Advance online publication. DOI: 10.1016/j.ijinfomgt.2020.102300

Melnykova, N., Shakhovska, N., ml, M. G., & Melnykov, V. (2019). Using big data for formalization of the patient's personalized data. *Procedia Computer Science*, *155*, 624–629. DOI: 10.1016/j.procs.2019.08.088

Meyer, E. (2014). *The culture map: Breaking through the invisible boundaries of global business*. Public Affairs Books.

Michailidis, E., & Banks, A. P. (2016). The relationship between burnout and risk-taking in workplace decision-making and decision-making style. *Work and Stress, 30*(3), 278–292. DOI: 10.1080/02678373.2016.1213773

Michie, S., Stralen, M. M., & West, R. (2011). The behaviour change wheel: A new method for characterising and designing behaviour change interventions. *Implementation Science : IS, 6*(1), 1–12. DOI: 10.1186/1748-5908-6-42

Microsoft. (2020). Transforming cash flow data with intelligent analytics. https://www.microsoft.com/insidetrack/blog/microsoft-transforms-cash-flow-data-into-an - intelligent-analytics-platform/

Microsoft. (2023). 2023 Work Trend Index Annual Report. https://www.microsoft.com/en- us/worklab/work-trend-index/will-ai-fix-work

Microsoft. (2023). 2023 Work Trend Index Report. https://assets- c4akfrf5b4d3f4b7.z01.azurefd.net/assets/2023/09/e3227681-b882-4050-b201-a631431ad2a5- WTI_Will_AI_Fix_Work_060723.pdf

Microsoft. (2024). 2024 Work Trend Index Annual Report. https://www.microsoft.com/en- us/worklab/work-trend-index/ai-at-work-is-here-now-comes-the-hard-part

Mikhailov, V. (2024). Revolutionizing Healthcare: The Use of Artificial Intelligence in Clinical Care. *International Journal of Advanced Engineering Technologies and Innovations, 10*(2), 718–735. https://ijaeti.com/index.php/Journal/article/view/528

Miller, C. A., & Dorneich, M. (2006). From Associate Systems to Augmented Cognition: 25 Years of User Adaptation in High Criticality Systems. . . *Arlington.*

Miller, E. K., & Cohen, J. D. (2001). An integrative theory of prefrontal cortex function. *Annual Review of Neuroscience, 24*(1), 167–202. DOI: 10.1146/annurev.neuro.24.1.167

Mitchell, T. M. (1997). *Machine Learning*. McGraw-Hill.

Moffat, K., Lacey, J., Zhang, A., & Leipold, S. (2016). The social license to operate: A critical review. *Forum (London), 89*(5), 477–488. DOI: 10.1093/forestry/cpv044

Mogaji, E., Viglia, G., Srivastava, P., & Dwivedi, Y. K. (2024). Is it the end of the technology acceptance model in the era of generative artificial intelligence? *International Journal of Contemporary Hospitality Management, 36*(10), 3324–3339. DOI: 10.1108/IJCHM-08-2023-1271

Mohri, M., Sivek, G., & Suresh, A. T. (2019, May). Agnostic federated learning. In *the International Conference on Machine Learning* (pp. 4615-4625). PMLR.

Moniz, A. (2021). *Beginning Azure cognitive services: data-driven decision making through artificial intelligence* (1st edition 2021). Apress., DOI: 10.1007/978-1-4842-7176-6

Monsell, S. (2021). "Control of mental processes." In Unsolved Mysteries of the Mind, pp. 93–148. DOI: DOI: 10.4324/9781315784960-4

Morgan, M. G., & Henrion, M. (1990). *Uncertainty (Morgan & Henrion, 1990): A Guide to Dealing with Uncertainty (Morgan & Henrion, 1990) in Quantitative Risk and Policy Analysis*. Cambridge University Press. DOI: 10.1017/CBO9780511840609

Mumali, F. (2022). Artificial Neural Network-Based Decision Support Systems in Manufacturing Processes: A Systematic Literature Review. *Computers & Industrial Engineering, 165*, 107964. Advance online publication. DOI: 10.1016/j.cie.2022.107964

Musa, H. G., Fatmawati, I., Nuryakin, N., & Suyanto, M. (2024). Marketing research trends using technology acceptance model (TAM): A comprehensive review of researches (2002–2022). *Cogent Business & Management, 11*(1), 2329375. Advance online publication. DOI: 10.1080/23311975.2024.2329375

Mushtaha, A., & Troyer, O. D. (2009). Cross-Culture and Website Design: Cultural Movements and Settled Cultural Variables. *Internationalization, Design and Global Development: Third International Conference, IDGD 2009, Held as Part of HCI International 2009*, 69–78. https://doi.org/DOI: 10.1007/978-3-642-02767-3_8

Mustafa Ayobami, R., Hameedat Bukola, O., Timothy Tolulope, O., Wilhelmina Afua, A., Onyeka Chrisanctus, O., & Adedoyin Tolulope, O. (2024). E-commerce and consumer behavior: A review of AI-powered personalization and market trends. *GSC Advanced Research and Reviews, 18*(3), 066-077. https://doi.org/DOI: 10.30574/gscarr.2024.18.3.0090

Muthukrishna, M., & Schaller, M. (2020). Are Collectivistic Cultures More Prone to Rapid Transformation? Computational Models of Cross-Cultural Differences, Social Network Structure, Dynamic Social Influence, and Cultural Change. *Personality and Social Psychology Review, 24*(2), 103–120. DOI: 10.1177/1088868319855783

Nakisa, B., Rastgoo, M. N., Rakotonirainy, A., Maire, F., & Chandran, V. (2018). Long short term memory hyperparameter optimization for a neural network based emotion recognition framework. *IEEE Access : Practical Innovations, Open Solutions, 6*, 49325–49338. DOI: 10.1109/ACCESS.2018.2868361

Nalla, L. N., & Reddy, V. M. (2024). AI-Driven Big Data Analytics for Enhanced Customer Journeys: A New Paradigm in E-Commerce. *2*(1), 719-740.

Nandi, S., Sarkis, J., Hervani, A. A., & Helms, M. M. (2021). Redesigning supply chains using blockchain-enabled circular economy and COVID-19 experiences. *Sustainable Production and Consumption*, *27*, 10–22. DOI: 10.1016/j.spc.2020.10.019 PMID: 33102671

Nardi, B. A. (1996). Activity theory and human-computer interaction. *Context and consciousness: Activity theory and human-computer interaction, 436*(1), 7-16.

Nazi, Z. A., & Peng, W. (2024). Large Language Models in Healthcare and Medical Domain: A Review. *Informatics (Basel)*, *11*(3), 57. Advance online publication. DOI: 10.3390/informatics11030057

Newell, A. (1992). Précis of Unified theories of cognition. . *Behavioral and Brain Sciences*, *15*(3), 425–437. DOI: 10.1017/S0140525X00069478

Newell, A., & Simon, H. A. (1972). *Human problem solving* (Vol. 104). Prentice-hall Englewood Cliffs.

Nikolaidis, S., Hsu, D., & Srinivasa, S. (2017). Human-robot mutual adaptation in collaborative tasks: Models and experiments. *The International Journal of Robotics Research*, *36*(5–7), 618–634. DOI: 10.1177/0278364917690593 PMID: 32855581

Nisha Angeline, M., Selvakumar, P., Gupta, M. K., Nazar, N., Elumalai, D., & Bhattacharya, S. (2025). Emerging Materials for Energy Storage and Environmental Applications. In Muduli, K., Moharana, B., Ales, S., & Biswal, D. (Eds.), *Using Computational Intelligence for Sustainable Manufacturing of Advanced Materials* (pp. 53–72). IGI Global Scientific Publishing., DOI: 10.4018/979-8-3693-7974-5.ch003

Niu, Y., Ying, L., Yang, J., Bao, M., & Sivaparthipan, C. B. (2021). Organizational business intelligence and decision making using big data analytics. *Information Processing & Management*, *58*(6), 102725. Advance online publication. DOI: 10.1016/j.ipm.2021.102725

Norman, D., & Euchner, J. (2023). Design for a Better World: A Conversation with Don Norman. *Research Technology Management*, *66*(3), 11–18. DOI: 10.1080/08956308.2023.2183015

Novet, J. (2017). *Google launches Cloud Dataprep, an embedded version of Trifacta*. Venture Beat.

O'Connor, J. (2021). Building greater transparency and accountability with the Violative View Rate. Google. https://blog.google/intl/en-in/company-news/technology/building-greater- transparency-and/

O'Dwyer, E., Pan, I., Charlesworth, R., Butler, S., & Shah, N. (2020). Integration of an energy management tool and digital twin for coordination and control of multi-vector smart energy systems. *Sustainable Cities and Society*, *62*, 102412.

Obermeyer, Z., Powers, B., Vogeli, C., & Mullainathan, S. (2019). Dissecting racial bias in an algorithm used to manage the health of populations. *Science*, *366*(6464), 447–453. DOI: 10.1126/science.aax2342 PMID: 31649194

OECD. (2019). Recommendation of the Council on Artificial Intelligence. OECD Legal Instruments. Retrieved from file:///Users/141420/Downloads/OECD-LEGAL-0449-en.pdf

ÓhÉigeartaigh, S. S., Whittlestone, J., Liu, Y., Zeng, Y., & Liu, Z.ÓhÉigeartaigh. (2020). Overcoming barriers to cross-cultural cooperation in AI ethics and governance. *Philosophy & Technology*, *33*(4), 571–593. DOI: 10.1007/s13347-020-00402-x

Okyere, E. (2021). How DeepMind's AI framework made Google Energy Efficient. Nural Research. https://www.nural.cc/deepmind-ai-framework/

Olawade, S. A., Zheng, Y., & Samuel, M. A. (2024). Mental health detection and treatment using artificial intelligence: A review of techniques and applications. *Journal of Advanced Health Informatics*, *14*(2), 120–134.

Ong, K., Mao, R., Satapathy, R., & Shirota Filho, R. (2024). Explainable natural language processing for corporate sustainability analysis. *Information Fusion*, *115*, 102726. DOI: 10.1016/j.inffus.2024.102726

Organisation for Economic Co-operation and Development (OECD). (2019). Artificial intelligence in society. OECD Publishing. https://www.oecd.org/en/publications/artificial-intelligence-in-society_eedfee77-en.html

Orie, R., Otabil, L., Agorua, J., & Iqbal, M. T. (2025). Enhanced IoT-based optimization for a hybrid power system in Cartwright, Labrador. *Energies*, *18*(7), 1566. Advance online publication. DOI: 10.3390/en18071566

Orzes, G., Moretto, A. M., Moro, M., Rossi, M., Sartor, M., Caniato, F., & Nassimbeni, G. (2020). The impact of the United Nations Global Compact on firm performance: A longitudinal analysis. *International Journal of Production Economics*, *227*, 107664. DOI: 10.1016/j.ijpe.2020.107664

Ostrau, C., Klarhorst, C., Thies, M., & Rückert, U. (2022). Benchmarking neuromorphic hardware and its energy expenditure. *Frontiers in Neuroscience*, *16*, 873935. https://doi.org/10.3389/fnins.2022.873935

Our World in Data. (2025). Annual global corporate investment in artificial intelligence, by type. https://ourworldindata.org/grapher/corporate-investment-in-artificial-intelligence-by-type

Ozdemir, D., Sharma, M., Dhir, A., & Daim, T. (2022). Supply chain resilience during the COVID-19 pandemic. *Technology in Society*, *68*, 101847. DOI: 10.1016/j.techsoc.2021.101847 PMID: 35075312

Page, S. (2008). *The Difference: How the Power of Diversity Creates Better Groups, Firms, Schools, and Societies - New Edition*. Princeton University Press., DOI: 10.1515/9781400830282

Pahl, A. K., Newnes, L., & McMahon, C. (2007). A generic model for creativity and innovation: Overview for early phases of engineering design. *Journal of Desert Research*, *6*(1/2), 5. DOI: 10.1504/JDR.2007.015561

Pan, D., Zheng, H., Xu, F., Ouyang, Y., Jia, Z., Wang, C., & Zeng, H. (2023). MSFR-GCN: A multi-scale feature reconstruction graph convolutional network for EEG emotion and cognition recognition. *IEEE Transactions on Neural Systems and Rehabilitation Engineering*, *31*, 3245–3254. DOI: 10.1109/TNSRE.2023.3304660 PMID: 37578922

Pang, Y., Zhang, M., Liu, Y., Li, X., Wang, Y., Huan, Y., Liu, Z., Li, J., & Wang, D. (2024). Large language model-based optical network log analysis using LLaMA2 with instruction tuning. *Journal of Optical Communications and Networking*, *16*(11), 1116–1132.

Papa, M., Carrassi, M., Muserra, A. L., & Wieczorek Kosmala, M. (2022). The impact of the European Union non financial information directive on environmental disclosure: Evidence from Italian environmentally sensitive industries. *Meditari Accountancy Research*, *30*(7), 87–120. DOI: 10.1108/MEDAR-03-2021-1247

Parasuraman, R., Sheridan, T. B., & Wickens, C. D. (2000). A model for types and levels of human interaction with automation. *IEEE Transactions on Systems, Man, and Cybernetics. Part A, Systems and Humans*, *30*(3), 286–297. DOI: 10.1109/3468.844354 PMID: 11760769

Patton, J. H., Stanford, M. S., & Barratt, E. S. (1995). Factor structure of the Barratt Impulsiveness Scale. *Journal of Clinical Psychology*, *51*(6), 768–774. DOI: 10.1002/1097-4679(199511)51:6<768::AID-JCLP2270510607>3.0.CO;2-1

Pavegen. (2025). What can Pavegen power? https://www.pavegen.com/what-can-pavegen-power

Petrides, M. (2005). Lateral prefrontal cortex: Architectonic and functional organization. *Philosophical Transactions of the Royal Society of London. Series B, Biological Sciences*, *360*(1456), 781–795. DOI: 10.1098/rstb.2005.1631

Pfizer. (2020, December 9). Statement on EMA cyberattack. https://www.pfizer.com/news/announcements/statement-ema-cyberattack

Pfizer. (2025). Artificial intelligence: On a mission to make clinical drug development faster and smarter. https://www.pfizer.com/news/articles/artificial_intelligence_on_a_mission_to_make_clinical_d rug_development_faster_and_smarter

Pizzi, S., Mastroleo, G., Venturelli, A., & Caputo, F. (2024). The digitalization of sustainability reporting processes: A conceptual framework. *Business Strategy and the Environment*, *33*(2), 1040–1050. DOI: 10.1002/bse.3544

Polanyi, M., & Sen, A. (2009). *The tacit dimension. With a new foreword by Amartya Sen*. The University Of.

Porcelli, A. J., & Delgado, M. R. (2017). Stress and decision making: Effects on valuation, learning, and risk-taking. *Current Opinion in Behavioral Sciences*, *14*, 33–39. DOI: 10.1016/j.cobeha.2016.11.015 PMID: 28044144

Prabhakaran, V., Qadri, R., & Hutchinson, B. (2022). Cultural incongruencies in artificial intelligence. arXiv preprint arXiv:2211.13069.

Pradhan, R. P., Samarakoon, S. M. R. K., Maradana, R. P., & Sahoo, P. (2024). Climate change disclosure and firm value in a frontier market: Exploring the determinants. *Natural Resources Forum*. https://doi.org/DOI: 10.1111/1477-8947.12462

Pu, L., Radics, R., Umar, M., Jeremiah, F., & Quan, Z. (2025). The potential of AI tools in shaping digital consumers' behavior: Investigating e-commerce engagement of Chinese Generation Z. *Asia Pacific Journal of Marketing and Logistics*, *37*(9), 2720–2737. Advance online publication. DOI: 10.1108/APJML-08-2024-1048

Pytel, K. (2025). Fuzzy logic applied to tunning mutation size in evolutionary algorithms. *Scientific Reports*, *15*(1), 1937. DOI: 10.1038/s41598-025-86349-5

Qadir, A., Mahum, R., El-Meligy, M. A., Ragab, A. E., AlSalman, A., & Hassan, H. (2024). An efficient deepfake video detection using robust deep learning. *Heliyon*, *10*(5), e25757. DOI: 10.1016/j.heliyon.2024.e25757

Raab, M., & Gigerenzer, G. (2015). The power of simplicity: A fast-and-frugal heuristics approach to performance science. *Frontiers in Psychology*, *6*. Advance online publication. DOI: 10.3389/fpsyg.2015.01672 PMID: 26579051

Rachmad, Y. E. (2024). *The Evolution of Consumer Behavior: Theories of Engagement, Influence, and Digital Interaction*. PT. Sonpedia Publishing Indonesia.

Raheel, A., Majid, M., Alnowami, M., & Anwar, S. M. (2020). Physiological sensors based emotion recognition while experiencing tactile enhanced multimedia. *Sensors (Basel)*, *20*(14), 4037. DOI: 10.3390/s20144037 PMID: 32708056

Rahevar, M., & Darji, C. S. (2024). The adoption of AI-driven Chatbots into a recommendation for e-commerce systems to targeted customer in the selection of product. *International Journal of Management. Economics and Commerce*, *1*(2), 128–137. DOI: 10.62737/m1vpdq75

Rahman, A., Debnath, T., Kundu, D., Islam, S., Airin Afroj Aishi, Sadia Sazzad, Sayduzzaman, M., & Band, S. S. (2024). Machine learning and deep learning-based approach in smart healthcare: Recent advances, applications, challenges and opportunities. 11(1), 58–109. .DOI: 10.3934/publichealth.2024004

Rajendran, S., Pan, W., Sabuncu, M. R., Chen, Y., Zhou, J., & Wang, F. (2024). Learning across diverse biomedical data modalities and cohorts: Challenges and opportunities for innovation. *Patterns (New York, N.Y.)*, *5*(2), 100913. Advance online publication. DOI: 10.1016/j.patter.2023.100913 PMID: 38370129

Rawte, V., Chakraborty, S., Pathak, A., Sarkar, A., Tonmoy, S. M. T. I., Chadha, A., Sheth, A., & Das, A. (2023). The Troubling Emergence of Hallucination in Large Language Models - An Extensive Definition, Quantification, and Prescriptive Remediations. EMNLP 2023 - 2023 Conference on Empirical Methods in Natural Language Processing, Proceedings, 2541–2573.

Razmus, W., Razmus, M., & Grabner-Kräuter, S. (2024). Joy and excitement in the purchase process: The role of materialism and brand engagement. *Journal of Product and Brand Management*, *33*(3), 330–344. DOI: 10.1108/JPBM-04-2022-3937

Rėklaitis, K., & Pilelienė, L. (2019). Principle Differences between B2B and B2C Marketing Communication Processes. *Management of Organizations: Systematic Research*, *81*(1), 73–86. DOI: 10.1515/mosr-2019-0005

Report, F. (2025). AI in Healthcare Market Size & Global Trends. https://www.fortunebusinessinsights.com/industry-reports/artificial-intelligence-in-healthcare-market-100534?utm_source=chatgpt.com

Rezvi, R. I., Rahman, K. O., Hasan, M. A., Nasrullah, F., Nusrat, N., Jishan, S. S., & Ahmed, S. (2025). How Digital Marketing Affiliates the Digital Stores: A Deep Dive into Shopify, Amazon, Walmart, and Other E-Commerce Giants. *Journal of Computer Science and Technology Studies*, *7*(2), 95–101. DOI: 10.32996/jcsts.2025.7.2.8

Ristevski, B., & Chen, M. (2018). Big data analytics in medicine and healthcare. *Journal of Integrative Bioinformatics, 2017*(0030), 1–5. DOI: 10.1515/jib-2017-0030 PMID: 29746254

Rivero, O. (2025). The role of artificial intelligence in strategic decision-making: Transforming managerial strategies in the digital age. *European Journal of Studies in Management and Business, 33*, 33–42. DOI: 10.32038/mbrq.2025.33.03

Rolls, E. T. (2023). Emotion, motivation, decision-making, the orbitofrontal cortex, anterior cingulate cortex, and the amygdala. *Brain Structure & Function, 228*(5), 1201–1257. DOI: 10.1007/s00429-023-02644-9 PMID: 37178232

Rolls, E. T. (2023). The orbitofrontal cortex, amygdala, reward value, emotion, and decision-making. *Neuroscience and Biobehavioral Reviews, 150*, 105066. Advance online publication. DOI: 10.1016/j.neubiorev.2023.105066

Rolnick, D., Donti, P. L., Kaack, L. H., Kochanski, K., Lacoste, A., Sankaran, K., Ross, A. S., Milojevic-Dupont, N., Jaques, N., Waldman-Brown, A., Luccioni, A. S., Maharaj, T., Sherwin, E. D., Mukkavilli, S. K., Kording, K. P., Gomes, C. P., Ng, A. Y., Hassabis, D., Platt, J. C., & Bengio, Y. (2023). Tackling climate change with machine learning. *ACM Computing Surveys, 55*(2), 1–96. DOI: 10.1145/3485128

Rooy, D. V. (2024). Human–machine collaboration for enhanced decision-making in governance. *Data & Policy, 6*, e60. DOI: 10.1017/dap.2024.72

Roy, K., Jaiswal, A., & Panda, P. (2019). Towards spike-based machine intelligence with neuromorphic computing. *Nature, 575*(7784), 607–617. DOI: 10.1038/s41586-019-1677-2

Ruan, Y., & Mezei, J. (2022). When do AI chatbots lead to higher customer satisfaction than human frontline employees in online shopping assistance? Considering product attribute type. *Journal of Retailing and Consumer Services, 68*, 103059. Advance online publication. DOI: 10.1016/j.jretconser.2022.103059

Rudin, C. (2019). Stop explaining black box machine learning models for high stakes decisions and use interpretable models instead. *Nature Machine Intelligence, 1*(5), 206–215. DOI: 10.1038/s42256-019-0048-x PMID: 35603010

Rudin, C., Chen, C., Chen, Z., Huang, H., Semenova, L., & Zhong, C. (2021). Interpretable machine learning: Fundamental principles and ten grand challenges. *Statistics Surveys, 16*(none), 1–85. DOI: 10.1214/21-SS133

Russell, S., & Norvig, P. (2021). *Artificial Intelligence: A Modern Approach* (4th ed.). Pearson.

Ryffel, T., Trask, A., Dahl, M., Wagner, B., Mancuso, J., Rueckert, D., & Passerat-Palmbach, J. (2018). A generic framework for privacy preserving deep learning. arXiv preprint arXiv:1811.04017.

Sabharwal, D., Sood, R. S., & Verma, M. (2022). Studying the relationship between artificial intelligence and digital advertising in marketing strategy. *Journal of Content. Community and Communication*, *16*(8), 118–126. DOI: 10.31620/JCCC.12.22/10

Sahoo, S. K., & Goswami, S. S. (2023). A comprehensive review of multiple criteria decision-making (MCDM) Methods: Advancements, applications, and future directions. *Decision Making Advances*, *1*(1), 25–48. DOI: 10.31181/dma1120237

Sajadi, A. (2024). Optimizing Data Input Interfaces Through User Error Analysis: A Statistical and HCI-Driven Approach. *International Journal of Advanced Human Computer Interaction*, *2*(2), 67–77. DOI: 10.5281/zenodo.10543292

Salama, E. S., El-Khoribi, R. A., Shoman, M. E., & Shalaby, M. A. W. (2021). A 3D-convolutional neural network framework with ensemble learning techniques for multi-modal emotion recognition. *Egyptian Informatics Journal*, *22*(2), 167–176. DOI: 10.1016/j.eij.2020.07.005

Saleh, R. A., & Zeebaree, S. R. M. (2025). Artificial Intelligence in E-commerce and Digital Marketing: A Systematic Review of Opportunities, Challenges, and Ethical Implications. *Asian Journal of Research in Computer Science*, *18*(3), 395–410. DOI: 10.9734/ajrcos/2025/v18i3601

Salih, A., Raisi-Estabragh, Z., Galazzo, I. B., Radeva, P., Petersen, S. E., Menegaz, G., & Lekadir, K. (2025). A Perspective on Explainable Artificial Intelligence Methods: SHAP and LIME. *Advanced Intelligent Systems*, *7*(1), 2400304. DOI: 10.1002/aisy.202400304

Samsuria, E., Mohd, S. A. M., Wahab, N. A., Romdlony, M. Z., Mohamad, S. Z. A., & Buyamin, S. (2025). An improved adaptive fuzzy-genetic algorithm based on local search for integrated production and mobile robot scheduling in job-shop flexible manufacturing system. *Computers & Industrial Engineering*, *204*, 111093. DOI: 10.1016/j.cie.2025.111093

Sarker, A. K., Azad, A. K., Rasul, M. G., & Doppalapudi, A. T. (2023). Prospect of green hydrogen generation from hybrid renewable energy sources: A review. *Energies*, *16*(3), 1556. Advance online publication. DOI: 10.3390/en16031556

Schoenbaum, G., Takahashi, Y., Liu, T. L., & McDannald, M. A. (2011). Does the orbitofrontal cortex signal value? *Annals of the New York Academy of Sciences*, *1239*(1), 87–99. DOI: 10.1111/j.1749-6632.2011.06210.x

Schwarzer, R., & Jerusalem, M. (1995). Generalized Self-Efficacy Scale. In Weinman, J., Wright, S., & Johnston, M. (Eds.), *Measures in health psychology: A user's portfolio* (pp. 35–37). NFER-Nelson.

Scott, L. D., Arslanian-Engoren, C., & Engoren, M. C. (2014). Association of sleep and fatigue with decision regret among critical care nurses. *American Journal of Critical Care: An Official Publication. American Journal of Critical Care, 23*(1), 13–23. DOI: 10.4037/ajcc2014191 PMID: 24382613

Scott, S. G., & Bruce, R. A. (1995). Decision-making style: The development and assessment of a new measure. *Educational and Psychological Measurement, 55*(5), 818–831. DOI: 10.1177/0013164495055005017

Sepúlveda, A., Castillo, F., Palma, C., & Rodriguez-Fernandez, M. (2021). Emotion recognition from ECG signals using wavelet scattering and machine learning. *Applied Sciences (Basel, Switzerland), 11*(11), 4945. DOI: 10.3390/app11114945

Sethumadhavan, R. P. P. S., Selvakumar, P., Kadu, S. G., Chandel, P. S., & Gupta, S. (2025). Environmental Sustainability and Employee Participation. In Şeker, C., Örücü, E., & Kaya, A. (Eds.), *Green Leadership Strategies for Increased Employee Engagement and Performance* (pp. 275–298). IGI Global Scientific Publishing., DOI: 10.4018/979-8-3373-2106-6.ch009

Shachter, R. D. (1986). Evaluating influence diagrams. Operations Research (Winston, 2004), 34(6), 871–882.

Sharafi, M., Yazdchi, M., Rasti, R., & Nasimi, F. (2022). A novel spatio-temporal convolutional neural framework for multimodal emotion recognition. *Biomedical Signal Processing and Control, 78*, 103970. DOI: 10.1016/j.bspc.2022.103970

Sheridan, T. B., & Verplank, W. L. (1978). *Human and Computer Control of Undersea Teleoperators*. MIT Man-Machine Systems Laboratory. DOI: 10.21236/ADA057655

Shmorgun, I., & Lamas, D. (2014). *Towards a Design Space for Ubiquitous Computing*. Springer International Publishing., DOI: 10.1007/978-3-319-07788-8_6

Shrestha, S. B., & Orchard, G. (2018). SLAYER: Spike layer error reassignment in time. In Bengio, S., Wallach, H., Larochelle, H., Grauman, K., Cesa-Bianchi, N., & Garnett, R. (Eds.), Vol. 31, pp. 1419–1428). Advances in Neural Information Processing Systems. Curran Associates.

Simon, H. A. (1955). A behavioral model of rational choice. *The Quarterly Journal of Economics, 69*(1), 99–118. DOI: 10.2307/1884852

Simon, H. A. (1997). *Models of Bounded Rationality: Empirically Grounded Economic Reason*. The MIT Press., DOI: 10.7551/mitpress/4711.001.0001

Singh, R. (2021). A Study of Artificial Intelligence and E-Commerce Ecosystem – A Customer's Perspective. *International journal of research in engineering, science and management, 4*(2), 78-87.

Singh, C., Dash, M. K., Sahu, R., & Kumar, A. (2024). Investigating the acceptance intentions of online shopping assistants in E-commerce interactions: Mediating role of trust and effects of consumer demographics. *Heliyon, 10*(3), e25031. DOI: 10.1016/j.heliyon.2024.e25031

Sohaib, O., Kang, K., & Miliszewska, I. (2021). Uncertainty Avoidance and Consumer Cognitive Innovativeness in E-Commerce. *Journal of Global Information Management, 27*(2), 59–77. DOI: 10.4018/JGIM.2019040104

Souchet, A. D., Philippe, S., Lourdeaux, D., & Leroy, L. (2022). Measuring Visual Fatigue and Cognitive Load via Eye Tracking While Learning with Virtual Reality Head-Mounted Displays: A Review. *International Journal of Human-Computer Interaction, 38*(9), 801–824. DOI: 10.1080/10447318.2021.1976509

Spiegelhalter, D. J., Dawid, A. P., Lauritzen, S. L., & Cowell, R. G. (1993). Bayesian (Spiegelhalter et al., 1993) analysis in expert systems. *Statistical Science, 8*(3), 219–247.

Sreen, N., Purbey, S., & Sadarangani, P. (2018). Impact of culture, behavior and gender on green purchase intention. *Journal of Retailing and Consumer Services, 41*, 177–189. DOI: 10.1016/j.jretconser.2017.12.002

Statista. (2023). Artificial intelligence (AI) adoption worldwide 2023, by industry and function. https://www.statista.com/statistics/1464584/ai-adoption-worldwide-industry-function/

Stevens Institute of Technology. (2025). Corporate education for Pfizer employees. https://www.stevens.edu/page-branding-landing/corporate-education-for-pfizer-employees

Stojčić, M., Zavadskas, E. K., Pamučar, D., Stević, Ž., & Mardani, A. (2019). Application of MCDM methods in sustainability engineering: A literature review 2008–2018. *Symmetry, 11*(3), 350. DOI: 10.3390/sym11030350

Strubell, E., Ganesh, A., & McCallum, A. (2019). Energy and policy considerations for deep learning in NLP. *Proceedings of the 57th Annual Meeting of the Association for Computational Linguistics*, 3645–3650. DOI: 10.18653/v1/P19-1355

Subasi, A., Tuncer, T., Dogan, S., Tanko, D., & Sakoglu, U. (2021). EEG-based emotion recognition using tunable Q wavelet transform and rotation forest ensemble classifier. *Biomedical Signal Processing and Control*, *68*, 102648. DOI: 10.1016/j.bspc.2021.102648

Sullivan, J. G., Vu, M. A. T., Stein, D. J., & Yazdanfar, A. (2022). Medial prefrontal cortex activation during third-party social interactions in infants. *Developmental Science*, *25*(5), e13245. Advance online publication. DOI: 10.1111/desc.13245

Sun, C., Shi, Z., Liu, X., & Ghose, A. (2025). The Effect of Voice AI on Digital Commerce. *Information Systems Research*, *36*(2), 1147–1166. Advance online publication. DOI: 10.1287/isre.2022.0140

Suta, A., Pintes, O., Molnár, P., Lukács, B., Kedves, L., & Tóth, Á. (2023). Sustainability and financial reporting in the XBRL era. *Chemical Engineering Transactions*, *107*, 577–582. DOI: 10.3303/CET23107097

Tabassam, A., Yaqoob, G., Cuong, V. H., Syed, M., Shahzadi, A., & Asghar, F. (2023). The Ethical Implication of Using Artificial Intelligence in Hiring and Promotion Decisions. 1(2). https://jomeri.org/index.php/jomeri/article/view/7

Tahir, H. A., Alayed, W., Hassan, W. U., & Haider, A. (2024). A Novel Hybrid XAI Solution for Autonomous Vehicles: Real-Time Interpretability Through LIME–SHAP Integration. *Sensors (Basel)*, *24*(21), 21. Advance online publication. DOI: 10.3390/s24216776 PMID: 39517672

Tang, W., Chen, S., Shao, J., Du, X., Wu, J., Zhang, X., Niu, Y., & Xue, C. (2025). Emerging Themes and Future Directions in Neurodesign and Human-Computer Interaction: A Systematic Review. *International Journal of Human-Computer Interaction*, *41*(2), 1652–1673. DOI: 10.1080/10447318.2024.2362486

Tang, Y. M., Chau, K. Y., Lau, Y.-y., & Zheng, Z. (2023). Data-Intensive Inventory Forecasting with Artificial Intelligence Models for Cross-Border E-Commerce Service Automation. *Applied Sciences (Basel, Switzerland)*, *13*(5), 3051. Advance online publication. DOI: 10.3390/app13053051

Tan, W. H., Chong, S., Fang, H.-W., Pan, K.-L., Mohamad, M., Lim, J. W., Tiong, T. J., Chan, Y. J., Huang, C.-M., & Yang, T. C.-K. (2021). Microbial Fuel Cell Technology—A Critical Review on Scale-Up Issues. *Processes (Basel, Switzerland)*, *9*(6), 985. DOI: 10.3390/pr9060985

Tate, L. (2024a). iLevyTate/SCANUE: 1.0.0-alpha (Version 1.0.0-alpha) [Computer software]. Zenodo. DOI: 10.5281/zenodo.14052759

Tate, L. (2024b). iLevyTate/scanue-v22: Zenodo Synchronization Second (Version 22.0) [Computer software]. Zenodo. DOI: 10.5281/zenodo.14510407

Tate, L. (2024c). iLevyTate/SCAN-Resources: 1.0.0-alpha (Version 1.0.0-alpha) [Dataset and code repository]. Zenodo. DOI: 10.5281/zenodo.140532

Tate, L. (2025a). STAC V1 implementation in iLevyTate/stac (Version 2.0.0.3) [Computer software, Jupyter Notebook]. Zenodo. DOI: 10.5281/zenodo.15867066

Tate, L., & Sanders, P. (2024). iLevyTate/SCAN: 1.0.0-alpha (Version 1.0.0-alpha) [Computer software]. Zenodo. DOI: 10.5281/zenodo.14052885

Tatineni, S. (2019). Ethical considerations in AI and data science: Bias, fairness, and accountability. *International Journal of Information Technology and Management Information Systems, 10*(1), 11–20.

Teece, D. J. (2007). Explicating dynamic capabilities: The nature and microfoundations of (sustainable) enterprise performance. *Strategic Management Journal, 28*(13), 1319–1350. DOI: 10.1002/smj.640

Thirunagalingam, A., & Selvakumar, P. (2025). Application of Automation and Robotics in the Oil and Gas Industry. In Abdullayev, V., & Khang, A. (Eds.), *Revolutionizing AI and Robotics in the Oil and Gas Industry* (pp. 229–244). IGI Global Scientific Publishing., DOI: 10.4018/979-8-3693-8156-4.ch011

Thomson, I., & Boutilier, R. G. (2011). Social license to operate. SME mining engineering handbook, 1, 1779-1796.

Tian, Y., & Frank, B. (2024). Optimizing live streaming features to enhance customer immersion and engagement: A comparative study of live streaming genres in China. *Journal of Retailing and Consumer Services, 81*, 103974. Advance online publication. DOI: 10.1016/j.jretconser.2024.103974

Toha, M., & Supriyanto, S. (2023). Factors influencing the consumer research process: Market target, purchasing behavior and market demand (literature review of consumer behavior). *Danadyaksa: Post Modern Economy Journal, 1*(1), 1–17. DOI: 10.69965/danadyaksa.v1i1.5

Toraman, S., & Dursun, Ö. O. (2021). GameEmo-CapsNet: Emotion Recognition from Single-Channel EEG Signals Using the 1D Capsule Networks. *TS. Traitement du Signal, 38*(6), 1689–1698. DOI: 10.18280/ts.380612

Tortoriello, M., Reagans, R., & McEvily, B. (2012). Bridging the knowledge gap: The influence of strong ties, network cohesion, and network range on the transfer of knowledge between organizational units. *Organization Science*, *23*(4), 1024–1039. DOI: 10.1287/orsc.1110.0688

Tran, N.-T., Trinh, V.-L., & Chung, C.-K. (2024). An Integrated Approach of Fuzzy AHP-TOPSIS for Multi-Criteria Decision-Making in Industrial Robot Selection. *Processes (Basel, Switzerland)*, *12*(8), 1723. DOI: 10.3390/pr12081723

Trompenaars, F., & Hampden-Turner, C. (2011). *Riding the waves of culture: Understanding diversity in global business*. Nicholas Brealey International.

Tubadji, A., Denney, T., & Webber, D. J. (2021). Cultural relativity in consumers' rates of adoption of artificial intelligence. *Economic Inquiry*, *59*(3), 1234–1251. DOI: 10.1111/ecin.12978

Turban, E., Sharda, R., & Delen, D. (2011). *Decision Support and Business Intelligence Systems* (9th ed.). Pearson.

Tversky, A., & Kahneman, D. (1973). Availability: A heuristic for judging frequency and probability. *Cognitive Psychology*, *5*(2), 207–232. DOI: 10.1016/0010-0285(73)90033-9

Tversky, A., & Kahneman, D. (1974). Judgment under Uncertainty: Heuristics and Biases. *Science*, *185*(4157), 1124–1131. DOI: 10.1126/science.185.4157.1124 PMID: 17835457

U.S. Food and Drug Administration (FDA). (2021). Guidance for industry: Software as a medical device (SaMD). Retrieved from https://www.fda.gov/

Umbrello, S., & Natale, S. (2024). Reframing Deception for Human-Centered AI. *International Journal of Social Robotics*, *16*(11), 103239. Advance online publication. DOI: 10.1007/s12369-024-01184-4

Unegbu, H. C. O., Yawas, D. S., Dan-asabe, B., & Alabi, A. A. (2025). Integrating renewable energy solutions in sustainable building projects: A case study of Nigerian urban centers. *Discover Civil Engineering*, *2*(1), 67. Advance online publication. DOI: 10.1007/s44290-025-00226-8

Upadhyaya, A. N., Debnath, P., Rani, P., Selvakumar, P., Manjunath, T. C., & Bhattacharya, S. (2025). AI-Based Disaster Prediction and Early Warning Systems. In Mobo, F. (Ed.), *AI and Emerging Technologies for Emergency Response and Smart Cities* (pp. 55–74). IGI Global Scientific Publishing., DOI: 10.4018/979-8-3693-9770-1.ch003

Van der Vegt, G. S., Essens, P., Wahlström, M., & George, G. (2015). Managing risk and resilience. *Academy of Management Journal, 58*(4), 971–980. DOI: 10.5465/amj.2015.4004

Van Noorden, R., & Perkel, J. M. (2023). AI and science: What 1,600 researchers think. *Nature, 621*(7980), 672–675. DOI: 10.1038/d41586-023-02980-0 PMID: 37758894

Vaswani, A., Shazeer, N., Parmar, N., Uszkoreit, J., Jones, L., Gomez, A. N., Kaiser, L., & Polosukhin, I. (2017). Attention is all you need. In von Luxburg, U., Guyon, I., Bengio, S., & Wallach, H. (Eds.), Vol. 30, pp. 5998–6008). Advances in Neural Information Processing Systems. Curran Associates.

Venkatesh, V., & Bala, H. (2008). Technology Acceptance Model 3 and a Research Agenda on Interventions. *Decision Sciences, 39*(2), 273–315. DOI: 10.1111/j.1540-5915.2008.00192.x

Venkatesh, V., & Davis, F. D. (2000). A Theoretical Extension of the Technology Acceptance Model: Four Longitudinal Field Studies. *Management Science, 46*(2), 186–204. DOI: 10.1287/mnsc.46.2.186.11926

Venkatesh, V., Morris, M. G., Davis, G. B., & Davis, F. D. (2003). User Acceptance of Information Technology: Toward a Unified View. *Management Information Systems Quarterly, 27*(3), 425–478. DOI: 10.2307/30036540

Venkatesh, V., Thong, J. Y. L., & Xu, X. (2012). Consumer Acceptance and Use of Information Technology: Extending the Unified Theory of Acceptance and Use of Technology. *Management Information Systems Quarterly, 36*(1), 157–178. DOI: 10.2307/41410412

Vetrivel, S. C., Sowmiya, K. C., Sabareeshwari, V., & Arun, V. P. (2024). Navigating the Digital Economy: The Crucial Role of Human-Computer Interaction. In *Social Reflections of Human-Computer Interaction in Education, Management, and Economics* (pp. 88–102). IGI Global., DOI: 10.4018/979-8-3693-3033-3.ch010

Vohra, K., Sinha, A. K., & Anand, A. (2024). Development of industry 4.0 based technology selection index using multi criteria decision making. *Operations Research, 58*(6), 5185–5209. DOI: 10.1051/ro/2024172

Vohs, K. D., Baumeister, R. F., Schmeichel, B. J., Twenge, J. M., Nelson, N. M., & Tice, D. M. (2008). Making choices impairs subsequent self-control: A limited-resource account of decision making, self-regulation, and active initiative. *Journal of Personality and Social Psychology, 94*(5), 883–898. DOI: 10.1037/0022-3514.94.5.883 PMID: 18444745

Vuong, Q. H. (2023). *Mindsponge theory.* Walter de Gruyter., DOI: 10.2478/9788367405157

Vygotsky, L. S., & Cole, M. (1978). *Mind in Society: Development of Higher Psychological Processes.* Harvard University Press.

Wachter, S., Mittelstadt, B., & Russell, C. (2017). Counterfactual explanations without opening the black box: Automated decisions and the GDPR. *SSRN, 31*(2), 841–887. DOI: 10.2139/ssrn.3063289

Wang, C., Ahmad, S. F., Bani Ahmad Ayassrah, A. Y. A., Awwad, E. M., Irshad, M., Ali, Y. A., Al-Razgan, M., Khan, Y., & Han, H. (2023). An empirical evaluation of technology acceptance model for Artificial Intelligence in E-commerce. *Heliyon, 9*(8), e18349. DOI: 10.1016/j.heliyon.2023.e18349

Wang, F., Wu, S., Zhang, W., Xu, Z., Zhang, Y., Wu, C., & Coleman, S. (2020). Emotion recognition with convolutional neural network and EEG-based EFDMs. *Neuropsychologia, 146*, 107506. DOI: 10.1016/j.neuropsychologia.2020.107506 PMID: 32497532

Wang, J. J., Miao, Z. H., Cui, F. B., & Liu, H. C. (2018). Robot evaluation and selection with entropy-based combination weighting and cloud TODIM approach. *Entropy (Basel, Switzerland), 20*(5), 349. DOI: 10.3390/e20050349

Wang, L., Huang, N., He, Y., Liu, D., Guo, X., Sun, Y., & Chen, G. (2025). Artificial Intelligence (AI) Assistant in Online Shopping: A Randomized Field Experiment on a Livestream Selling Platform. *Information Systems Research*, isre.2023.0103. Advance online publication. DOI: 10.1287/isre.2023.0103

Wang, S. (2025). Public Perceptions of Artificial Intelligence in 20 Countries: Assessing Individual-and Country-Level Factors. *Cross-Cultural Research*, •••, 10693971251336803. DOI: 10.1177/10693971251336803

Wang, X., Ma, Y., Cammon, J., Fang, F., Gao, Y., & Zhang, Y. (2023). Self-supervised EEG emotion recognition models based on CNN. *IEEE Transactions on Neural Systems and Rehabilitation Engineering, 31*, 1952–1962. DOI: 10.1109/TNSRE.2023.3263570 PMID: 37015115

Wang, X., & Yin, M. (2021). "Are explanations helpful? A comparative study of the effects of explanations in AI-assisted decision-making." *Proceedings of the 26th International Conference on Intelligent User Interfaces (IUI '21)*, pp. 26–36. DOI: DOI: 10.1145/3397481.3450650

Wang, X., Zhang, J., He, C., Wu, H., & Cheng, L. (2023). A novel emotion recognition method based on the feature fusion of single-lead EEG and ECG signals. *IEEE Internet of Things Journal, 11*(5), 8746–8756. DOI: 10.1109/JIOT.2023.3320269

Wang, Y., Tian, W., Xu, J., Tian, Y., Xu, C., Ma, B., Hao, Q., Zhao, C., & Liu, H. (2023). Wearable wireless dual-channel EEG system for emotion recognition based on machine learning. *IEEE Sensors Journal, 23*(18), 21767–21775. DOI: 10.1109/JSEN.2023.3303441

Wani, T. M., Gunawan, T. S., Qadri, S. A. A., Kartiwi, M., & Ambikairajah, E. (2021). A comprehensive review of speech emotion recognition systems. *IEEE Access : Practical Innovations, Open Solutions*, *9*, 47795–47814. DOI: 10.1109/ACCESS.2021.3068045

Warrier, U., Singh, P., Jien, C. W., Kee, D. M. H., Yi, G. Z., Jiann, T. W., Liang, T. Y., Sb, G., Nair, S., Nair, R. K., Lokhande, S. D., & Ganatra, V. (2021). Factors that Lead Amazon.com to A Successful Online Shopping Platform. *International Journal of Tourism and Hospitality in Asia Pasific*, *4*(1), 7–17. DOI: 10.32535/ijthap.v4i1.1017

Wei, B., Huang, J., Ren, R., & Hua, M. (2025). Virtual Shopping Assistant, Real User Experience: The Impact of Co-creation and Experiential Value on Consumer Purchase Intention. *Lecture Notes in Computer Science*, *15769*, 214–233. Advance online publication. DOI: 10.1007/978-3-031-93861-0_14

Wei, J., Wang, X., Schuurmans, D., Bosma, M., Ichter, B., Xia, F., Chi, E., Le, Q., & Zhou, D. (2022). Chain-of-thought prompting elicits reasoning in large language models. *Advances in Neural Information Processing Systems*, *35*, 24824–24837.

Wemm, S. E., & Wulfert, E. (2017). Effects of Acute Stress on Decision Making. *Applied Psychophysiology and Biofeedback*, *42*(1), 1–12. DOI: 10.1007/s10484-016-9347-8 PMID: 28083720

Wien, A. H., & Peluso, A. M. (2021). Influence of human versus AI recommenders: The roles of product type and cognitive processes. *Journal of Business Research*, *137*, 13–27. DOI: 10.1016/j.jbusres.2021.08.016

Wilden, R., Gudergan, S. P., Nielsen, B. B., & Lings, I. (2013). Dynamic capabilities and performance: Strategy, structure and environment. *Long Range Planning*, *46*(1-2), 72–96. DOI: 10.1016/j.lrp.2012.12.001

Wiles, J. (2022). 2022 Gartner Hype Cycle. Gartner. https://www.gartner.com/en/articles/what- s-new-in-artificial-intelligence-from-the-2022-gartner-hype-cycle

Williamson, S. M., & Prybutok, V. (2024). The Era of Artificial Intelligence Deception: Unraveling the Complexities of False Realities and Emerging Threats of Misinformation. *Information (Basel), 15*(6), 299. DOI: 10.3390/info15060299

Wilson, B., Lakshmanan, K., Dix, A., Rahat, A., & Roach, M. (2025). A Crossroads for Hybrid Human-Machine Decision-Making. In Meo, R., & Silvestri, F. (Eds.), *Machine Learning and Principles and Practice of Knowledge Discovery in Databases* (Vol. 2134, pp. 316–322). Springer Nature Switzerland., DOI: 10.1007/978-3-031-74627-7_24

Winston, W. L. (2004). *Operations Research (Winston, 2004): Applications and Algorithms* (4th ed.). Cengage Learning.

Wirtz, B. W. (2024). *Digital business and electronic commerce: Strategy, business models and technology*. Springer Nature., DOI: 10.1007/978-3-031-50289-7

Wixom, B. H., & Ross, J. W. (2017). How to monetize your data. *MIT Sloan Management Review, 58*(3).

Wong, S., Yeung, J. K.-W., Lau, Y.-Y., & Kawasaki, T. (2023). A case study of how Maersk adopts cloud-based blockchain integrated with machine learning for sustainable practices. *Sustainability, 15*(9). *Sustainability (Basel), 7305*(9), 7305. Advance online publication. DOI: 10.3390/su15097305

World Design Council. (2025, July 14). *Case Study: How Tesla Used Design Thinking to Redefine the Automotive Experience.* https://worlddesigncouncil.org/blog/case-study-how-tesla-used-design-thinking-to-redefine-the-automotive-experience/

Wuchty, S., Jones, B. F., & Uzzi, B. (2007). The increasing dominance of teams in production of knowledge. *Science, 316*(5827), 1036–1039. DOI: 10.1126/science.1136099 PMID: 17431139

Wu, Z., Aw, E. C. X., Tan, G. W. H., & Ooi, K. B. (2025). Speak and shop! Transforming retail experience with AI voice assistants. *Internet Research.* Advance online publication. DOI: 10.1108/INTR-02-2024-0149

Xie, J., Nozawa, W., & Managi, S. (2023). Do investors incorporate financial materiality? Remapping environmental information for carbon-sensitive investment. *Corporate Social Responsibility and Environmental Management, 30*(6), 2881–2899. DOI: 10.1002/csr.2524

Xu, K., Zhou, H., Zheng, H., Zhu, M., & Xin, Q. (2024). Intelligent Classification and Personalized Recommendation of E-commerce Products Based on Machine Learning. *arXiv preprint arXiv:2403.19345.*

Yalla, S. P., Subudhi, P. S., & Ramachandaramurthy, V. K. (2022). Topological review of hybrid RES-based multi-port converters. *IET Renewable Power Generation*, *16*(6), 1087–1106. DOI: 10.1049/rpg2.12356

Yang, C. Y., Wang, C. Y., & Pao, C. L. (2025). The impact of negative emotions on product purchase decisions: A study of watch design. *BMC Psychology*, *13*(1), 447. DOI: 10.1186/s40359-025-02775-0

Yang, X., Chen, A., PourNejatian, N., Shin, H. C., Smith, K. E., Parisien, C., Compas, C., Martin, C., Costa, A. B., & Flores, M. G. (2023). A large language model for electronic health records. npj. *Digital Medicine*, *5*(1).

Yoesoep Edhie, R.Yoesoep Edhie Rachmad. (2022). Social Media Marketing Mediated Changes In Consumer Behavior From E-Commerce To Social Commerce. *International Journal of Economics and Management Research*, *1*(3), 227–242. DOI: 10.55606/ijemr.v1i3.152

Yoon, K. (1987). A reconciliation among discrete compromise solutions. *The Journal of the Operational Research Society*, *38*(3), 277–286. DOI: 10.1057/jors.1987.44

Younus, S., Rasheed, F., & Zia, A. (2015). Identifying the Factors Affecting Customer Purchase Intention. *Global Journal of Management and Business Research*, *15*(2), 8–13.

Yu, C., & Wang, M. (2022). Survey of emotion recognition methods using EEG information. *Cognitive Robotics*, *2*, 132–146. DOI: 10.1016/j.cogr.2022.06.001

Zenke, F., & Vogels, T. P. (2021). The remarkable robustness of surrogate gradient learning for instilling complex function in spiking neural networks. *Nature Machine Intelligence*, *3*(1), 76–87. DOI: 10.1038/s42256-020-00286-3

Zhang, D. C., Highhouse, S., & Nye, C. D. (2019). Development and validation of the General Risk Propensity Scale (GRiPS). *Journal of Behavioral Decision Making*, *32*(2), 152–167. DOI: 10.1002/bdm.2102

Zhang, F., Li, X., & Wu, Y. (2023). Footstep-powered floor tile: Design and evaluation of an electromagnetic frequency up-converted energy harvesting system enhanced by a cylindrical Halbach array. *Sustainable Energy Technologies and Assessments*, *60*, 103571. Advance online publication. DOI: 10.1016/j.seta.2023.103571

Zhang, L., Xia, B., Wang, Y., Zhang, W., & Han, Y. (2023). A fine-grained approach for EEG-based emotion recognition using clustering and hybrid deep neural networks. *Electronics (Basel)*, *12*(23), 4717. DOI: 10.3390/electronics12234717

Zhang, Z., & Hyland, K. (2023). The Role of Digital Literacy in Student Engagement with Automated Writing Evaluation (AWE) Feedback on Second Language Writing. *Computer Assisted Language Learning*. Advance online publication. DOI: 10.1080/09588221.2023.2256815

Zhao, L., Feng, X., Feng, X., Qin, B., & Liu, T. (2023). Length extrapolation of transformers: A survey from the perspective of position encoding. arXiv preprint arXiv:2312.17044.

Zheng, A., Zhang, J., Wang, R., & Poutievski, L. (2023). *How Traffic Analytics Shapes Traffic Engineering, Topology Engineering, and Capacity Planning of Jupiter. In 2023 Optical Fiber Communications Conference and Exhibition*. OFC., DOI: 10.1364/OFC.2023.W3A.2

Zhou, M., Duan, N., Liu, S., & Shum, H.-Y. (2020). Progress in Neural NLP: Modeling, Learning, and Reasoning. *Engineering (Beijing)*, *6*(3), 275–290. DOI: 10.1016/j.eng.2019.12.014

Zhou, R., Chen, L., & Yu, K. (2024, May). Is LLM a reliable reviewer? a comprehensive evaluation of LLM on automatic paper reviewing tasks. In *Proceedings of the 2024 joint international conference on computational linguistics, language resources and evaluation (LREC-COLING 2024)* (pp. 9340-9351).

Zhou, X., Jin, K., Shang, Y., & Guo, G. (2020). Visually interpretable representation learning for depression recognition from facial images. *IEEE Transactions on Affective Computing*, *11*(3), 542–555. DOI: 10.1109/TAFFC.2018.2828819

Zorzetti, M., Signoretti, I., Salerno, L., Marczak, S., & Bastos, R. (2022). Improving agile software development using user-centered design and lean startup. *Information and Software Technology*, *141*, 106718. Advance online publication. DOI: 10.1016/j.infsof.2021.106718

About the Contributors

Zian Shah Kabir is an experienced information systems specialist with a background in the telecommunication industry for +17 years. Zian completed his PhD in information systems engineering under the Faculty of Engineering & IT at UTS. He is serving as a casual academic and worked as a research assistant on a digital health project. Before attending the PhD program, he attained a master's in information and telecommunication engineering and a bachelor's degree in electrical and electronic engineering. Zian has a long tenure track and professional experience in the telecommunication industry. He served as a Deputy Director at the Bangladesh Telecommunication Regulatory Commission (BTRC). he is an affiliated member of the Institutions of Engineers Bangladesh (IEB), KAIST GCC (Commercialization Center), International Council on Systems Engineering (INCOSE), Systems Engineering Society of Engineering (SESA), Australian Computer Society (ACS), and Commonwealth Engineers Council.

***

Javaria Ajoon holds an M.S. in Computer Science (2009) from COMSATS University Islamabad, Pakistan. She is currently a Subject Specialist in Information Technology at the Elementary and Secondary Education Department, Khyber Pakhtunkhwa, Pakistan. An independent researcher with a keen interest in Artificial Intelligence and Machine Learning, she is pursuing opportunities to undertake a Ph.D. and contribute to advancements in computing and technology education.

Nivashini G is a dedicated faculty from the Department of Information Technology at Karpagam Academy of Higher Education, located in Coimbatore, Tamil Nadu, India. With a strong interest in technology and innovation, she is passionate about exploring emerging trends in the IT field and developing solutions that contribute to the digital world. Hailing from the vibrant city of Coimbatore (PIN code: 641021), Nivashini is known for her commitment to learning, problem-solving skills, and enthusiasm for continuous growth in both academic and professional pursuits. Her goal is to build a successful career in the field of information technology and make a meaningful impact in the tech industry.

Qazi Mudassar Ilyas received his Ph.D. in Information and Communication Engineering from Huazhong University of Science and Technology, China, in 2005. Over the years, he has held academic positions at several leading institutions in Pakistan, including Ghulam Ishaq Khan Institute, the University of the Punjab, and COMSATS Institute of Information Technology . He is currently serving as an Associate Professor at King Faisal University in Saudi Arabia. His research spans artificial intelligence, machine learning, knowledge management, and information retrieval. He has also contributed to academic leadership, curriculum development, and quality assurance initiatives across multiple institutions.

Ziaul Islam is currently serving as an Assistant Engineer at West Zone Power Distribution PLC (WZPDCL) which is an enterprise of Bangladesh Power Development Board (BPDB). WZPDCL is responsible for power distribution in the western zone of Bangladesh where this organization is committed to advancing the efficiency and reliability of power distribution. Md. Ziaul Islam completed his Master of Science (M.Sc.) in Electrical and Electronic Engineering (EEE) from Bangladesh University of Engineering and Technology (BUET) and his Bachelor of Science (B.Sc.) in EEE from Ahsanullah University of Science and Technology (AUST). With a strong research background in electrical energy, power systems and renewable technologies, Md. Ziaul Islam focuses on innovative solutions to modern energy challenges, particularly in the integration of renewable sources and optimization of power systems. His academic training and professional experience reflect a dedication to improving energy infrastructure and promoting sustainable technological advancements.

Meera K L is a finance professional with over 20+ years of experience in industry and academia. Currently, she is working as an Associate Professor at Christ Academy Institute of Advanced Studies, Bangalore. She has held leadership roles, including IQAC Coordinator and HOD at Dayananda Sagar Business Academy. A Cambridge-trained educator, she adopts innovative, student-centered teaching methods. Her research focuses on investment avenues for middle-class investors, with publications in ABDC, Scopus, and Web of Science journals. She believes in the transformative power of education to inspire growth.

Kyeong Kang is an academic and course coordinator for Bachelor of Science in IT (IS Management major)) at University of Technology Sydney. She is an expert in

Innovative Digital Culture in E-service System Design particularly in E-commerce and E-government. Her research focus is not only limited to the system innovation but also towards positive impacts on people and society. She has completed several industry and internationally collaborated research projects. She has published over 20 journal articles and book chapters. She has edited a book in E-commerce. Her research is widely published and cited.

Ben Kennedy is an Artificial Intelligence researcher affiliated with Capitol Technology University in Laurel, Maryland. His research focuses on biologically inspired modular architectures, notably the Synthetic Cognitive Augmentation Network (SCAN), emphasizing adaptive learning, advanced cognitive agent integration, and human-centered AI collaboration. Kennedy actively develops sophisticated experimental platforms designed to enhance cognitive augmentation, align AI agents with human cognitive processes, and ensure ethical integration within complex human-AI systems.

M. R. Arun completed his BE, ME and PhD from Anna University, Chennai. He completed PG Diploma (PGDEEM) from Annamalai University, M.Sc. Counselling Psychology from TNOU University, M.A. Public Administration from Madras University and B.Lib.Sc. from Alagappa University. He is in the teaching profession for the past 17 years, published around 101 papers in Journals, Patents and Conferences, Active participant crossing more than 150 events and internationally networked through various Professional bodies. He is skilled and technically proficient in various simulation software associated to Research in Engineering. He organized and coordinated around 15 events in working institution. Authored 12 Books, 15 Book Chapters and he has been technically integrated in various international journals as an Editorial Board Member and Reviewer. His area of interest includes Video Processing, Remote Automation, Wireless Communication, Quantum Computing, Machine Learning & Artificial Intelligence.

T.C. Manjunath was born on 6th February 1967. He did his B.E. from RV College of Engg. in 1989 (B'lore Univ), M.E. in Robotics & Automation from Govt.'s LD College of Engg., (Gujarat Univ., Ahmedabad) in 1992 with FCD-First Rank & PhD in Systems & Control Engg. from IIT Bombay in the year 2007, got best research scholar award for the entire IITB. He is currently working as the Dean Research (R & D) and as Professor in Dept. of Computer Science & Engg. with specialization in IoT, Cyber Security, Blockchain Technology in Rajarajeswari College of Engineering, Bengaluru, Karnataka. He has got a teaching (academic), research & administrative experience of more than 35+ years (10 years as Principal & 8 years as HoD, 1 year as Dean Research & rest as faculty) and has published more than 1500+ research papers in various national & international journals & conferences, with 175+ patents, 100+ book-chapters, 25 awards, 100s of Keynote talks and also published 25 text books in various field of science, engineering & technology, completed projects worth Rs. 9 Crores till date & guided 10 PhD scholars in VTU. He is a member of 30 Professional Society Chapters and is a fellow of IE, IETE, AMIEE, Sr. Member of IEEE and a chartered engineer.

Keerthana N V is an Assistant Professor with 8 Years of experience in teaching and 1-year experience in research (IUCAA LIGO Data Centre, Pune). Her area of interest is Network Security, Blockchain Security, and Machine Learning Algorithms. She is pursuing PhD at Anna University in Security in Blockchain. She has a working experience in High-Performance Computing (HTCondor) in LIGO INDIA Project. She has written many research articles in Network Security and assisted Scientists in analyzing Datasets in HTCondor.

P.Sathyasutha, currentlyworking as an Assistant Professor in the Department of Computer Science and Engineering at Gnanamani College of Technology,Namakkal, Tamil Nadu. She has completed her B.E. in Information Technology, M.E. in Computer Science and Engineering and pursuing Ph.D. in Information and Communication Engineering, under Anna University, Chennai. She has 18 years of Academic, Administration, Teaching, and Researchexperience.She is a highly accomplished professional in the field of Computer Science and Engineering, specializing in Artificial Intelligence. She has authored numerous articles published in various journals and actively supervises B.E and M.E students undertaking their final year projects. Additionally, she holds an active life membership in the Professional Society of ISTE. Her research interests encompass Networking, Data Analytics, Artificial Intelligence, Machine Learning and theory of computing. She actively participates in college-wide activities and demonstrates strengths in maintaining confidence, positive attitude, effective teamwork and commitment.

P. Selvakumar is currently an Associate Professor in the Department of Science and Humanities at Nehru Institute of Technology, Coimbatore, Tamil Nadu, India. Prior to joining Nehru Institute of Technology, he held positions as an Assistant Professor in the Department of Science and Humanities various engineering colleges in Coimbatore, Tamil Nadu, India, spanning from 2011 to 2023. Dr.P. Selvakumar's academic and research interests lie primarily in the field of phytochemistry. He has made significant contributions to his field, with an impressive record that includes 30 journal publications, 12 patents, and 20 conference papers. Additionally, he has organized 5 seminars and conferences, further demonstrating his commitment to academic and scientific exchange. Throughout his career, Dr. P.Selvakumar has been recognized for his dedication and excellence in teaching, having been honored with several awards, including the Best Teacher Awards. He holds a prominent position in the academic community, contributing actively to research.

Sancheeta Pugalia is a Lecturer at the School of Professional Practice and Leadership, Faculty of Engineering and IT, University of Technology Sydney (UTS), Australia. Holding a Doctor of Philosophy in Women Entrepreneurship from the University of Technology Sydney, a Master of Science by Research from the Indian Institute of Technology Madras, and a Bachelor of Engineering in Industrial Engineering, she combines theoretical expertise with practical application.

R. Loganathan received B.E degree in Computer Science and Engineering from Annamalai University, Annamalainagar, Tamil Nadu, India in 2007 and the M.E degree in Computer Science and Engineering from Annamalai University, Annamalainagar, Tamil Nadu, India in 2012. i am Currently pursuing Ph.D degree in SRM University, Chennai in the area of Cyber Security. I have teaching experience of 13 years. At present i am working as an Assistant Professor in Cyber Security department at Paavai Engineering College, Namakkal.

Undabatla Rambabu is an Associate Professor in the Department of MBA at Lakireddy Bali Reddy College of Engineering, Mylavaram, with over 16 years of teaching and research experience. He specializes in Managerial Economics, Financial Management, Security Analysis, and Financial Accounting. A Ph.D. holder in Commerce & Business Administration from Acharya Nagarjuna University, he has published extensively in reputed journals, presented at national and international conferences, and received multiple awards for academic excellence, including the NPTEL Discipline Star Award and Best Faculty Award for 2021–2023.

Vandana N. Revankar is an educator, researcher, and academic leader with over two decades of experience across the fields of Aviation, Insurance, Education, and Management. She holds a Ph.D. in Marketing and is UGC-NET qualified. Currently serving as Assistant Professor at St. Francis College, Bengaluru, she has contributed extensively to curriculum design, academic advisory, and student enrichment in Management and Aviation education.She has authored textbooks, published research papers in reputed journals, and presented in national and international conferences. With her expertise in bridging academia and industry, Dr. Revankar continues to inspire learners and shape the future of Management, Tourism and Aviation studies.

José De Jesús Reyes-Sánchez, Ph.D. in senior management, has six master's degrees, in education, in management, professional soccer technique, in physical preparation, in professional soccer, in physical rehabilitation sciences, in sports sciences, as well as in coaching and integral well-being, who is a research professor at the TECNM Instituto Tecnológico Superior De Fresnillo, with more than 20 years of teaching experience, PRODEP profile and coordinator of the academic body with vast experience in conferences and workshops, currently novelist and writer.

Samkit J. Shah, Ph.D. candidate in Mechanical & Automobile Engineering at A. D. Patel Institute of Technology, CVM University, specializes in advanced internal combustion engine performance through variable compression ratio (VCR) and exhaust gas recirculation (EGR) integration. His doctoral work investigates synergistic VCR–EGR strategies to optimize brake thermal efficiency and emissions in spark-ignition engines. He holds an M.E. in Internal Combustion & Automobile Engineering (8.84 CGPA, GTU, 2015) and a B.E. in Mechanical Engineering (7.71 CGPA, GTU, 2012). His early research explored supercharging techniques for SI engines and continuously variable transmissions, culminating in two authored monographs: Hybrid Electric Vehicle Technology and Supercharging of S.I. Engine. With over 30 peer-reviewed publications in journals such as SAE, ISNT, IJMTER and IJRSI, Samkit's contributions span hybrid drivetrains, transmission design, and alternative fuels. He serves on editorial boards for IJGET and IJTET and reviews for the International Journal of Automobile Engineering Research and Development. In academia, he has led research projects and mentored graduate students at Parul Institute of Engineering & Technology and ITM Universe Institute. His professional affiliations include SAE India and the Society of Plastic Engineers, reflecting his interdisciplinary approach to sustainable vehicle systems.

Mahendrakumar Subramaniam received the Bachelors degree in Electronics andCommunication Engineering in 2006, the Masters degree in Communication Systemsin 2009 and received Ph.D - Information and Communication Engineering in 2019from Anna University, Chennai, India. He is currently working as Professor inthe department of Electronics and Communication Engineering, Velalar College ofEngineering and Technology since 2009. He is an Anna University recognized Ph.Dsupervisor and eight scholars are pursuing Ph.D under his supervision. Hisresearch interests include Antenna, Microwave, Wireless Network, MAC andNetwork Protocols. He has published 32 papers in international journals. He isa life member of ISTE and IEICE.

T. Nithya presently serves as an Assistant Professor in the Department of Information Technology at Velalar College of Engineering and Technology in Erode, Tamil Nadu. She earned his B.Tech. in II, followed by an M.E. in Computer Science and Engineering from Anna University, Chennai. With an extensive 18 years of experience in academia, Administration, Teaching, and Research, she is a highly accomplished professional in the field of Information Technology, specializing in Data Analysis and Data Mining. She has published more than 20 research papers in National and International journals and presented more than 50 papers in National and International conferences across India. She published 2 Indian Patents. She actively supervises students undertaking their final year projects at Anna University Chennai. Additionally, she holds an active life membership in the Professional Society of ISTE. Her research interests encompass Data Analysis, Machine Learning and Deep Learning.

V.Chandrasekaran received B.E Degree in Electronics and Communication Engineering from Kongu Engineering College, Perundurai in 2001 and M.Tech Degree in Communication Systems from SASTRA University in 2004. From 2004 to 2006, he worked as a Lecturer in the Department of ECE, Maharaja Engineering College, Avinashi. He has completed Doctoral Programme from Anna University in 2014. From 2006 to till date, he worked in Velalar College of Engineering and Technology, Erode. Currently he is working as a Professor and Head in the Department of Medical Electronics, Velalar College of Engineering and Technology, Erode, Tamilnadu, India. He is an Anna University recognized Ph.D supervisor and 9 scholars are pursuing and 3 scholars completed Ph.D under his supervision. He is a life member of IETE, ISTE, ISSE and BMESI. He has published several papers in National and International Journals and Conferences, patents, book chapters and books. His academic interests include Wireless Networks,5G Networks, Signal processing, Media Access Control and Networking Protocols.

Index

A

Adaptive Heuristics 127
AI 1, 2, 3, 4, 5, 6, 7, 8, 9, 10, 11, 12, 13, 14, 15, 16, 17, 18, 19, 20, 22, 23, 24, 25, 29, 33, 35, 42, 45, 48, 52, 53, 54, 57, 62, 63, 65, 69, 76, 77, 78, 79, 81, 83, 84, 85, 86, 87, 88, 89, 90, 91, 92, 93, 94, 95, 96, 97, 98, 99, 100, 101, 102, 103, 104, 105, 106, 107, 108, 109, 110, 111, 113, 114, 115, 116, 117, 118, 126, 134, 137, 138, 139, 141, 155, 163, 168, 169, 183, 184, 185, 186, 187, 188, 189, 190, 191, 192, 193, 194, 195, 196, 197, 198, 199, 200, 201, 202, 203, 204, 205, 206, 207, 208, 209, 210, 211, 212, 213, 214, 215, 216, 217, 243, 244, 245, 247, 250, 253, 254, 255, 256, 257, 258, 259, 260, 261, 262, 263, 264, 265, 266, 267, 268, 269, 270, 272, 273, 274, 275, 276, 277, 279, 288, 293, 299, 300, 302, 303, 305, 307, 344, 371, 373
AI application 259
Artificial Intelligence 1, 2, 5, 21, 22, 23, 25, 33, 38, 42, 48, 52, 53, 55, 57, 76, 78, 80, 81, 85, 86, 88, 89, 107, 108, 109, 110, 111, 154, 156, 183, 212, 213, 214, 215, 216, 217, 240, 244, 245, 254, 255, 256, 261, 267, 268, 269, 270, 272, 273, 274, 275, 276, 277, 298, 299, 350, 351, 359, 369, 371, 372, 373

C

case study 47, 56, 80, 81, 153, 172, 173, 179, 180, 183, 191, 192, 193, 212, 215, 233, 235, 257, 293, 305, 306, 374, 377
Cognitive analytics 339, 340, 341, 342, 343, 344, 360, 364, 366, 367, 368, 369, 378
Cognitive Biases 7, 9, 36, 63, 244
consumer behavior 155, 167, 245, 247, 248, 250, 252, 253, 254, 255, 261, 264, 270, 274, 276, 277, 331
Conversational AI 213
cultural adaptation 122, 140, 261, 265

D

data-driven decision making 90, 103, 109, 111
Data-driven decision-making 1, 4, 5, 6, 85, 86, 87, 88, 89, 90, 91, 92, 93, 94, 95, 96, 100, 101, 102, 103, 104, 105, 111, 341, 349, 351, 378
Decision-making 1, 2, 3, 4, 5, 6, 7, 8, 9, 10, 11, 12, 13, 14, 15, 17, 18, 20, 27, 28, 29, 30, 31, 32, 33, 34, 35, 36, 37, 38, 39, 40, 41, 43, 44, 47, 48, 49, 50, 54, 55, 56, 57, 60, 62, 63, 65, 66, 67, 70, 71, 72, 74, 75, 76, 77, 78, 79, 80, 82, 83, 84, 85, 86, 87, 88, 89, 90, 91, 92, 93, 94, 95, 96, 97, 98, 99, 100, 101, 102, 103, 104, 105, 111, 114, 116, 117, 118, 123, 124, 130, 138, 139, 142, 146, 149, 150, 151, 153, 154, 155, 156, 160, 163, 164, 165, 166, 171, 172, 175, 180, 183, 184, 185, 186, 187, 188, 189, 190, 191, 192, 194, 199, 203, 207, 208, 209, 210, 211, 212, 213, 215, 217, 219, 220, 221, 222, 223, 224, 225, 226, 228, 232, 235, 237, 238, 240, 241, 243, 244, 245, 247, 249, 250, 251, 252, 253, 256, 257, 261, 262, 263, 264, 265, 269, 295, 297, 308, 332, 336, 339, 341, 343, 344, 345, 348, 349, 351, 354, 358, 362, 365, 378
Decision Making 1, 2, 3, 4, 5, 6, 7, 8, 10, 13, 16, 17, 18, 19, 20, 21, 22, 24, 25, 29, 30, 34, 52, 53, 55, 56, 57, 59, 60, 61, 62, 64, 65, 66, 67, 68, 69, 70, 72, 74, 75, 76, 78, 81, 82, 83, 85, 87, 90, 93, 94, 97, 98, 103, 107, 109, 111, 143, 153, 171, 183, 189, 191, 220, 222, 241, 257, 269
Decision Making in Human-Computer Interaction 59

Decision Support Systems 2, 7, 10, 13, 16, 17, 21, 41, 57, 69, 70, 72, 76, 77, 78, 82, 187, 203, 238, 245, 264
decision trees 15, 39, 40, 95, 153, 154, 155, 160, 162, 163, 164, 166, 168, 171, 176
Digital commerce 243, 244, 245, 246, 247, 248, 250, 252, 253, 254, 255, 256, 257, 258, 259, 260, 261, 262, 263, 264, 265, 271, 275

E

EEGNet 307, 308, 309, 310, 311, 317, 318, 319, 326, 327, 328, 329, 330, 331, 332, 334
Electroencephalogram 38, 53, 58, 334, 335, 337
Emerging technologies 57, 62, 82, 83, 264, 339, 340, 341, 342, 343, 364, 368, 378
Energy Efficiency 114, 117, 118, 119, 131, 133, 134, 135, 289, 293, 294, 300
ESG reporting 339, 365, 366, 367, 376, 378
ethical considerations 43, 61, 67, 103, 119, 187, 217, 256, 258
expectation-execution-mitigation framework 87, 93, 95, 111
Expert Judgment 157

F

Footstep Power Generators 279, 280, 281, 283, 291, 299, 302
fuzzy logic 219, 220, 221, 222, 227, 228, 234, 239, 240, 241

G

genetic algorithm 227, 228, 234, 240, 241

H

hallucination 85, 86, 87, 89, 90, 91, 92, 93, 94, 95, 100, 101, 102, 104, 105, 108, 109, 110, 111, 185, 214
Healthcare 1, 2, 3, 4, 5, 6, 7, 8, 9, 10, 11, 12, 13, 14, 15, 16, 17, 18, 19, 20, 21, 22, 23, 24, 25, 30, 41, 42, 46, 60, 69, 70, 78, 79, 95, 98, 107, 108, 109, 114, 139, 153, 155, 160, 162, 163, 174, 176, 185, 186, 187, 191, 192, 199, 201, 202, 203, 204, 205, 206, 207, 208, 209, 210, 211, 301, 307, 308, 331, 332, 334, 342, 365, 375
Hybrid Power 279, 280, 281, 287, 288, 290, 294, 296, 297, 298, 299, 301, 303, 305

I

influence diagrams 153, 154, 155, 163, 164, 166, 167, 168, 169, 176, 178, 181
Innovative Decision Making 1, 2, 3, 4, 6, 7, 8, 10, 16, 17, 18, 19, 20, 24, 25, 111
Innovative Decision-Making Support Framework 1, 3

M

Machine learning 3, 4, 5, 6, 11, 18, 19, 21, 22, 23, 40, 41, 52, 54, 56, 76, 78, 79, 86, 88, 91, 105, 107, 155, 156, 162, 169, 184, 185, 189, 190, 195, 200, 201, 205, 214, 217, 256, 277, 288, 295, 299, 308, 309, 310, 311, 316, 329, 330, 336, 349, 363, 373, 377, 378
MCDM 83, 219, 220, 221, 222, 224, 225, 226, 234, 238, 239, 240, 241
Multinational Companies 183, 191, 192, 207, 209, 250

N

Neural network 82, 308, 309, 329, 334, 335, 336, 337
Neuromorphic Computing 113, 114, 116, 134, 137, 139, 140, 142
Noise-to-Power Conversion 284

O

Organizational resilience 339, 341, 342, 343, 347, 353, 357, 364, 368, 372, 373, 374, 378

P

Personalized AI 20, 113, 117
Psychometric Alignment 113, 137

R

reliability engineering 220, 222
risk analysis 9, 172
Robot selection 219, 220, 221, 222, 226, 228, 233, 239, 240, 241

S

simulation 19, 42, 70, 73, 76, 131, 136, 141, 157, 158, 162, 171, 174, 181, 312, 356
Spiking Neural Networks 116, 141, 143
stochastic modeling 153, 154, 155, 170, 171, 174, 175, 176, 179, 180
strategic planning 72, 122, 154, 162, 164, 189
Sustainable AI 191
Sustainable Energy 280, 289, 292, 293, 297, 298, 306

T

technology acceptance 187, 213, 245, 247, 253, 254, 255, 257, 259, 260, 261, 262, 263, 264, 267, 273, 276
Thermoelectric Harvesters 279, 280, 281, 284, 286, 287, 288, 291, 292, 293, 299, 302

U

uncertainty 9, 28, 39, 40, 53, 56, 63, 68, 71, 125, 138, 149, 153, 154, 155, 156, 158, 160, 163, 165, 166, 171, 172, 178, 180, 181, 186, 194, 196, 201, 202, 222, 223, 244, 251, 253, 262, 263, 267, 275, 340, 354, 360, 361
Unconventional 279, 280, 281, 282, 283, 285, 286, 289, 290, 291, 293, 294, 295, 296, 297, 298, 299, 300, 301, 302, 303
user experience 3, 15, 60, 61, 64, 67, 69, 75, 129, 243, 245, 252, 277, 311, 330

www.ingramcontent.com/pod-product-compliance
Lightning Source LLC
LaVergne TN
LVHW062327300926
849161LV00007B/1607